Lecture Notes in Computer Scie

T0238088

Commenced Publication in 1973
Founding and Former Series Editors:
Gerhard Goos, Juris Hartmanis, and Jan van Leeuwen

Editorial Board

Constantinos Dovrolis (Ed.)

Passive and Active Network Measurement

6th International Workshop, PAM 2005
Boston, MA, USA, March 31 – April 1, 2005
Proceedings

 Springer

Volume Editor

Constantinos Dovrolis
Georgia Institute of Technology, College of Computing
801 Atlantic Drive, Atlanta, GA 30332, USA
E-mail: dovrolis@cc.gatech.edu

Library of Congress Control Number: 2005922930

CR Subject Classification (1998): C.2, C.4

ISSN   0302-9743
ISBN-10   3-540-25520-6 Springer Berlin Heidelberg New York
ISBN-13   978-3-540-25520-8 Springer Berlin Heidelberg New York

Springer is a part of Springer Science+Business Media

springeronline.com

© Springer-Verlag Berlin Heidelberg 2005
Printed in Germany

Typesetting: Camera-ready by author, data conversion by Scientific Publishing Services, Chennai, India
Printed on acid-free paper   SPIN: 11414100   06/3142   5 4 3 2 1 0

Preface

Welcome to the 6th International Workshop on Passive and Active Measurement, held in Boston, Massuchusetts. PAM 2005 was organized by Boston University, with financial support from Endace Measurement Systems and Intel.

PAM continues to grow and mature as a venue for research in all aspects of Internet measurement. This trend is being driven by increasing interest and activity in the field of Internet measurement. To accommodate the increasing interest in PAM, this year the workshop added a Steering Committee, whose members will rotate, to provide continuity and oversight of the PAM workshop series.

PAM plays a special role in the measurement community. It emphasizes pragmatic, relevant research in the area of network and Internet measurement. Its focus reflects the increasing understanding that measurement is critical to effective engineering of the Internet's components. This is clearly a valuable role, as evidenced by the yearly increases in the number of submissions, interest in, and attendance at PAM.

PAM received 84 submissions this year. Each paper was reviewed by three or four Program Committee (PC) members during the first round. Papers that received conflicting scores were further reviewed by additional PC members or external reviewers (typically two). After all reviews were received, each paper with conflicting scores was discussed extensively by its reviewers, until a consensus was reached. The PC placed particular emphasis on selecting papers that were fresh and exciting research contributions. Also, strong preference was given to papers that included validation results based on real measurements. Eventually, out of the 84 submissions, 24 were accepted as full papers and 12 were accepted as poster presentations. The accepted papers cover a wide range of measurement areas, from topology and routing to wireless network and application measurements. Also, the PAM 2005 program shows the truly diverse and international identity of the network measurement community; the 36 accepted contributions are from 13 different countries in North and South America, Asia, Europe, and Oceania.

This year's conference was held in Boston's Back Bay, a lively neighborhood in the heart of the city. Our location is close to many of the best attractions Boston has to offer, and we hope that attendees found an opportunity to enjoy some of them.

The workshop depended on the support of many people, whom we wish to thank. The PC members worked long and hard to evaluate papers, write reviews, and come to consensus on paper acceptance; this work took place under tight deadlines.

We would like to thank Boston University for their support. Special thanks go to Manish Jain at Georgia Tech, who prepared the conference webpage, ran the online review system (CyberChair), and helped significantly in preparing the final proceedings. Many thanks also to George Smaragdakis at Boston University for helping with the conference website and with the local arrangements. And we are grateful to Ellen Grady at Boston University for her work in local arrangements, financial management, and registration. The PC Chair wishes to acknowledge the authors of CyberChair, the software used to manage the review process, for making their system publicly available.

We are very grateful to Endace Measurement Systems and to Intel for financial support that allowed us to keep registration costs low.

Finally, our thanks to all authors of submitted papers, to the speakers, and to all the participants for making PAM 2005 a success!

We hope attendees enjoyed the PAM 2005 workshop, and had a pleasant stay in Boston.

Mark Crovella (General Chair)
Constantinos Dovrolis (PC Chair)

Chadi Barakat, Nevil Brownlee, Mark Crovella,
Constantinos Dovrolis, Ian Graham, and Ian Pratt
(Steering Committee)

Organization

PAM 2005 was organized by Boston University.

Program Committee

Suman Banerjee (University of Wisconsin-Madison)
Chadi Barakat (INRIA)
Supratik Bhattacharyya (Sprint ATL)
Andre Broido (CAIDA)
Nevil Brownlee (University of Auckland)
Neal Cardwell (Google)
kc Claffy (CAIDA)
Les Cottrell (SLAC)
Mark Crovella (Boston University)
Christophe Diot (Intel Research)
Constantine Dovrolis (Georgia Tech)
Avi Freedman (Akamai)
Ian Graham (Endace)
Khaled Harfoush (North Carolina State University)
Loki Jorgenson (Apparent Networks)
Simon Leinen (Switch)
David Meyer (Cisco Systems)
Dina Papagiannaki (Intel Research)
Ian Pratt (University of Cambridge)
Matt Roughan (University of Adelaide)
Kave Salamatian (CNRS/LIP6)
Anees Shaikh (IBM T.J. Watson)
Patrick Thiran (EPFL)
Alex Tudor (Agilent Labs)
Steve Uhlig (Université catholique de Louvain)
Darryl Veitch (University of Melbourne)

Additional Reviewers

Olivier Bonaventure	Marina Fomenkov	Bradley Huffaker
Manish Jain	Dima Krioukov	Matthew Luckie
Priya Mahadevan	Jiri Navratil	Cristel Pelsser
Bruno Quoitin	John Reumann	Ulrich Schmid
Kay Sripanidkulchai	Paul Vixie	Brett Watson

Sponsoring Institutions

Endace Measurement Systems (www.endace.com)
Intel Corporation (www.intel.com)

Table of Contents

Section 4: Topology Measurements

Section 5: Wireless Network Measurements

Section 6: Monitoring Facilities

Section 7: Routing and Traffic Engineering Measurements

Section 8: Spectroscopy and Bandwidth Estimation

Section 9: Poster Session

On the Impact of Bursting on TCP Performance

Ethan Blanton[1] and Mark Allman[2]

[1] Purdue University
eblanton@cs.purdue.edu
[2] ICSI/ICIR
mallman@icir.org

Abstract. Periodically in the transport protocol research community, the idea of introducing a *burst mitigation strategy* is voiced. In this paper we assess the prevalence and implications of bursts in the context of real TCP traffic in order to better inform a decision on whether TCP's congestion control algorithms need to incorporate some form of burst suppression. After analyzing traffic from three networks, we find that bursts are fairly rare and only large bursts (of hundreds of segments) cause loss in practice.

1   Introduction

Within the transport protocol research community, an idea that crops up periodically is that of introducing *bursting mitigation* into the standard congestion control algorithms. Transport protocols can naturally send line-rate bursts of segments for a number of reasons (as sketched below). Studies have shown that these line-rate bursts can cause performance problems by inducing a rapid build-up of queued segments at a bottleneck link, and ultimately dropped segments and a reduced transmission rate when the queue is exhausted.

In this paper we focus on the study of *micro-bursts* and exclude *macro-bursts* from consideration. A micro-burst is a group of segments transmitted at line rate in response to a single event (usually the receipt of an acknowledgment). A macro-burst, on the other hand, can stretch across larger time scales. For instance, while using the slow start algorithm [APS99], TCP[1] increases the congestion window (and therefore the transmission rate) exponentially from one round-trip to the next. This is an increase in the macro-burstiness of the connection. The micro-burstiness, however, is unaffected as TCP sends approximately 2–3 segments per received acknowledgment (ACK) throughout slow start (depending on whether the receiver employs delayed ACKs [Bra89, APS99]).

An example of naturally occurring bursting behavior is given in [Hay97], which shows that TCP connections over long-delay satellite links with advertised windows precisely tuned to the appropriate size for the delay and bandwidth of the network path suffer from burst-induced congestion when loss occurs.

---

[1] The measurements and discussions presented in this paper are in terms of TCP, but also apply to SCTP [SXM+00] and DCCP's [KHF04] CCID 2 [FK04], since they use similar congestion control techniques.

C. Dovrolis (Ed.): PAM 2005, LNCS 3431, pp. 1–12, 2005.
© Springer-Verlag Berlin Heidelberg 2005

Ideally, a TCP connection is able to send both retransmissions and new data segments during a loss recovery phase [FF96]. However, if there is no room in the advertised window, new segments cannot be sent during loss recovery. Upon exiting loss recovery (via a large cumulative ACK), TCP's window will slide and a line-rate burst of segments will be transmitted. [Hay97] shows that this burst — which is roughly half the size of the congestion window before loss — can result in an overwhelmed queue at the bottleneck link, causing further loss and additional performance sacrifice. [Hay97] also shows this bursting situation to apply to a number of TCP variants (Reno [APS99], NewReno [Hoe96, FF96], FACK [MM96], etc.). Finally, [Hay97] shows that bursts *can impact TCP performance*, but the experiments outlined are lab-based and offer no insight into how often the given situation arises in the Internet. In this paper we assess the degree to which these micro-bursting situations arise in the wild in an attempt to inform a decision as to whether TCP should mitigate micro-bursting. While out of scope for this paper we note that [AB04] compares a number of burst mitigation techniques. In addition to advertised window constraints discussed above, micro-bursts can be caused by several other conditions, including (but not limited to):

- **ACK loss.** TCP uses a cumulative acknowledgment mechanism that is robust to ACK loss. However, ACK loss causes TCP's window to slide by a greater amount with less frequency, potentially triggering longer than desired bursts in the process.
- **Application Layer Dynamics.** Ultimately, the application provides TCP with a stream of data to transmit. If the application (for whatever reason) provides the data to TCP in a bursty fashion then TCP may well transmit micro-bursts into the network. Note: an operating system's socket buffer provides a mechanism that can absorb and smooth out some amount of application burstiness, especially in bulk transfer applications. However, the socket buffer does not always help in applications that asynchronously obtain data to send over the network.
- **ACK Reordering.** Reordered ACKs[2] cause an ACK stream that appears similar to a stream containing ACK loss. If a cumulative ACK "passes" ACKs transmitted earlier by the endpoint, then the later ACK (which now arrives earlier) triggers the transmission of a micro-burst, while the earlier ACKs (arriving later) will be thrown away as "stale".

The causes of bursting discussed above are outlined in more detail in [JD03] and [AB04]. Also note that the causes of bursts are not TCP variant specific, but rather apply to all common TCP versions (Reno, NewReno, SACK, etc.).

[JD03] also illustrates the impact of micro- and macro-bursts on aggregate network traffic. In particular, [JD03] finds that these source-level bursts create

---

[2] Reordered data segments can also cause small amounts of bursting, if the reordering is modest. However, if the reordering is too egregious then false loss recovery will be induced, which is a different problem from bursting. For a discussion of the issues caused by data segment reordering, see [BA02, ZKFP03].

scaling in short timescales and can cause increased queuing delays in intermediate nodes along a network path. In contrast, in this paper we concentrate on
characterizing bursts and determine the frequency of bursts. We then use the
analyzed data to inform a discussion on whether it behooves TCP to prevent
bursts from a performance standpoint, as proposed in the literature (e.g., in
[FF96, HTH01]).

We offer several contributions in this paper after outlining our measurement
methodology in § 2. First, we characterize observed, naturally occurring micro-
bursts from three networks in § 3. Next, we investigate the implications of the
observed micro-bursts in § 4. Finally, in § 5 we conclude with some preliminary
discussion into the meaning of the results from § 3 and § 4 as they relate to
the question of whether a burst mitigation mechanism should be introduced
into TCP.

2   Measurement Methodology

First, we define a "burst" for the remainder of the paper as a sequence of at
least 4 segments sent between two successive ACKs (i.e., a "micro-burst"). The
"magic number" of 4 segments comes from the specification of TCP's congestion
control algorithms [APS99]. On each ACK during slow start, and roughly once
every 1–2 round-trip times (RTT) in congestion avoidance, TCP's algorithms
call for the transmission of 2–3 segments at line-rate. Therefore, micro-bursts of
3 or fewer segments have been deemed reasonable in ideal TCP and are common
in the network. We consider bursts of more than 3 segments to be "unexpected",
in that they are caused by network and application dynamics rather than the
specification of the congestion control algorithms. That is not to say that TCP is
in violation of the congestion control specification in these circumstances — just
that outside dynamics have caused TCP to deviate from the envisioned sending
pattern. These unexpected bursts are the impetus for the various proposals to
mitigate TCP's burstiness, and therefore they are the focus of our study. We do
note that [AFP02] allows TCP to transmit an initial 4 segment burst when the
maximum segment size is less than 1096 bytes. However, this is not taken into
account in our definition of a burst since it is a one-time only allowance.

In principle, the above definition of a burst is sound. However, in analyzing
the data we found a significant vantage point problem in simply using the number
of segments that arrive between two successive ACKs. As shown in [Pax97] there
is a general problem in TCP analysis with matching particular ACKs with the
packets they liberate — even when the monitoring point is the end host involved
in the TCP connection. However, when the monitoring point is not the end
host the problem is exacerbated. Table 1 illustrates the problem by showing
the different order of events observed inside the TCP stack of the end host
and at our monitor point. In the second column of this example, two ACKs
arrive at the end host and each trigger the transmission of two data segments,
as dictated by TCP's sliding window mechanism. However, the third column
shows a different (and frequently observed) story from the monitor's vantage

Table 1. Example of vantage point problem present in our datasets

Event Number	End Host	Monitor
0	ACK_{n+1}	ACK_{n+1}
1	$DATA_{m+1}$	ACK_{n+2}
2	$DATA_{m+2}$	$DATA_{m+1}$
3	ACK_{n+2}	$DATA_{m+2}$
4	$DATA_{m+3}$	$DATA_{m+3}$
5	$DATA_{m+4}$	$DATA_{m+4}$

point. In this case, the monitor observes both ACKs before observing any of the data segments and subsequently notes all four data segments transmitted. Using the notion sketched above, these four data segments would be recorded as a burst when in fact they were not. This scenario can, for example, be caused by ACK compression [Mog92]. If the ACKs are compressed before the monitor such that they arrive within t seconds, where t is less than the round-trip time (RTT) between the monitor and the end host, r, then the situation illustrated in table 1 will be found in the traces. An even thornier problem occurs when a group of compressed ACKs arrives over an interval a bit longer than r. In this case, the overlap between noting ACKs and data packets makes it nearly impossible to untangle the characteristics of the bursts (or, even their presence).

We cope with this problem by *accumulating* the ACK information. For instance, in table 1 since two ACKs without any subsequent data segments are recorded our analysis allows up to 6 data segments to be sent before determining a burst occurred. This heuristic does not always work. For instance, if 6 data segments were observed between two subsequent ACKs in a trace file there is no way to conclusively determine that 3 data segments were sent per ACK. The case when the first ACK triggered 2 data segments and the second ACK triggered 4 data segments (a burst) is completely obfuscated by this heuristic. Another problem is that ACKs could conceivably be lost between the monitor and the end host which would likewise cause the analysis to mis-estimate the bursting characteristic present on the network. In our analysis, if we note more than $3N$ segments sent in response to N ACKs we determine a burst has been transmitted. The length of this burst is simply recorded as the number of segments noted. In other words, we do not attempt to ascribe some of the data segments to each ACK. This is surely an overestimate of the size of the burst. However, as will be shown in the following sections, this small vantage point problem is unlikely to greatly impact the results because the results show that the difference between bursts of size M and bursts of size $M \pm x$ for some small value of x (e.g., 1–10) is negligible. Therefore, while the numbers reported in this paper are slight mis-estimates of the true picture of bursting in the network we believe the *insights* are solid.

To assess the prevalence and impact of micro-bursting, we gathered four sets of packet traces from three different networks. We analyze those connections involving web servers on the enterprise network. That is, we focus on local web

Table 2. Dataset characteristics

Dataset	Start	Duration	Servers	Clients (/24s)	Conns.	Bogus
Anon	7/24/03	≈26 hours	1,202	5,319 (4,541)	295,019	5,955 (2.0%)
LBNL	10/22/03	≈11 hours	947	22,788 (19,689)	196,085	2,362 (1.2%)
ICSI$_1$	1/4/04	≈14 days	1	24,752 (21,571)	223,906	221 (0.1%)
ICSI$_2$	9/18/04	≈14 days	1	23,956 (20,874)	198,935	114 (0.1%)

servers' sending patterns, rather than the sending patterns of remote servers that are responding to local web clients' requests. The characteristics of the four trace files used in our study are given in table 2. The first trace, denoted Anon, consists of roughly 26 hours of web traffic recorded near web servers at a biology-related research facility that asked not to be identified. The tracing architecture is, however, outlined in [MHK+03]. The second trace represents roughly 11 hours of web server traffic at the Lawrence Berkeley National Laboratory (LBNL) in Berkeley, CA, USA. The final two datasets represent requests to a single web server at the International Computer Science Institute (ICSI), also in Berkeley, during two different two week periods in 2004.

These packet traces are analyzed with a custom-written tool called *conninfo*, which analyzes the data-carrying segments sent by web servers on the enterprise network. *Conninfo* mainly tracks the number of data segments sent between two subsequent pure acknowledgment (ACK) segments as sketched above. In addition to recording micro-burst sizes, *conninfo* also records which (if any) segments within a burst are retransmitted. Finally, *conninfo* records several ancillary metrics such as the total data transfer size, the duration of the connection, etc.

Conninfo attempts to process each connection in the dataset. However, as indicated in the last column of table 2, a small fraction of connections were removed from each dataset. These connections exhibit strange behavior that *conninfo* either does not or cannot understand; for example, several "connections" (which are perhaps some sort of attack or network probe) consist of a few random data segments with noncontiguous sequence numbers. As the table shows, the fraction of connections removed from further analysis is relatively small and, therefore, we do not believe this winnowing of the datasets biases the overall results presented in this paper.

3 Characterizing Bursts

In this section we provide a characterization of the bursts observed in the traces we studied. Figure 1 shows the distributions of burst sizes in each of the datasets in terms of both segments, bytes and time. Figure 1(a) shows that the distribution of burst sizes when measured in terms of segments is similar across all datasets. In addition, the figure shows that over 90% of the bursts are less than 15 segments in length. Figure 1(b) shows the burst size in terms of bytes per burst. This distribution generally follows from the segment-based distribution if 1500 byte segments are assumed. While the LBNL and ICSI datasets are similar

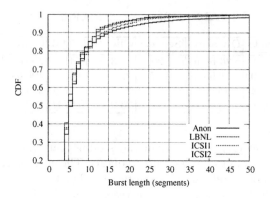

(a) Segments

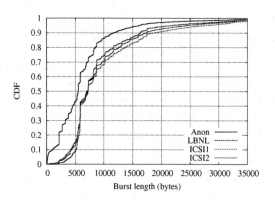

(b) Bytes

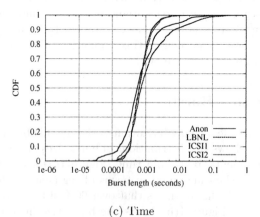

(c) Time

Fig. 1. Distribution of burst sizes

in terms of byte-based burst size, the Anon distribution indicates smaller bursts. Since we did not observe smaller bursts in the Anon dataset when measuring in terms of segments, it appears that the segment sizes used in the Anon network are generally smaller than at LBNL and ICSI. We generated packet size distributions for all networks and there is a clear mode of 20% in the Anon dataset at 576 bytes that is not present in either the ICSI or LBNL datasets. Otherwise, the distributions of packet sizes are roughly the same, thus explaining the discrepancy in figure 1.

Figure 1(c) shows the distribution of the amount of elapsed time covered by each burst. This plot confirms that the vast majority of bursts happen within a short (less than 10 msec) window of time. This is essentially a double-check that our methodology of checking data between subsequent ACKs and our analysis tool are both working as envisioned. While we found bursts that encompass a fairly long period of time (over a second) these are the exception rather than the rule and upon close examination of the time-sequence plots these look to be artifacts of network dynamics mixing with application sending patterns that are difficult to systematically detect. Therefore, we believe that our analysis techniques are overall sound.

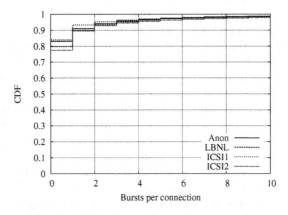

Fig. 2. Distribution of bursts per connection

Next we turn our attention to the prevalence of bursts. Figure 2 shows the distribution of the number of bursts per connection in our four datasets. The figure shows that the burst prevalence is roughly the same across datasets. As shown, over 75% of the connections across all the datasets experienced no bursts of 4 or more segments. However, note that many of the connections that did not burst could not because of the limited amount of data sent or because TCP's congestion window never opened far enough to allow 4 or more segments to be transmitted.

Next we look beyond the on-the-wire nature of bursts and attempt to determine the *root cause* of the bursts. Table 3 shows the determined causes of the

Table 3. Percentage of bursts triggered by the given root cause

Dataset	Bursts	Initial Window	Exit Loss Recovery	Stretch ACKs	Window Opening	App. Pattern	Unknown
Anon	274,880	1.8	0.2	26.3	5.0	17.0	49.6
LBNL	187,176	0.9	0.3	22.9	3.1	32.8	40.0
ICSI$_1$	165,023	6.4	0.7	23.5	4.8	24.0	40.6
ICSI$_2$	228,063	4.2	5.1	22.4	4.5	23.3	45.1

bursts found in each dataset. First, the second column of the table shows that each dataset contains a wealth of bursts. Next, the third column of the table shows that 1–6% of the bursts are observed in the initial window of data transmission. The fourth column shows a similarly small amount of bursting caused by the sender being limited by the advertised window during loss recovery and then transmitting a burst upon leaving loss recovery (when a large amount of the advertised window is freed by an incoming ACK). These first two causes of loss account for a small fraction of the bursts, but the fraction does vary across datasets. We have been unable to derive a cause for this difference and so ascribe it to the heterogeneous nature of the hosts, operating systems, routers, etc. at the various locations in the network.

The fifth column in table 3 shows that roughly 20–25% of the bursts are caused by stretch ACKs (acknowledgments that newly ACK more than 2 times the number of bytes in the largest segment seen in the connection). Stretch ACKs arrive for a number of reasons. For instance, some operating systems generate stretch ACKs [Pax97] in the name of economy of processing and bandwidth. In addition, since TCP's ACKs are cumulative in nature simple ACK loss can cause stretch ACKs to arrive. Finally, ACK reordering can cause stretch ACKs due to an ACK generated later passing an earlier ACK. The origin of each stretch ACK is therefore ambiguous given our limited vantage point, and hence we did not try to untangle the possible root causes.

The sixth column represents a somewhat surprising bursting cause that we did not expect. From the server's vantage point we observe ACKs arriving from the web client that acknowledge the data transmitted as expected but that do not free space in the advertised window — and, hence, do not trigger further data transmission when the sender is constrained by the advertised window. When an ACK that opens advertised window space finally does arrive a burst of data is transmitted. This phenomenon happens in modest amounts (3–5% of bursts) in all the datasets we examined.

The seventh column in the table shows the percentage of bursts caused by the application's sending pattern. We expected this cause of bursts to be fairly low since our mental model of web transfers is that objects are pushed onto the network as fast as possible. However, 17–33% of the bursts happened after all transmitted data was acknowledged and no other bursting scenario explained the burst, indicating that the data was triggered by the application rather than being clocked out by the incoming ACKs. This could be explained by a persistent HTTP connection that did not use pipelining [FGM$^+$97] — or, which was kept open while the user was viewing a given web page and then re-used to fetch another item from the same server.

Finally, the last column of the table is the most troubling in that it indicates that we could not accurately determine the cause of 40–50% of the bursts across all the datasets. Part of the future work in this area will be to develop additional techniques to determine why this bursting is happening. However, the problem is daunting in that we examined a large number of time-sequence plots for connections containing the unknown burst causes and at times we could not figure out why the burst happened ourselves — let alone design a heuristic to detect it!

4   Implications of Bursts

In this section we explore the implications of the bursting uncovered in the last section on the TCP connections themselves. It is beyond our scope (and data) to evaluate the implications the bursting has on competing traffic and the network itself. Figure 3 shows the probability of losing at least one segment in

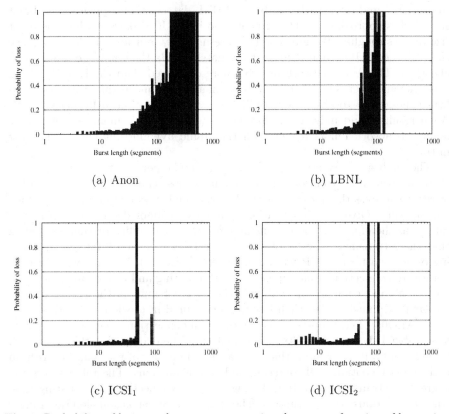

(a) Anon

(b) LBNL

(c) ICSI$_1$

(d) ICSI$_2$

Fig. 3. Probability of losing at least one segment in a burst as a function of burst size (in segments)

Table 4. Retransmission rates observed inside and outside bursts

Dataset	Conns.	Bursts	Burst Loss Rate (%)	Non-Burst Loss Rate (%)
Anon	4,233	69,299	70.9	41.2
LBNL	5,685	45,282	23.0	22.9
$ICSI_1$	4,805	39,832	16.1	14.5
$ICSI_2$	8,201	72,069	26.6	20.5

a burst as a function of the burst size (in segments) for each of our datasets. The figure shows that for modest burst sizes (tens of segments or less) that the probability of losing a segment from the burst is fairly low (roughly less than 5%). As the burst size increases, the likelihood of experiencing a drop within a burst also increases. Bursts on the order of hundreds of segments in our datasets are clearly doomed to overwhelm intervening routers and experience congestion. One interesting note is in the shape of the plots. The Anon dataset shows a fairly smooth ramp-up in the probability of loss in a burst as the burst size increases. However, in both the LBNL and ICSI datasets there is a clear point at which the chances of losing at least one segment in a burst jumps from less than 5% to over 20% and often to 100%. In the LBNL dataset this happens when burst size reaches approximately 60 segments and in the ICSI dataset when the burst size reaches roughly 50 segments. These results may indicate a maximum queue size at or near LBNL and ICSI that ultimately limits the burst size that can be absorbed. The Anon network may be more congested than the ICSI or LBNL networks and therefore the chances of a non-empty queue vary with time and hence the ability to absorb bursts likewise varies over time. Alternatively, the Anon results could indicate the presence of active queue management, whose ability to absorb bursts depends on the traffic dynamics at any given point in time.

The analysis above assesses the question of whether bursts cause *some* loss. Next we focus our attention on the amount of loss caused by bursts. In other words we address the question: is the loss rate higher when bursting than when not bursting? Given the information at hand we cannot determine precise loss rates and therefore use the retransmission rate as an indicator (understanding that the retransmission rate and the loss rate can be quite different depending on loss patterns, TCP variant used, etc. [AEO03]). We first winnow each dataset to connections that experience both bursting and retransmissions to allow for a valid comparison. This has the effect of making the reported rates *appear* to be much higher than loss rates measured in previous network studies (e.g., [AEO03]) because we are disregarding all connections that experienced no loss. Table 4 shows the results of our analysis. When comparing this table with tables 2 and 3 it is apparent that only a small minority of the connections from each dataset contain both bursting and retransmissions. The table shows the aggregate retransmission rate to be higher when connections are bursting than when connections are not bursting. The change in the retransmission rate ranges from nearly non-existent in the LBNL dataset to a roughly 75% increase in the Anon dataset. The large increase in the Anon network agrees with the results presented above that the network is generally more congested and the bottleneck queue closer to the drop point than the other networks we studied. Therefore, bursts cause a large increase in the loss rates experienced in this network while the other networks were better able to absorb the bursts.

5   Conclusions and Future Work

The work in this paper is focused on the impact of bursting on TCP connections themselves. From the above preliminary data analysis we note that micro-bursts are not frequent in TCP connections — with over 75% of the connections in the three networks studied showing no bursting. When bursting does occur, burst sizes are predominantly modest with over 90% of the bursts are less than 15 segments across the datasets we studied. Furthermore, in these modest bursts the probability of experiencing loss within the burst is small (generally less than 5% across datasets). However, bursts of hundreds of segments do occur and such large bursts nearly always experience some loss. We analyzed the cause of bursts and found the two predominant known causes of bursting to be the reception of stretch ACKs and application sending patterns. Unfortunately, our analysis techniques also failed to find the cause of 40–50% of the bursts we observed. An area for future work will be to further refine the analysis to gain further insights into these unclassified bursts (however, as described in § 3 this is a challenging task). Finally, we find an increase in the loss rate experienced within bursts with the loss rate experienced outside of bursts. The increase ranged from slight to approximately 75% depending on the network in question.

A key piece of future work is in understanding how the results given in [JD03] relate to those given in this paper. That is, the preliminary results of this paper indicate that micro-bursting is not likely to hurt performance, while [JD03] shows that the network impact of bursting is non-trivial. Before applying a mitigation to TCP to smooth or reduce bursts it would be useful to correlate the network issues found in [JD03] with specific bursting situations. For instance, if only particular kinds of bursting are yielding the scaling behavior noted in [JD03] then mitigating only those bursting situations may be a desirable path forward.

Acknowledgments

Andrew Moore and Vern Paxson provided the Anon and LBNL datasets, respectively. Sally Floyd, Vern Paxson and Scott Shenker provided discussions on this study. The anonymous reviewers provided good feedback on the submission of this paper and their comments improved the final product. This work was partially funded by the National Science Foundation under grant number ANI-0205519. Our thanks to all!

References

[AB04]   Mark Allman and Ethan Blanton. Notes on Burst Mitigation for Transport Protocols. December 2004. Under submission.

[AEO03]   Mark Allman, Wesley Eddy, and Shawn Ostermann. Estimating Loss Rates with TCP. *ACM Performance Evaluation Review*, 31(3), December 2003.

[AFP02]   Mark Allman, Sally Floyd, and Craig Partridge. Increasing TCP's Initial Window, October 2002. RFC 3390.

[APS99]   Mark Allman, Vern Paxson, and W. Richard Stevens. TCP Congestion Control, April 1999. RFC 2581.

[BA02]   Ethan Blanton and Mark Allman. On Making TCP More Robust to Packet Reordering. *ACM Computer Communication Review*, 32(1):20–30, January 2002.

[Bra89]   Robert Braden. Requirements for Internet Hosts – Communication Layers, October 1989. RFC 1122.

[FF96]   Kevin Fall and Sally Floyd. Simulation-based Comparisons of Tahoe, Reno, and SACK TCP. *Computer Communications Review*, 26(3), July 1996.

[FGM+97]   R. Fielding, Jim Gettys, Jeffrey C. Mogul, H. Frystyk, and Tim Berners-Lee. Hypertext Transfer Protocol – HTTP/1.1, January 1997. RFC 2068.

[FK04]   Sally Floyd and Eddie Kohler. Profile for DCCP Congestion Control ID 2: TCP-like Congestion Control, November 2004. Internet-Draft draft-ietf-dccp-ccid2-08.txt (work in progress).

[Hay97]   Chris Hayes. Analyzing the Performance of New TCP Extensions Over Satellite Links. Master's thesis, Ohio University, August 1997.

[Hoe96]   Janey Hoe. Improving the Start-up Behavior of a Congestion Control Scheme for TCP. In *ACM SIGCOMM*, August 1996.

[HTH01]   Amy Hughes, Joe Touch, and John Heidemann. Issues in TCP Slow-Start Restart After Idle, December 2001. Internet-Draft draft-hughes-restart-00.txt (work in progress).

[JD03]   Hao Jiang and Constantinos Dovrolis. Source-Level IP Packet Bursts: Causes and Effects. In *ACM SIGCOMM/Usenix Internet Measurement Conference*, October 2003.

[KHF04]   Eddie Kohler, Mark Handley, and Sally Floyd. Datagram Control Protocol (DCCP), November 2004. Internet-Draft draft-ietf-dccp-spec-09.txt (work in progress).

[MHK+03]   Andrew Moore, James Hall, Christian Kreibich, Euan Harris, and Ian Pratt. Architecture of a Network Monitor. In *Passive & Active Measurement Workshop 2003 (PAM2003)*, April 2003.

[MM96]   Matt Mathis and Jamshid Mahdavi. Forward Acknowledgment: Refining TCP Congestion Control. In *ACM SIGCOMM*, August 1996.

[Mog92]   Jeffrey C. Mogul. Observing TCP Dynamics in Real Networks. In *ACM SIGCOMM*, pages 305–317, 1992.

[Pax97]   Vern Paxson. Automated Packet Trace Analysis of TCP Implementations. In *ACM SIGCOMM*, September 1997.

[SXM+00]   Randall Stewart, Qiaobing Xie, Ken Morneault, Chip Sharp, Hanns Juergen Schwarzbauer, Tom Taylor, Ian Rytina, Malleswar Kalla, Lixia Zhang, and Vern Paxson. Stream Control Transmission Protocol, October 2000. RFC 2960.

[ZKFP03]   Ming Zhang, Brad Karp, Sally Floyd, and Larry Peterson. RR-TCP: A Reordering-Robust TCP with DSACK. In *Proceedings of the Eleventh IEEE International Conference on Networking Protocols (ICNP)*, November 2003.

A Study of Burstiness in TCP Flows

Srinivas Shakkottai[1], Nevil Brownlee[2], and kc claffy[3]

[1] Department of Electrical and Computer Engineering,
University of Illinois at Urbana-Champaign, USA
sshakkot@uiuc.edu
[2] CAIDA, University of California at San-Diego, USA
and Department of Computer Science,
The University of Auckland, New Zealand
nevil@auckland.ac.nz
[3] Cooperative Association for Internet Data Analysis,
University of California at San-Diego, USA
kc@caida.org

Abstract. We study the burstiness of TCP flows at the packet level. We
aggregate packets into entities we call "flights". We show, using a simple
model of TCP dynamics, that delayed-acks and window dynamics would
potentially cause flights at two different timescales in a TCP flow— the
lower at the order of 5-10 ms (sub-RTT) and the higher at about 10 times
this value (order of an RTT seen by the flow). The model suggests that
flight sizes would be small at the lower timescale, regardless of the network environment. The model also predicts that the network conditions
required for the occurrence of flights at the larger timescale are either
large buffers or large available bandwidths — both of which result in
a high bandwidth delay product environment. We argue that these two
conditions indicate that the TCP flow does not operate in a congestion
control region , either because the source of traffic is unaware of congestion or because there is so much bandwidth that congestion control
is not required. We verify our model by passive Internet measurement.
Using the trace files obtained, we collect statistics on flights at the two
timescales in terms of their frequency and size. We also find the dependence of the sizes and frequency of flights on the Internet environment in
which they occurred. The results concur strongly with our hypothesis on
the origins of flights, leading us to the conclusion that flights are effective
indicators of excess resource in the Internet.

1 Introduction

TCP is the dominant protocol in today's Internet. It has been observed [1, 2]
that TCP sometimes sends packets in the form of deterministic aggregations.
The timescale at which this phenomenon occurs is at the RTT level, which
indicates that we should study it at the packet level in individual flows. We
consider the steady state characteristics of TCP at a packet level and investigate
the frequency with which TCP flows have recognizable structure that we can

C. Dovrolis (Ed.): PAM 2005, LNCS 3431, pp. 13–26, 2005.
© Springer-Verlag Berlin Heidelberg 2005

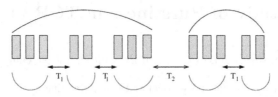

Fig. 1. Illustration of two aggregation levels. Packets may be aggregated into flights at different time scales. At the lower time scale we see five flights, while at the higher time scale we see two

label *flight behavior*. Fig. 1 shows a sequence of thirteen packets and we observe deterministic behavior of packet aggregates at two time scales.

Definition 1. *A small time scale flight (STF) is a sequence of packets whose inter-arrival times differ by at most 'T' percent, where 'T' is a fixed threshold value.*

At the smaller time scale we look at inter-arrival times between single packets; if the inter-arrival times are nearly identical then we say that the packets belong to a single STF. However, observing packets at such a fine resolution obscures the temporal relations that might exist between aggregations of packets. In other words, there may be deterministic behavior between the STFs themselves. In the figure, there are two groups of STFs, within which STFs have nearly identical inter-arrival times.

Definition 2. *A large time scale flight (LTF) is a sequence of aggregations of packets whose inter-arrival times differ by at most 'T' percent, where 'T' is a fixed threshold value.*

By our definition, aggregations of STFs with nearly identical inter-arrival times are defined to be LTFs. We recognize that the terms "small" and "large" are relative. Both terms are with respect to the RTT seen by a flow. The inter-arrival times between packets of an STF are on the order of 5-10 milliseconds (sub-RTT), while the inter-arrival times between STFs are on the order of 40-1000 milliseconds (order of RTT seen by the flow).

Flight behavior of TCP has been a matter of considerable debate. In fact there is not even a standard terminology for the phenomenon; other names for flight-like phenomena are *bursts* [3] and *rounds* [4], where "bursts" usually describe phenomena similar to our STFs and "rounds" usually describe phenomena similar to our LTFs. While modeling TCP flows some authors simply assume the flight nature of TCP [4, 5]. As far as we know, there are no published statistics on flight behavior, and no studies investigating the correlation of flight occurrence with the Internet environment in which TCP operates. Also, there do not seem to be any algorithms for identifying the structure of TCP flows — the method used in the only other work we are aware of in the area [6], is dependent on visually classifying flows.

1.1   TCP Model

Two facets of TCP design could potentially lead to flights, each one at a different time scale.

1. Since many TCP implementations [1, 2] implement *delayed-acks*, a host may send multiple packets for every *ack* it receives. Implementations of delayed-acking vary in terms of the maximum delay (200-500 ms). Many implementations also require that there be a maximum of one outstanding un-acked packet, nominally leading to acknowledgment of alternate packets. Transmission of such packets back to back at source could result in the observation of STFs at the measurement point if the network delays are relatively constant.

2. TCP follows a window-based congestion control mechanism with self-clocking, i.e., the window size changes and packets are transmitted only when acknowledgments are received. If acknowledgments are received with relatively constant inter-arrival times, it would give rise to STFs being sent with similar inter-arrival times, i.e., LTFs.

Another phenomenon that may occur is that of constant-rate flows.

Definition 3. *A constant-rate flow (CRF) is a large TCP flow in which aggregations of two or three packets are observed with nearly identical spacing between the aggregations.*

From the definition of LTFs, it is clear that CRFs are nothing but large LTFs, where we say that a flow is large if it has over 30 packets. Other names for such flows are "rate-limited flows" and "self-clocked flows" [6].

From the above discussion, the origin of of STFs lies in the fact that delayed-acks acknowledge a small sequence of packets (often alternate packets) resulting in the back-to-back transmission of a small number of packets at the source. It seems clear, therefore, that STFs would naturally be of small size regardless of the network environment that the TCP flow in which they occur sees.

However, the question arises: what network environment would be conducive to LTF behavior? We conjecture that LTFs of large size can exist only in high bandwidth-delay product (BDP) regimes. The reason is that as long as no drops occur, TCP increases its window size by some value depending on whether it is in slow-start or congestion avoidance. Only if the network is able to absorb all the packets in the congestion window of a TCP flow will acks be received at deterministic times at the source, leading to transmission of packets at deterministic times. The absorption may take place in two ways:

1. Suppose that the buffer sizes are large in the path of a flow and bandwidth is limited. Then, regardless of congestion window size, the actual throughput is bandwidth constrained. The large buffer size in effect absorbs the packets and delays them so that the source does not see any drops. TCP is unable to estimate the available bandwidth as it is blinded by the large buffer.
   Fig. 2 depicts the case where there is a large buffer between the source and destination. We have assumed, for illustration purposes, that the delay is large enough to ensure that every packet is acked inspite of the delayed-ack

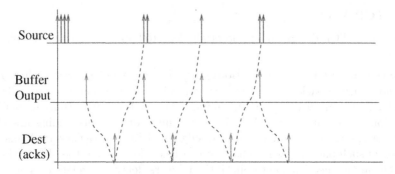

Fig. 2. Illustration of how large buffers in a bandwidth constrained path of a TCP flow lead to LTFs. The congestion window at source gradually increases, but since the buffer absorbs excess packets, the source does not know of the bandwidth constraint

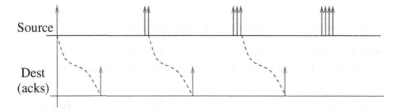

Fig. 3. Illustration of how a large bandwidth and medium delay results in flights in the slow-start phase of TCP. Large bandwidth implies that the source can increase the congestion window to a large size without drops occurring. In this case flights are indicative that congestion control is irrelevant since the network has a large available bandwidth

implementation. The source is in congestion avoidance phase and reception of an ack could result in either the source transmitting one packet or an increase in window size with the source transmitting two packets. The source never loses a packet and assumes that excess bandwidth is available. So the window size continuously increases. TCP is thus blind to congestion in this scenario.

2. Another possible scenario is when bandwidth is high and delay is moderate. In such a case the link absorbs the packets, and large windows of packet aggregations proceed through the network. There is no congestion in the network and TCP congestion control is not required. This scenario is illustrated in Fig. 3 in the slow start phase. We could draw a similar diagram for the congestion-avoidance phase.

We summarize our main hypotheses and the conjectures that we make in Table 1.

Table 1. Summary of our main hypothesis and the conjectures based on them

Hypothesis 1	STFs arise due to the implementation of delayed-acks.
Conjecture 1	The size of STFs are on the order of two or three packets
Conjecture 2	The frequency of STFs is independent of the network environment.
Hypothesis 2	LTFs arise due to window dynamics of TCP.
Conjecture 3	LTFs could be of large size (potentially several hundred packets)
Conjecture 4	The frequency of LTFs increases with increasing BDP.

1.2   Flights as Indicators of Excess Resource

Why should we study flights? What are they good for? Let us consider the question in detail. Two assumptions that network designers traditionally make are:

1. Link capacities are low and many users contend for their use. The expected load is close to the capacity of the links. Hence the tremendous volume of research on the "single bottleneck scenario".
2. To handle demands close to the capacity, buffer sizes should be of the order of the bandwidth-delay product of the link.

Usually such design gives rise to recommendations for large buffer sizes, which in turn has given rise to high bandwidth infrastructure with huge buffer capacities. If the usage assumptions were correct, neither of our two scenarios for flight existence would exist, congestion control would be relevant, and the resource on the Internet would be utilized at high efficiency. On the other hand, the presence of flights is a symptom that we have over designed the Internet — there are enormous resources, in terms of buffer sizes or link capacities, being shared by remarkably few users. In other words, flights are a symptom that TCP congestion control is having no effect, either due to hiding of congestion by buffers, or because there is so much bandwidth that the packets sail through the network. Consistent with the above is the fact that observations of packets on 10 Mb/s Ethernet (for example those in the packet sequence plots in [1, 2]) show clear flight behavior.

1.3   Main Results

We use three different packet traces, all from OC-48 ($\approx 2.5Gb/s$) links, and call them BB1-2002, BB2-2003 and Abilene-2002 [7]. Together these packet traces represent a high diversity of IP addresses, applications, geographic locations and access types. For instance, the *BB1-2002* trace shows about 30% of bytes destined for locations in Asia, with flows sourced from about 15% of all global autonomous systems (AS). The *BB2-2003* has even higher diversity with flows from about 24% of all global ASs. The *Abilene-2002* trace has a large fraction of non-web traffic. Since all three traces give nearly identical results, we provide graphs from only one trace: BB1-2002.

We summarize our main results as follows:

1. We propose a simple threshold-based algorithm, which robustly identifies the different time scale aggregation levels.
2. We verify our hypothesis of two distinct phenomena — delayed acks and window dynamics — giving rise to two classes of packet behavior by studying the statistics of each aggregation level.
3. We show how the algorithm naturally leads to a method of identifying CRFs as large LTFs.
4. We further confirm Hypothesis 1 — delayed acks causing STFs — by verifying Conjectures 1 and 2 — that STF sizes are on the order of two to three packets and are independent of network conditions such as round trip time (RTT), bandwidth and BDP. The observation on the size of STFs illustrates that the source transmits a small number (usually 2 or 3) of back to back packets resulting in an STF at the point of measurement.
5. We verify the Conjecture 4 — high BDP regimes permitting LTFs — by studying the variation in LTF lengths as a function of BDP and showing that LTFs that have a much larger number of packets occur at higher BDPs.
6. Finally, using the statistics on LTFs of large size, we verify Conjecture 3 — LTFs can be of large size — and conclude that currently about 12-15% of flows over thirty packets in length in the traces we study are not responding to congestion control, either because the they are unaware of congestion or because there is no congestion on their paths.

2   Algorithms

In this section we describe the algorithms we use for identification of flights. We first consider the case of identifying STFs. Consider a sequence of packets p_1, p_2, p_3, with inter-arrival times (IATs) δ_1 and δ_2 between the first and second pairs of packets, respectively. Then we consider the ratio $g(\delta_1, \delta_2) = |\frac{\delta_2 - \delta_1}{\delta_1}|$. We decide whether a packet belongs to a particular STF depending on whether

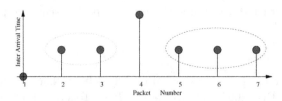

Fig. 4. Illustration of how we find STFs. We group packets $1 - 3$ together as a 2-inter-arrival time unit flight, and so on. The large gap between packets 3 and 4 appears as a singleton

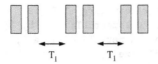

Fig. 5. A sequence of packet pairs in which STFs do not capture complete information

$g > T$ or $g \leq T$, where T is a threshold value. We use the IAT as a measure of the scale of the flight and call the units *IAT units (IATU)*. Thus a flight of 1 IATU means that the observed IAT was different from the preceding and following IATs. An IATU of 2 would mean that two successive IATs were nearly identical. Fig. 4 depicts two STFs, the first of size 2 IATU and the second of size 3 IATU. Our STF detection algorithm is as follows:

```
1. Start with IAT=constant, flight size=1;
2. Compare previous IAT with current IAT and
     calculate g (as defined above).
3. If g is within threshold then
     increment flight size by 1;
   else if flight size >1 start a new
   flight of size 0;
   else start a new flight of size 1;
4. Set previous IAT <- current IAT;
```

The 'else if' line in our above algorithm means that an out-of-threshold IAT indicates the end of a STF, but a sequence of out-of-threshold IATs indicates consecutive 1-IATU STFs. The singleton shown in Fig. 4 indicates such behavior. Our STFs may therefore have 2 packets (1 IATU), 3 packets (2 IATU), 4 packets (3 IATU) and so on. Of course, a 1 IATU STF simply means that at the low time scale, the algorithm did not observe any deterministic behavior.

However, the situation illustrated in Fig. 5 might occur. Here we see a sequence of packet pairs, which are identified by the above algorithm as distinct STFs. But there deterministic behavior between packet pairs at a larger time scale. We would like to have an algorithm that would identify such behavior and aggregate all six packets as an LTF.

We observe that deterministic behavior in the larger time scale can potentially occur only when the STF algorithm reports that the current IAT is different from the previous one (if not, the current packet would be part of the current STF and we update the STF size by one and proceed to the next packet). Also, since we are interested in large timescales, we need to know if the current IAT is larger than the previous IAT. So, if we keep the most recent large IAT in memory, we may compare it to the current IAT and check if they are within a threshold of each other. If they are, then we update the size of the current LTF by one. We merely need to add the following lines to Step 3 of the LTF algorithm:

```
3. (continued) If current IAT > previous IAT then
     compare current IAT with most recent large IAT;
```

```
If g(current IAT, most recent large IAT) is
 within threshold then increment LTF size by 1;
else if LTF size>1 start a new LTF of size 0;
else start a new LTF of size 1;
Set most recent large IAT  <- current IAT;
```

Looking at Fig. 5 again, we see that the above extension would result in the identification of the packet sequence as one LTF as desired. We remark at this point that the choice of threshold value does not seem to be critical to the algorithm. The reason for this observation is that the timescales of IATs for STFs and LTFs are different. As mentioned in the introduction, the typical IATs between packets of an STF are 5-10 ms, whereas the IATs between aggregations of an LTF are about 10 times this. In our analysis we used several values of T ($\frac{1}{16}, \frac{1}{4}, \frac{1}{2}$, 1, 2, 4 and 8) with nearly identical results.

3   Frequency and Size of Flights

In this section we show that our flight detection algorithm is successful and also illustrate the fact that considering two aggregation levels of packets yields a clear picture of TCP behavior. We ran the algorithm with different threshold values on the packet traces and show only some illustrative graphs here.

We first consider the statistics of STFs in Fig. 6 and Fig. 7. Recall that the unit of flight size is IATU. We can convert IATU in STFs into packets by recalling that a 1 IATU STF is a packet whose leading and trailing IATs were different, a

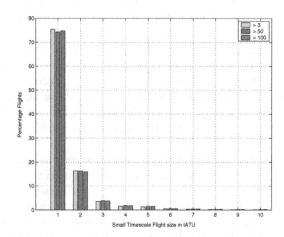

Fig. 6. Small timescale flight size distribution in IATU for BB1-2002. The left column in each set of bars is for flows greater than 3 packets in length; the middle for those greater than 50; and the right column is for flows greater than 100 packets in length. We notice that flights are usually small (in terms of IATU and hence in packets) irrespective of the number of packets in the flow

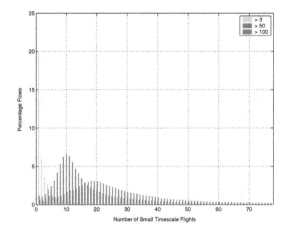

Fig. 7. Number of small timescale flights for trace BB1-2002 on a per flow basis. The left histogram is for flows greater than 3 packets in length, the middle for those greater than 50 and the right histogram is for flows greater than 100 packets in length. We notice that STFs are more common in flows with a larger number of packets

2 IATU STF has three packets and so on. Fig. 6 shows the distribution of STF sizes. We see that regardless of the number of packets in the flow, STF sizes are usually quite small — a size 7 IATU occurs in less than 1% of the STFs. Also, we may easily calculate that the mean STF size is 2.5 packets regardless of the number of packets in the flow.

Fig. 7 shows the distributions of number of STFs on a per flow basis. We see that STFs are much more common in flows with a large number of packets.

We consider flight behavior at the timescale of the RTT seen by the flows in Fig. 8 and Fig. 9. As we have just seen, the STFs of which the LTFs are composed are an average of 2.5 packets in length. We may thus get an estimate of the number of packets in an LTF by multiplying its IATU size by this number. Fig. 8 shows the size distribution of LTFs. The statistics are quite different from the STF size distribution that we analyzed earlier. Flights are much more common at the larger timescale. The graph follows a distribution that is proportional to $\frac{1}{LTF\ size}$. Thus, even at this timescale, decay of flight sizes is fairly quick. Finally, we plot the distributions of number of LTFs on a per flow basis in Fig. 9. We see that as with STFs, LTFs are much more common in flows with a large number of packets.

We draw the following conclusions from the flight statistics observed above:

1. Our initial hypotheses from our model of TCP were that there would be two distinct aggregation levels at different timescales caused by delayed acks and TCP window dynamics. The hypotheses are borne out by the fact that we usually see short STFs, normally consisting of two or three packets, indicating delayed acks. We also see much larger LTFs indicative of windows of packets transmitted in pairs and triplets (i.e., as STFs) with similar spacings between the aggregations.

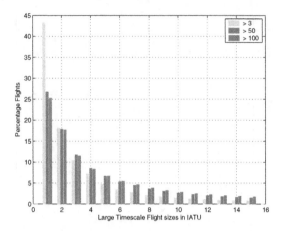

Fig. 8. Large timescale flight size distribution in IATU for BB1-2002. The left column in each set of bars is for flows greater than 3 packets in length; the middle for those greater than 50; and the right column is for flows greater than 100 packets in length. We notice that the LTF size distribution varies proportionally to $\frac{1}{LTF\ size}$

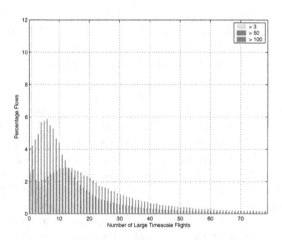

Fig. 9. Number of large timescale flights for trace BB1-2002 on a per flow basis. The left histogram is for flows greater than 3 packets in length, the middle for those greater than 50 and the right histogram is for flows greater than 100 packets in length. We notice that LTFs are more common in flows with a larger number of packets

2. Over 75% of flows having over 50 packets contain LTFs. We identify fairly large LTFs of up to 16 IATUs, i.e., with an average of 40 packet or more, thus verifying Conjecture 3 — that LTFs could potentially be large (and hence be identified as CRFs). What we observe in such cases are aggregations of two or three packets being transmitted at a constant rate. Thus, our algorithm offers a simple means of identifying CRFs. If we consider a flow to be a CRF

if it has over 30 packets in equally spaced aggregations, then about 12-15% of flows are constant-rate flows. These flows are clearly not limited by PC clock speed, as Brownlee and Claffy also observe in [8].

3. From the statistics on the number of flights seen in flows we conclude that many flows are composed of fairly small deterministic packet aggregations at the large timescale, which indicates that the congestion window in these flows grows only up to 10-12 packets before feedback from the network causes it to reduce. Thus, large time scale structure is lost with the growth of TCP congestion windows.

4   Relationship with Network Environment

We now study the relation between flights and the characteristics of the path that a flow traverses, namely the round trip time (RTT), the bandwidth and the BDP. Our usage of the term RTT is to indicate the entire path delay inclusive of queuing delays. We measure RTT by the syn-ack method whose validity has been largely established in [9].

We first consider STFs. We already know that the majority of them are two packets in size, irrespective of flow size. We would like to know if any network characteristics affect STFs larger than two packets. If their origin has to do with delayed acks, i.e., the source constrains them to be small, then the variation of their occurrence with the network parameters should not be significant. In Fig. 10, we show a probability histogram ($\times 100\%$) of STFs larger than two packets in size in different RTT regimes. We see that the chance of seeing an STF larger than two packets is about 1% regardless of the RTT.

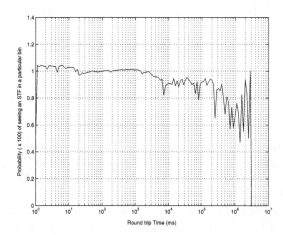

Fig. 10. Fraction of traffic having STFs larger than two packets in each RTT regime for BB1-2002. The variation is about 0.4% indicating the independence of STFs and RTT

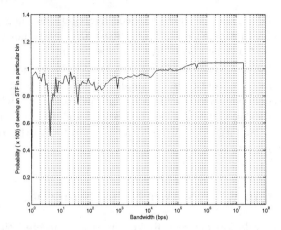

Fig. 11. Fraction of traffic having STFs in each bandwidth regime for BB1-2002. The flat nature indicates that the probability of seeing an STF is independent of bandwidth

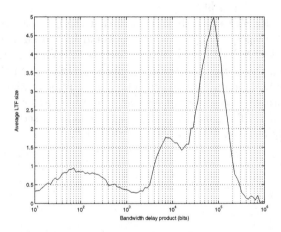

Fig. 12. Variation of LTF sizes as a function of BDP for BB1-2002. The plot peaks at high BDP. Note that the number of data points drops sharply after the peak, and hence the graph ends here

We next examine the relation between STFs and bandwidth (Fig. 11). By 'bandwidth', we mean the total bytes transferred divided by the lifetime of the flow. Here too the probability ($\times 100\%$) of seeing at least one STF larger than two packets in size for any particular bandwidth is nearly constant at 1%. We also analyzed the frequency of seeing at least one STF with more than two packets with regard to BDP and found that the probability ($\times 100\%$) of this event too was about 1%, with a variation of less than 0.5%.

We now study the variation of LTF sizes as a function of BDP, which we show in Fig. 12. We see that on average, LTF sizes are higher at higher BDP. The graph peaks at 10 kb with an average LTF size of 5 (i.e. most flows in this regime had LTFs consisting of of 12 or more packets). The number of available points is small after the peak and the graph dips sharply. The above facts support Conjecture 4 — high BDP is conducive to LTFs being large — since it means that the network has the capacity to absorb the large windows of packets.

5   Conclusion

We studied deterministic temporal relations between TCP packets using several packet traces, which were from different backbone fibers and represent a large fraction of ASs and prefixes. Such traces give an indication of Internet traffic characteristics, since phenomena occurring at the edges are reflected in temporal relations between packets at the measurement point. We studied aggregation of packets at two time scales (of the order of 5-10 ms and 50-1000 ms) in order to verify our hypothesis that two distinct facets of TCP structure should give rise to two different types of temporal relations. In doing so we proposed a simple threshold algorithm for identification of flights. Our TCP model predicted that high BDP environments would be conducive to CRFs. Such an environment could exist only if the network were over-provisioned — either in terms of large buffers or large bandwidth — and hence CRFs are indicative of excess resource in the network. Through statistics on flight sizes and frequency, we verified Conjectures 1 and 3 — that STFs should be short and LTFs can be long. We then verified Conjectures 2 and 4 — that STFs should not depend on the network environment, whereas LTFs should be benefited by large BDPs — by studying the correlations between flights and different network parameters. We thus showed that Hypothesis 1 and 2 — delayed-acks giving rise to STFs and window dynamics giving rise to LTFs — are valid. We concluded that about 12 – 15% of Internet flows in our traces do not operate in a congestion control region. In the future we would like to study how the occurrence of flights changes over the years both on the backbone as well as access links, to understand flights as indicators of excess network resource. Such a study would give us an idea of whether congestion on the Internet has been increasing or decreasing over time.

Acknowledgment

The authors would like to acknowledge Andre Broido (CAIDA) for his participation in many of the discussions that led to this work. This research was funded in part by NSF grant ANI-0221172.

References

1. Stevens, R.: TCP/IP illustrated, Vol.1, Addison-Wesley (1994)
2. Paxson, V.E.: Measurements and Analysis of End-to-End Internet Dynamics. PhD dissertation, University of California, Lawrence Berkeley National Laboratory (1997)

3. Sarvotham, S., Riedi, R., Baraniuk, R.: Connection-level analysis and modeling of network traffic. In: Proceedings of IMW 2001. (2001)
4. Padhye, J., Firoiu, V., Towsley, D., Krusoe, J.: Modeling TCP throughput: A simple model and its empirical validation. In: Proceedings of ACM SIGCOMM '98. (1998)
5. Zhang, Y., Breslau, L., Paxson, V., Shenker, S.: On the Characteristics and Origins of Internet Flow Rates. In: Proceedings of ACM SIGCOMM. (2002)
6. Downey, A.: TCP Self-Clocking. Technical Report TR-2003-002, Olin College, Computer Science (2003)
7. Abilene: Trace obtained from the NLANR PMA webpage (URL below) (2002) http://pma.nlanr.net/Traces/long/ipls1.html.
8. Brownlee, N., Claffy, K.C.: Understanding Internet Streams: Dragonflies & Tortoises. IEEE Communications Magazine (2002)
9. Aikat, J., Kaur, J., Smith, F., Jeffay, K.: Variability in TCP round-trip times. In: Proceedings of IMW. (2003)

On the Stationarity of TCP Bulk Data Transfers

Guillaume Urvoy-Keller

Institut Eurecom, 2229, route des crêtes,
06904 Sophia-Antipolis, France
urvoy@eurecom.fr

Abstract. While the Internet offers a single best-effort service, we remark that (i) core backbones are in general over provisioned, (ii) end
users have increasingly faster access and (iii) CDN and p2p solutions
can mitigate network variations. As a consequence, the Internet is to
some extent already mature enough for the deployment of multimedia
applications and applications that require long and fast transfers, e.g.
software or OS updates. In this paper, we devise a tool to investigate
the stationarity of long TCP transfers over the Internet, based on the
Kolomogorov-Smirnov goodness of fit test. We use BitTorrent to obtain
a set of long bulk transfers and test our tool. Experimental results show
that our tool correctly identify noticeable changes in the throughput of
connections. We also focus on receiver window limited connections to
try to relate the stationarity observed by our tool to typical connection
behaviors.

1 Introduction

The current Internet offers a single best-effort service to all applications. As a
consequence, losses and delay variations are managed by end-hosts. Applications
in the Internet can be classified into two classes: elastic applications, e.g. web or
e-mail, that can tolerate throughputs and delays variations; and real time applications, that are delay sensitive (e.g. voice over IP) or throughput sensitive (e.g.
video-on-demand). With respect to the above classification, a common belief is
that the current Internet with its single best-effort service requires additional
functionality (e.g. DiffServ, MPLS) to enable mass deployment of real-time applications. Still, a number of facts contradict, or at least attenuate, this belief:
(i) recent traffic analysis studies have deemed the Internet backbone ready to
provide real-time services [15]; (ii) the fraction of residential users with high
speed access, e.g. ADSL or cable, increases rapidly; (iii) network-aware coding
schemes, e.g. mpeg4-fgs [7], combined with new methods of transmission like
peer-to-peer (p2p) techniques, e.g. Splitstream [5], have paved the way toward
the deployment of real-time applications over the Internet.

The above statements have lead us to investigate the variability of the service provided by the Internet from an end connection point of view. As TCP is
carrying most of the bytes in the Internet [10], our approach is to concentrate on
long lived TCP connections. Bulk data transfers represent a significant portion

C. Dovrolis (Ed.): PAM 2005, LNCS 3431, pp. 27–40, 2005.

of the current Internet traffic load, especially with p2p applications [2]. By analyzing bulk data transfers, we expect to better understand the actual interaction
between TCP and the Internet. This is important for future applications [1] and
also for CDN providers that rely on migrating traffic on the "best path" from
central to surrogate servers [8]. CDN providers generally rely on bandwidth estimation tools, either proprietary or public tools [12] to perform path selection.
However, the jury is still out on the stationarity horizon provided by such tools,
i.e. how long will the estimation provided by the tool remain valid or at least
reasonable. In the present work, we propose and evaluate a tool that should help
solving these issue. The rest of this paper is organized as follows. In Section
2, we review the related work. In Section 3, we present our dataset. In Section
4, we present our tool to extract stationarity periods in a given connection. In
Section 5, we discuss results obtained on our dataset. Conclusions and future
work directions are presented in Section 6.

2   Related Work

Mathematically speaking, a stochastic process $X(t)$ is stationary if its statistical
properties (marginal distribution, correlation structure) remain constant over
time.

Paxson et al. [17] have studied the stationarity of the throughput of short
TCP connections (transfers of 1Mbytes) between NIMI hosts. The major difference between this work and the present work is that we consider long bulk data
transfer (several tens of minutes) and our dataset is (obviously) more recent
with hosts with varying access capacity, whereas NIMI machines consistently
had good Internet connectivity. Other studies [4, 13] have concentrated on the
non stationarity observed on high speed link with a high number of aggregated
flows. They studied the time scales at which non stationarity appears and the
causes behind it. Also, recently, the processing of data streams has emerged as
an active domain in the database research community. The objective is to use
database techniques to process on-line stream at high speed (e.g. Internet traffic on a high speed link). In the data stream context, detection of changes is a
crucial task [14, 3].

3   Dataset

Our objective is to devise a tool to assess the stationarity of TCP bulk data
transfers. To check the effectiveness of the tool, we need to gather samples, i.e.
long TCP transfers, from a wide set of hosts in the Internet. A simple way to
attract traffic from a variety of destinations around the world is to use a p2p
application. As we are interested in long data transfers, we used BitTorrent,

---

[1] Our focus in the present work is on throughput, which is an important QoS metrics for some multimedia applications, e.g. VoD, but arguably not all multimedia
applications, a typical counter-example being VoIP.

a popular file replication application [11]. A BitTorrent session consists in the replication of a single large file on a set of peers. BitTorrent uses specific algorithms to enforce cooperations among peers. The data transfer phase is based on the swarming technique where the file to be replicated is broken into chunks (typical chunk size is 256 kbytes) that peers exchange with one another. The BitTorrent terminology distinguishes between peers involved in a session that have not yet completed the transfer of the file, which are called *leechers* and peers that have already completed the transfer, which are called *seeds*. Seeds remain in the session to serve leechers. Connections between peers are permanent TCP connections. Due to the BitTorrent algorithms [11], a typical connection between two hosts is a sequence of on periods (data transfers) and off periods (where only keep-alive messages are transfered). Figure 1, where y axis values are one second throughputs samples, depicts a typical one way connection of approximately 14 hours with clear on and off phases.

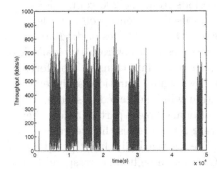

Fig. 1. A typical (one-way) BitTorrent connection

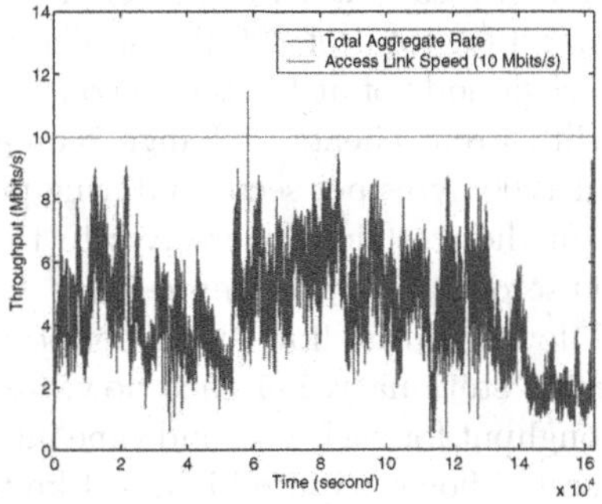

Fig. 2. Aggregate rate of the BitTorrent application during the experiment

The dataset we have collected consists of connections to about 200 peers that were downloading (part of) the file (latest Linux Mandrake release) from a seed located at Eurecom. More precisely, a tcpdump trace of 10 Gbytes was generated during a measurement period of about 44 hours. While the 200 connections are all rooted at Eurecom, the 10 Mbits/s access link of Eurecom should not constitute a shared bottleneck for two reasons. First, with BitTorrent, a client (leecher or seed) does not send to all its peers simultaneously but only to 4 of them, for sake of efficiency. Second, the total aggregate throughput remains in general far below the 10 Mbits/s as shown in figure 2 while the average traffic generally observed on this link (to be added to the traffic generated by our BitTorrent client to obtain the total offered load for the link) exhibits an average rate around 1 Mbits/s with a peak rate below 2 Mbits/s.

To illustrate the diversity of these 200 peers, we have used the maxmind service (http://www.maxmind.com/) to assess the origin country of the peers. In table 1, we ranked countries based on the peers that originate from each of them. Unsurprisingly, we observe a lot of US peers (similar observation was made in [11]

Table 1. Origin countries of the 200 peers

Country	# peers	Country	# peers	Country	# peers	Country	# peers
US	87	NL	4	BR	2	YU	1
UK	24	DE	3	LT	2	BE	1
CA	14	AU	3	CN	1	AT	1
FR	12	PE	3	NO	1	ES	1
IT	8	AE	3	SI	1	CH	1
SE	8	CL	2	TW	1		
PL	7	PT	2	CZ	1		

for a similar torrent, i.e. Linux Redhat 9.0) while the other peers are distributed over a wide range of 27 countries (see http://encyclopedia.thefreedictionary.com/ ISO%203166-1 for the meaning of the abbreviations used in table 1).

Our objective is to study long bulk data transfers in the Internet. To obtain meaningful samples, we extracted the on periods from the 200 connections, resulting in a total of 399 flows. The algorithm used to identify off-periods is to detect periods of at least 15 seconds where less than 15 kbytes of data are sent, as BitTorrent clients exchange keep-alive messages at a low rate (typically less than 1000 bytes per second) during periods where no data transfer is performed. We further restricted ourselves to the 184 flows whose duration is higher than 1600 seconds ($\sim$ 26.6 minutes), for reasons that will be detailed in section 4. We call flow or initial flow an on-period and stationary flow a part of a flow that is deemed stationary. For each flow, we generate a time series that represents the throughput for each 1 second time interval. The average individual throughput of these 184 flows is quite high, 444 kbits/s. Overall, these flows correspond to the transfer of about 50 Gbytes of data over a cumulated period of about 224 hours (the flows of duration less than 1600 seconds represent about 14 Gbytes of data). Due to its size, we cannot claim that our dataset is representative of the bulk transfers in the Internet. It is however sufficiently large to demonstrate the effectiveness of our tool. It also shows that BitTorrent is a very effective application to collect long TCP transfers from a variety of hosts in terms of geographical location and access link speed (even if it is unlikely to observe clients behind modem lines, as downloading large file behind a modem line is unrealistic).

4   Stationarity Analysis Tool

4.1   Kolmogorov-Smirnov (K-S) Test

Given two i.i.d samples $X_1(t)_{t \in \{1,...n\}}$ and $X_2(t)_{t \in \{1,...n\}}$, the Kolmogorov-Smirnov test enables us to determine whether the two samples are drawn from the same distributions or not. The test is based on calculating the empirical cumulative distribution functions of both samples and evaluating the absolute maximum difference $D_{\max}$ between these two functions. The limit distribution of $D_{\max}$ under the null hypothesis (X_1 and X_2 drawn from the same distribution) is known and thus $D_{\max}$ is the statistics the test is built upon. In the sequel of this paper, we used the matlab implementation of the K-S test with 95% confidence levels.

4.2   K-S Test for Change Point Detection

Our objective is to detect stationary regions in time series, or equivalently to detect change points (i.e. border points between stationary regions). We used the K-S test to achieve this goal. Previous work already used the K-S test to detect changes [9, 3], though not in the context of traffic analysis.

The basic idea behind our tool is to use two back-to-back windows of size w sliding along the time series samples and applying the K-S test at each shift of the windows. If we assume a time series of size n, then application of the K-S test leads to a new binary time series of size $n - 2w$, with value zero whenever the null hypothesis could not be rejected and one otherwise. The next step is to devise a criterion to decide if a '1' in the binary time series corresponds to a false alarm or not. Indeed, it is possible to show that even if all samples originate from the same underlying distribution, the K-S test (or any other goodness of fit test [16]) can lead to spurious '1' values. The criterion we use to deem detection of a change point is that at least $w_{min} \approx \frac{w}{2}$ consecutive ones must be observed in the binary time series. w_{min} controls the sensitivity of the algorithm. The intuition behind setting w_{min} to a value close to $\frac{w}{2}$ is that we expect the K-S test to almost consistently output '1' from the moment when the right-size window contains about 25% of points from the "new" distribution (the distribution after the change point) up to the moment when the left-size window contains about 25% of points from the "old" distribution. In practice, a visual inspection of some samples revealed that using such values for w_{min} allows to correctly detect obvious changes in the time series. Figure 3 presents an example on one of our TCP flows time series (aggregated at a 10 seconds time scale - see next section for details) along with the scaled binary time series output by the tool and the change points (vertical bars). This example illustrates the ability of the test to isolate stationary regions. Note also that the output of the binary time series that represents the output of the K-S test for each window position (dash line in figure 3) exhibits a noticeable consistency. This is encouraging as oscillations in the output of the test would mean that great care should be taken in the design of the change point criterion. As this is apparently not the case, we can expect our simple criterion (w_{min} consecutive '1' values to detect a change) to be effective.

4.3   K-S Test in the Presence of Correlation

We want to apply the K-S change point tool described in the previous section to detect changes in the throughput time series described in section 3. However, we have to pay attention that, due to the close loop nature of TCP, consecutive one-second throughputs samples are correlated[2]. If all samples are drawn from

---

[2] While correlation and independence are not the same, we expect that removing correlation will be sufficient in our context to obtain some almost independent samples.

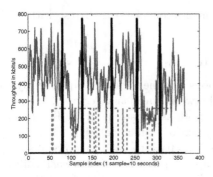
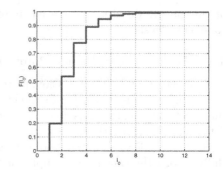

Fig. 3. Initial time series (thin line), binary time series (dash line) and change points (thick bars)

Fig. 4. Cumulative distri. function of l_0

the same underlying distribution, a simple heuristic to build an uncorrelated time series out of a correlated time series is to (i) compute the auto-correlation function of the initial time series, (ii) choose a lag l_0 at which correlation is close enough to zero and (iii) aggregate the initial time series over time intervals of size l_0. Specifically, let $X(t)_{t\in\{1,...n\}}$ be the initial time series. Its auto-correlation function is $AC(f) = \frac{\sum_{i=1}^{n-f} \bar{X}(i+f)\bar{X}(i)}{n\sigma_{\bar{X}}^2}$, where $\bar{X}(t) \triangleq X(t) - E[X]$ and $\sigma_{\bar{X}}^2$ is the variance of $\bar{X}$. $AC(f)$ measures the amount of correlation between samples located at positions t and $t + f$. If ever the time series is i.i.d., then $|AC(f)|$ should be upper bounded by $\frac{2}{\sqrt{n}}$ for $f > 1$ [6]. For a correlated time series, we can choose l_0 such that $\forall f > l_0, |AC(f)| \leq \frac{2}{\sqrt{n}}$. We then generate the aggregate time series $Y(t)_{t\in\{1,...\lceil\frac{n}{l_0}\rceil\}}$ where $Y(t) = \frac{\sum_{u=t\times l_0+1}^{(t+1)\times l_0} X(u)}{l_0}$. This method is however not applicable to our TCP time series as changes in the network conditions prevent us from assuming the same underlying distributions over the whole duration of a flow.

To overcome this difficulty and be able to use the K-S test, we aggregate each time series at a fixed value of $l_0 = 10$. This means that we average the initial time series over intervals of 10 seconds. As the average throughput of the flows is 444 kbits/s, an average flow will send more than 400 packets (of size 1500 bytes) in a 10 second time interval, which is reasonably large enough for a TCP connection to have lost memory of its past history (e.g. to have fully recovered from a loss). To assess the level of correlation that persists in the time series after aggregation at the 10 second time scale, we have computed, for each stationary interval obtained with our tool, the autocorrelation function of the process in this interval. We then derive the lag l_0 after which the autocorrelation function remains (for 95% of the cases) in the interval $\left[-\frac{2}{\sqrt{n}}, \frac{2}{\sqrt{n}}\right]$. Figure 4 represents the cumulative distribution function of l_0. We notice that about 95% of the l_0 values are below 5, which indicates that the "remaining" correlation is of short term kind only.

Table 2. Change point detection tool performance in the presence of correlation

a	w	w_{min}	% of cases with one detection in [450, 550]	Average number of detections
0.2	40	15	100	2.4
0.2	40	40	100	1
0.2	80	15	100	2.4
0.2	80	40	100	1.2
0.5	40	15	100	5.6
0.5	40	40	100	1.1
0.5	80	15	100	4.5
0.5	80	40	100	1.9
0.9	40	15	100	14.6
0.9	40	40	99	6.5
0.9	80	15	89.9	8
0.9	80	40	90.7	5.6

Based on the result of figure 4, one could however still argue that we should continue further the aggregation of the time series for which the correlation is apparently too large, say for $l_0 \geq 3$. Note however that the choice of the time scale at which one works directly impacts the separation ability of the K-S test. Indeed, as we use windows of w samples, a window corresponds to a time interval of $10 \times w$ seconds, and we won't be able to observe stationary periods of less than $10 \times w$ seconds. For example, the results presented in section 5 are obtained with $w = 40$, which means that we won't be able to observe stationary periods of less than 400 seconds (~ 6.7 minutes). Thus, there exists a trade-off between the correlation of the TCP throughput time series that calls for aggregating over large time intervals and the separation ability of the test that calls for having as much small windows as possible.

A second reason why we have chosen to aggregate at a fixed 10 second time scale value is that we expect our tool to be robust in the presence of short term correlation. We investigate this claim in the next section, on synthetic data, where we can tune the amount of correlation. While by no means exhaustive, this method allows us to obtain insights on the behavior of K-S test in the presence of correlation.

4.4   Test of the Robustness of the Tool with Synthetic Data

We consider a first-order auto-regressive process X with $X(t) = aX(t - 1) + Z(t), \forall t\{1, \ldots n\}$ where Z is a purely random process with a fixed distribution. We choose two distributions for Z (leading to Z_1 and Z_2) to generate two samples $X_1(t)$ and $X_2(t)$. We then form the compound vector $[X_1(t)X_2(t)]$ and apply the K-S change point test. We can vary the a parameter to tune the amount of correlation and test how the K-S change point test behaves. Specifically, we consider $a \in \{0.2, 0.5, 0.9\}$ as these values roughly correspond to l_0 values (as defined in the previous section) equal respectively to 2, 5 and 20. With respect

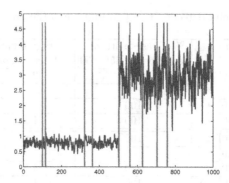

Fig. 5. Sample trajectory of $[X_1 X_2]$ with the detected change points (vertical bars) for $w = 40$ and $w_{min} = 15$

to the results presented in figure 4, we expect the K-S test to behave properly for $a \leq 0.5$ (i.e. $l_0 \leq 5$). In table 2, we present results obtained when $Z_1(t)$ and $Z_2(t)$ are derived from normal distributions with respective means and variances $(0.4, 0.3)$ and $(1.5, 1.5)$ where a given sample $Z_1(t)$ (resp. $Z_2(t)$) is obtained by averaging 10 independent samples drawn from the normal distribution with parameters $(0.4, 0.3)$ (resp. $(1.5, 1.5)$). The main idea behind this averaging phase is to smooth $X_1(t)$ and $X_2(t)$ in a similar fashion that the throughput samples are smoothed at a 10 second time scale in the case of our BitTorrent dataset. As the transition between $X_1(t)$ and $X_2(t)$ is sharp thanks to the difference in mean between Z_1 and Z_2, we expect that the change point tool will correctly detect it. Now, depending on the correlation structure, it might happen that more change points are detected. This is reflected by the results presented in table 2, where for different values of a, w and w_{min}, we compute over 1000 independent trajectories, the average number of detections made by the algorithm (without false alarm, we should obtain 1) and the percentage of cases for which a change is detected in the interval $[450, 550]$ that corresponds to the border between $X_1(t)$ and $X_2(t)$ in the compound vector $[X_1(t)X_2(t)]$, each vector having a size of 500 samples. When the latter metric falls below 100%, it indicates that the correlation is such that our tool does not ncessarily notice the border between $X_1(t)$ and $X_2(t)$ any more. From table 2, we note that such a situation occurs only for $a = 0.9$. Also, when the amount of correlation increases, the average number of points detected increases dramatically, as the correlation structure of the process triggers false alarms as illustrated by the trajectory depicted in figure 5. For a given w value, increasing the treshold w_{min} helps reducing the rate of false. Note that while the results obtained here on synthetic data seem to be better for a criterion $w_{min} = 40$, we used $w_{min} = 15$ on our dataset as it was giving visually better results. A possible reason is the small variance of the throughput time series as compared to the corresponding mean for our dataset. More generally, we note that tuning w and w_{min} is necessary to tailor the tool to the specific needs of a user or an application.

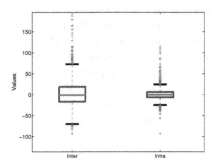 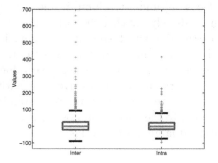

Fig. 6. Boxplot representation of the inter jump in mean (left side) and intra jump in mean (right side)

Fig. 7. Boxplot representation of the inter jump in standard deviation(left side) and intra jump in standard deviation (right side)

4.5   Empirical Validation on Real Data

For the results obtained in this section and the rest of the paper, we used $w = 40$ as special care must be taken when using the K-S test for smaller values [16]. Also, we consider $w_{\min} = 15$ as it visually gives satisfying results on our dataset. In addition, to obtain meaningful results, we restrict the application of the tool to time series with at $4 \times w$ samples (the tool will thus output at least $2 \times w$ results), i.e. to flows that last at least 1600 seconds.

Our change point analysis tool can be easily validated with synthetic data. However, we need to further check whether the results obtained on real traces are reasonable or not. We thus applied our tool on our 184 flows to obtain 818 stationary flows. To assess the relevance of the approach, we proceeded as follows: for any two neighboring stationary flows from the same flow, we compute their means μ_1 and μ_2 and their standard deviations σ_1 and σ_2. We then compute the "jump in mean" $\Delta_\mu = \frac{\mu_2 - \mu_1}{\mu_1} \times 100$ and "jump in standard deviation" $\Delta_\sigma = \frac{\sigma_2 - \sigma_1}{\sigma_1} \times 100$. We then break each stationary flows into two sub-flows of equal size and compute their means μ_1^i and μ_2^i and standard deviations σ_1^i and σ_2^i ($i = 1, 2$). We can then define jumps in means and standard deviations between two sub-flows of a given stationary flow. The latter jumps are called intra jumps while the jumps between stationary flows are called inter jumps. The idea behind these definitions is to demonstrate that the distributions of intra jumps are more concentrated around their mean value than the distributions of inter jumps. To compare those distributions, we used boxplot representations. A boxplot of a distribution is a box where the upper line corresponds to the 75 percentile $\hat{p}_{0.75}$ of the distribution, the lower line to the 25 percentile $\hat{p}_{0.25}$ and the central line to the median. In addition, the $\hat{p}_{0.25} - 1.5 \times IQR$ and $\hat{p}_{0.75} + 1.5 \times IQR$ values ($IQR = \hat{p}_{0.75} - \hat{p}_{0.25}$ is the inter quantile range, which captures the variability of the sample) are also graphed while the samples falling outside these limits are marked with a cross. A boxplot allows to quickly compare two distributions and to assess the symmetry and the dispersion of a given distribution. In figure 6, we plotted the boxplots for the inter jump in mean (left side) and intra jump in mean (right side). From these representations, we immediately see that the intra jump distribution is thinner than the inter jumps distribution which complies with our initial intuition. Note also that the means of the inter and intra jump distributions are close to zero as the Δ_μ definition can result in positive or negative values and it is quite reasonable that overall, we observe as much positive as negative jumps. Figure 7 depicts the boxplots for the inter and intra jumps in standard deviations. The results are somehow similar to the ones for jumps in mean although less pronounced and more skewed toward large positive values.

5   Results on the BitTorrent Dataset

5.1   Stationary Periods Characterization

As stated in the previous section, the K-S change point tool has extracted 818 stationary flows out of the 184 initial flows. This means that, on average, a flow is cut into 4.45 stationary flows. Figure 8 represents the cumulative distribution functions (cdf) of the duration of stationary and initial flows. Stationary flows have an average duration of 16.4 minutes while initial flows have an average duration of 73 minutes.

Figure 9 represents the cumulative distribution functions of throughputs of the stationary and initial flows. Overall, stationary flows tend to exhibit larger throughputs than initial flows. Indeed, the mean throughput of stationary flows is 493.5 kbits/s as compared to 444 kbit/s for the initial ones. This discrepancy is an indication that the K-S change point test is working properly as it extracts from the initial flows stationary periods where the throughputs significantly differ from the mean throughput of the flow. The cdfs differ at the end because whenever the K-S test exhibits a small period (relative to the flow it is extracted from) with high throughput, it will become one sample for the cdf of stationary flows, whereas it might have little impact for the corresponding sample for the cdf of the initial flows (if the high throughput part only corresponds to a small fraction of the initial flow).

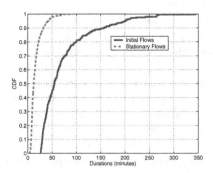 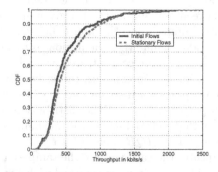

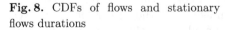

Fig. 8. CDFs of flows and stationary flows durations

Fig. 9. CDFs of flows and stationary flows throughputs

Using our tool, we can also investigate transitions between consecutive stationary periods. The left boxplot of figure 6 allows us to look globally at transitions between stationary periods. From this figure, we can observe that most of the changes result in jumps of the mean value that are less than 20% in absolute values. This is encouraging for applications that can tolerate such changes in their observed throughput since they can expect to experience quite long stable periods, typically several tens of minutes (at least in the context of our dataset). However, a lot of values fall outside the plus or minus $1.5 \times IQR$ interval, meaning that some transitions are clearly more sharp than others.

5.2   The Case of Receiver Window Limited Connections

In a effort to relate the stationarity observed by our tool to the intrinsic characteristics of the connections, we considered the case of receiver window limited flows. A receiver window limited flow is a flow whose throughput is limited by the advertised window of the receiver. The motivation behind this study is that as receiver window limited flows are mostly constrained by some end hosts characteristics (the advertised window of the receiver), they should exhibit longer stationary periods than other flows. Indeed, the intuition is that those other flows have to compete "more" for resources along their path with side traffic, which should affect their throughput, leading to change points.

We first have to devise a test that flags receiver window limited flows. We proceed as follows. For each flow, we generate two time series with a granularity of 10 seconds. The first time series, $Adv(t)$ represents the advertised window of the receiver while the second one, $Out(t)$ accounts for the difference between the maximum unacknowledged byte and the maximum acknowledged byte. The second time series provides an estimate of the number of outstanding bytes on the path at a given time instant. The $Out(t)$ time series is accurate except during loss periods. Note that the computation of $Out(t)$ is possible since our dataset was collected at the sender side, as the Eurecom peer in the BitTorrent session was acting as a seed during the measurement period. A flow is then flagged receiver window limited if the following condition holds:

$$\frac{\sum_{t=1}^{N} 1_{Adv(t)-3 \times MSS \leq Out(t) \leq Adv(t)}}{N} \geq 0.8$$

where N is the size of the two time series and MSS is the maximum segment size of the path. The above criterion simply states that 80% of the time, the estimated number of outstanding packets must lie between the advertised window minus three MSS and the advertised window. By choosing a treshold of 80%, we expect to be conservative.

Application of the test on our dataset leads us to flag about 13.7% of the flows as receiver window limited. The next issue is to choose the non window limited flows. We adopt the following criterion:

$$\frac{\sum_{t=1}^{N} 1_{Out(t) \leq Adv(t)-3 \times MSS}}{N} \geq 0.9.$$

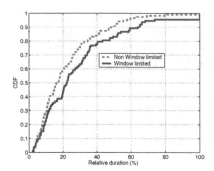
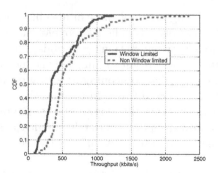

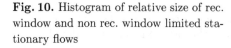

Fig. 10. Histogram of relative size of rec. window and non rec. window limited stationary flows

Fig. 11. Histogram of rec. window and non rec. window limited stationary flows throughputs

Applying the above criterion, we obtained about 14.4% of non receiver window limited flows. A straightforward comparison of the durations of the stationary flows extracted from the flows of the two families (receiver window limited and non receiver limited) is misleading as the duration of their respective connections is different. We thus use two other metrics. First, we compute the number of stationary flows into which a flow is cut in each family. We obtain that the receiver window limited flows are on average cut into 3.5 stationary flows while non receiver window limited flows are cut into 4.5 stationary flows. The second metric we consider is the relative size, in percentage, of the stationary flows with respect to the size of flow they are extracted from for the two familied. Figure 10 represents the cumulative distribution functions of the percentages for the two families. From this figure, we observe that receiver window limited stationary flows are relatively larger than non receiver window limited ones in most cases. Also, in figure 11, we plot the cumulative distributions of the throughput of the stationary flows for both families. We conclude from figure 11 that receiver window limited stationary flows exhibit significantly smaller throughputs values than non receiver window limited ones. This might mean that receiver limited flows correspond to paths with larger RTT than non receiver window limited ones, as this would prevent these flows from achieving high throughput values. This last point as well as our definition of window limited flows (we only considered around 28% of the flows of our dataset to obtain those results) would clearly deserve more investigation.

6   Conclusion and Outlook

Internet Traffic analysis becomes a crucial activity, e.g. for ISPs to do troubleshooting or for content providers and researchers that are willing to devise new multimedia services in the Internet. Once information on some path has been collected, its needs to be analyzed. The first step is to divide traces into somewhat homogeneous period and to flag anomalies. In this paper, we concentrate on the analysis of the service perceived by long TCP connections in the Internet. We have developed a change point analysis tool that extracts stationary periods within connections. We follow a non parametric approach and based our tool on the Kolmogorov-Smirnov goodness of fit test. We validated our change point tool in various ways on synthetic and operational datasets. Overall, the tool manages to correctly flag change points as long as little correlation persists at the time scale at which it is applied. We worked at the 10 second time scale, which is a reasonable time scale for some multimedia applications such as VoD. We also focused on receiver window limited connections to relate the stationarity observed by our tool to typical connection behaviors.

As future work, we intent to pursue in this direction by correlating the stationarity periods with some other network events like RTT variations or loss rates. We would also like to study the extent to which our tool could be used in real time and to investigate how it could be tailored to the need of some specific applications. It is also necessary to compare our tool with some other change point techniques [1].

Acknowledgment

The author is extremely grateful to the anonymous reviewers for their valuable comments and to M. Siekkinen for the trace collection and time series extraction.

References

1. M. Basseville and I. V. Nikiforov, *Detection of Abrupt Changes - Theory and Application*, Prentice-Hall, Inc. ISBN 0-13-126780-9, 1993.
2. N. Ben Azzouna, F. Clerot, C. Fricker, and F. Guillemin, "Modeling ADSL traffic on an IP backbone link", *Annals of Telecommunications*, December 2004.
3. S. Ben-David, J. Gehrke, and D. Kifer, "Detecting changes in data streams", In *Proceedings of the 30th International Conference on Very Large Databases*, 2004.
4. J. Cao, W. S. Cleveland, D. Lin, and D. X. Sun, "On the nonstationarity of Internet traffic", In *Proceedings of the 2001 ACM SIGMETRICS international conference on Measurement and modeling of computer systems*, pp. 102–112, ACM Press, 2001.
5. M. Castro, P. Druschel, A.-M. Kermarrec, A. Nandi, A. Rowstron, and A. Singh, "SplitStream: High-bandwidth multicast in a cooperative environment", In *Proceedings of SOSP'03*, New York, USA, October 2003.
6. C. Chatfield, *The analysis of time series - An introduction*, Chapman & Hall, London, UK, 1996.
7. P. De Cuetos, P. Guillotel, K. Ross, and D. Thoreau, "Implementation of Adaptive Streaming of Stored MPEG-4 FGS Video over TCP", In *International Conference on Multimedia and Expo (ICME02)*, August 2002.
8. J. Dilley, B. Maggs, J. Parikh, H. Prokop, and R. Sitaraman, and B. Weihl, "Globally distributed content delivery", *Internet Computing, IEEE*, pp. 50–58, Sept.-Oct 2002.
9. H. Eghbali, "K-S Test for Detecting Changes from Landsat Imagery Data", *IEEE Trans Syst., Man & Cybernetics*, 9(1):17–23, January 1979.

10. M. Fomenkov, K. Keys, D. Moore, and k claffy, "Longitudinal study of Internet traffic from 1998-2003", Cooperative Association for Internet Data Analysis - CAIDA, 2003.

11. M. Izal, G. Urvoy-Keller, E. Biersack, P. Felber, A. Al Hamra, and L. Garcés-Erice, "Dissecting BitTorrent: Five Months in a Torrent's Lifetime", In *Passive and Active Measurements 2004*, April 2004.

12. M. Jain and C. Dovrolis, "End-to-end available bandwidth: measurement methodology, dynamics, and relation with TCP throughput", *IEEE/ACM Transactions on Networking*, 11(4):537–549, 2003.

13. T. Karagiannis and et al., "A Nonstationary Poisson View of Internet Traffic", In *Proc. Infocom 2004*, March 2004.

14. B. Krishnamurthy, S. Sen, Y. Zhang, and Y. Chen, "Sketch-based change detection: methods, evaluation, and applications", In *IMC '03: Proceedings of the 3rd ACM SIGCOMM conference on Internet measurement*, pp. 234–247, ACM Press, 2003.

15. A. Markopoulou, F. Tobagi, and M. J. Karam, "Assessing the quality of voice communications over Internet backbones", *IEEE/ACM Transactions on Networking*, 11:747–760, October 2003.

16. S. Siegel and N. J. Castellan, *Nonparametric statistics for the Behavioral Sciences*, McGraw-Hill, 1988.

17. Y. Zhang, V. Paxson, and S. Shenker, "The Stationarity of Internet Path Properties: Routing, Loss, and Throughput", ACIRI, May 2000.

Toward the Accurate Identification of Network Applications

Andrew W. Moore[1,*] and Konstantina Papagiannaki[2]

[1] University of Cambridge
andrew.moore@cl.cam.ac.uk
[2] Intel Research, Cambridge
dina.papagiannaki@intel.com

Abstract. Well-known port numbers can no longer be used to reliably identify network applications. There is a variety of new Internet applications that either do not use well-known port numbers or use other protocols, such as HTTP, as wrappers in order to go through firewalls without being blocked. One consequence of this is that a simple inspection of the port numbers used by flows may lead to the inaccurate classification of network traffic. In this work, we look at these inaccuracies in detail. Using a full payload packet trace collected from an Internet site we attempt to identify the types of errors that may result from port-based classification and quantify them for the specific trace under study. To address this question we devise a classification methodology that relies on the full packet payload. We describe the building blocks of this methodology and elaborate on the complications that arise in that context. A classification technique approaching 100% accuracy proves to be a labor-intensive process that needs to test flow-characteristics against multiple classification criteria in order to gain sufficient confidence in the nature of the causal application. Nevertheless, the benefits gained from a content-based classification approach are evident. We are capable of accurately classifying what would be otherwise classified as unknown as well as identifying traffic flows that could otherwise be classified incorrectly. Our work opens up multiple research issues that we intend to address in future work.

1   Introduction

Network traffic monitoring has attracted a lot of interest in the recent past. One of the main operations performed within such a context has to do with the identification of the different applications utilising a network's resources. Such information proves invaluable for network administrators and network designers. Only knowledge about the traffic mix carried by an IP network can allow efficient design and provisioning. Network operators can identify the requirements of

* Andrew Moore thanks the Intel Corporation for its generous support of his research fellowship.

C. Dovrolis (Ed.): PAM 2005, LNCS 3431, pp. 41–54, 2005.

different users from the underlying infrastructure and provision appropriately. In addition, they can track the growth of different user populations and design the network to accommodate the diverse needs. Lastly, accurate identification of network applications can shed light on the emerging applications as well as possible mis-use of network resources.

The state of the art in the identification of network applications through traffic monitoring relies on the use of well known ports: an analysis of the headers of packets is used to identify traffic associated with a particular port and thus of a particular application [1, 2, 3]. It is well known that such a process is likely to lead to inaccurate estimates of the amount of traffic carried by different applications given that specific protocols, such as HTTP, are frequently used to relay other types of traffic, e.g., the NeoTeris VLAN over HTTP product. In addition, emerging services typically avoid the use of well known ports, e.g., some peer-to-peer applications. This paper describes a method to address the accurate identification of network applications in the presence of packet payload information[1]. We illustrate the benefits of our method by comparing a characterisation of the same period of network traffic using ports-alone and our content-based method.

This comparison allows us to highlight how differences between port and content-based classification may arise. Having established the benefits of the proposed methodology, we proceed to evaluate the requirements of our scheme in terms of complexity and amount of data that needs to be accessed. We demonstrate the trade-offs that need to be addressed between the complexity of the different classification mechanisms employed by our technique and the resulting classification accuracy. The presented methodology is not automated and may require human intervention. Consequently, in future work we intend to study its requirements in terms of a real-time implementation.

The remainder of the paper is structured as follows. In Section 2 we present the data used throughout this work. In Section 3 we describe our content-based classification technique. Its application is shown in Section 4. The obtained results are contrasted against the outcome of a port-based classification scheme. In Section 5 we describe our future work.

2   Collected Data

This work presents an application-level approach to characterising network traffic. We illustrate the benefits of our technique using data collected by the high-performance network monitor described in [5].

The site we examined hosts several Biology-related facilities, collectively referred to as a *Genome Campus*. There are three institutions on-site that employ

---

[1] Packet payload for the identification of network applications is also used in [4]. Nonetheless, no specific details are provided by [4] on the implementation of the system thus making comparison infeasible. No further literature was found by the authors regarding that work.

Table 1. Summary of traffic analysed

	Total Packets	Total MBytes
Total	573,429,697	268,543
	As percentage of Total	
TCP	94.819	98.596
ICMP	3.588	0.710
UDP	1.516	0.617
OTHER	0.077	0.077

about 1,000 researchers, administrators and technical staff. This campus is connected to the Internet via a full-duplex Gigabit Ethernet link. It was on this connection to the Internet that our monitor was placed. Traffic was monitored for a full 24 hour, week-day period and for both link directions.

Brief statistics on the traffic data collected are given in Table 1. Other protocols were observed in the trace, namely IPv6-crypt, PIM, GRE, IGMP, NARP and private encryption, but the largest of them accounted for fewer than one million packets (less than 0.06%) over the 24 hour period and the total of all OTHER protocols was fewer than one and a half million packets. All percentage values given henceforth are from the total of UDP and TCP packets only.

3   Methodology

3.1   Overview of *Content-Based* Classification

Our content-based classification scheme can be viewed as an iterative procedure whose target is to gain sufficient confidence that a particular traffic stream is caused by a specific application. To achieve such a goal our classification method operates on traffic flows and not packets. Grouping packets into flows allows for more-efficient processing of the collected information as well the acquisition of the necessary context for an appropriate identification of the network application responsible for a flow. Obviously, the first step we need to take is that of aggregating packets into flows according to their 5-tuple. In the case of TCP, additional semantics can also allow for the identification of the start and end time of the flow. The fact that we observe traffic in both directions allows classification of all nearly flows on the link. A traffic monitor on a unidirectional link can identify only those applications that use the monitored link for their datapath.

One outcome of this operation is the identification of unusual or peculiar flows — specifically *simplex* flows. These flows consist of packets exchanged between a particular port/protocol combination in only one direction between two hosts. A common cause of a simplex flow is that packets have been sent to an invalid or non-responsive destination host. The data of the simplex flows were not discarded, they were classified — commonly identified as carrying worm and

Table 2. Methods of flow identification

Identification Method	Example
I Port-based classification (only)	—
II Packet Header (including I)	*simplex* flows
III Single packet signature	Many worm/virus
IV Single packet protocol	IDENT
V Signature on the first KByte	P2P
VI first KByte Protocol	SMTP
VII Selected flow(s) Protocol	FTP
VIII (All) Flow Protocol	VNC, CVS
IX Host history	Port-scanning

virus attacks. The identification and removal of simplex flows (each flow consisting of between three and ten packets sent over a 24-hour period) allowed the number of unidentified flows that needed further processing to be significantly reduced.

The second step of our method iteratively tests flow characteristics against different criteria until sufficient certainty has been gained as to the identity of the application. Such a process consists of nine different identification sub-methods. We describe these mechanisms in the next section. Each identification sub-method is followed by the evaluation of the acquired certainty in the candidate application. Currently this is a (labour-intensive) manual process.

3.2 Identification Methods

The nine distinct identification methods applied by our scheme are listed in Table 2. Alongside each method is an example application that we could identify using this method. Each one tests a particular property of the flow attempting to obtain evidence of the identity of the causal application.

Method **I** classifies flows according to their port numbers. This method represents the state of the art and requires access only to the part in the packet header that contains the port numbers. Method **II** relies on access to the entire packet header for both traffic directions. It is this method that is able to identify simplex flows and significantly limit the number of flows that need to go through the remainder of the classification process. Methods **III** to **VIII** examine whether a flow carries a well-known signature or follows well-known protocol semantics. Such operations are accompanied by higher complexity and may require access to more than a single packet's payload. We have listed the different identification mechanisms in terms of their complexity and the amount of data they require in Figure 1. According to our experience, specific flows may be classified positively from their first packet alone. Nonetheless, other flows may need to be examined in more detail and a positive identification may be feasible

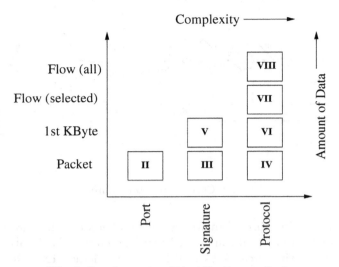

Fig. 1. Requirements of identification methods

once up to 1 KByte of their data has been observed[2]. Flows that have not been classified at this stage will require inspection of the entire flow payload and we separate such a process into two distinct steps. In the first step (Method **VII**) we perform full-flow analysis for a subset of the flows that perform a control-function. In our case FTP appeared to carry a significant amount of the overall traffic and Method **VII** was applied only to those flows that used the standard FTP control port. The control messages were parsed and further context was obtained that allowed us to classify more flows in the trace. Lastly, if there are still flows to be classified, we analyse them using specific protocol information attributing them to their causal application using Method **VIII**.

In our classification technique we will apply each identification method in turn and in such a way that the more-complex or more-data-demanding methods (as shown in Figure 1) are used only if no previous signature or protocol method has generated a match. The outcome of this process may be that (i) we have positively identified a flow to belong to a specific application, (ii) a flow appears to agree with more than one application profile, or (iii) no candidate application has been identified. In our current methodology all three cases will trigger manual intervention in order to validate the accuracy of the classification, resolve cases where multiple criteria have generated a match or inspect flows that have not matched any identification criteria. We describe our validation approach in more detail in Section 3.4.

The successful identification of specific flows caused by a particular network application reveals important information about the hosts active in our trace.

---

[2] The value of 1 KByte has been experimentally found to be an upper bound for the amount of packet information that needs to be processed for the identification of several applications making use of signatures. In future work, we intend to address the exact question of what is the necessary amount of payload one needs to capture in order to identify different types of applications.

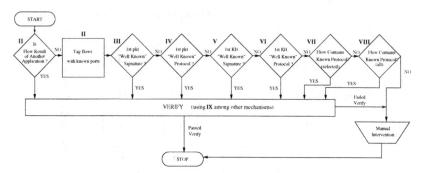

Fig. 2. Classification procedure

Our technique utilises this information to build a knowledge base for particular host/port combinations that can be used to validate future classification by testing conformance with already-observed host roles (Method **IX**). One outcome of this operation is the identification of hosts performing port scanning where a particular destination host is contacted from the same source host on many sequential port numbers. These flows evidently do not belong to a particular application (unless port scanning is part of the applications looked into). For a different set of flows, this process validated the streaming audio from a pool of machines serving a local broadcaster.

Method **IX** can be further enhanced to use information from the host name as recorded in the DNS. While we used this as a process-of-last-resort (DNS names can be notoriously un-representative), DNS names in our trace did reveal the presence of an HTTP proxy, a Mail exchange server and a VPN endpoint operating over a TCP/IP connection.

3.3   Classification Approach

An illustration of the flow through the different identification sub-methods, as employed by our approach, is shown in Figure 2. In the first step we attempt to reduce the number of flows to be further processed by using context obtained through previous iterations. Specific flows in our data can be seen as "child" connections arising from "parent" connections that precede them. One such example is a web browser that initiates multiple connections in order to retrieve parts of a single web page. Having parsed the "parent" connection allows us to immediately identify the "child" connections and classify them to the causal web application.

A second example, that has a predominant effect in our data, is passive FTP. Parsing the "parent" FTP session (Method **VIII**) allows the identification of the subsequent "child" connection that may be established toward a different host at a non-standard port. Testing whether a flow is the result of an already-classified flow at the beginning of the classification process allows for the fast characterisation of a network flow without the need to go through the remainder of the process.

If the flow is not positively identified in the first stage then it goes through several additional classification criteria. The first mechanism examines whether a flow uses a well-known port number. While port-based classification is prone to error, the port number is still a useful input into the classification process because it may convey useful information about the identity of the flow. If no well-known port is used, the classification proceeds through the next stages. However, even in the case when a flow is found to operate on a well-known port, it is tagged as well-known but still forwarded through the remainder of the classification process.

In the next stage we test whether the flow contains a known signature in its first packet. At this point we will be able to identify flows that may be directed to well-known port numbers but carry non-legitimate traffic as in the case of virus or attack traffic. Signature-scanning is a process that sees common use within Intrusion Detection Systems such as *snort* [6]. It has the advantage that a suitable scanner is often optimised for string-matching while still allowing the expression of flexible matching criteria. By scanning for signatures, applications such as web-servers operating on non-standard ports may be identified.

If no known signature has been found in the first packet we check whether the first packet of the flow conveys semantics of a well-known protocol. An example to that effect is IDENT which is a single packet IP protocol. If this test fails we look for well-known signatures in the first KByte of the flow, which may require assembly of multiple individual packets. At this stage we will be able to identify peer-to-peer traffic if it uses well known signatures. Traffic due to SMTP will have been detected from the port-based classification but only the examination of the protocol semantics within the first KByte of the flow will allow for the confident characterisation of the flow. Network protocol analysis tools, such as *ethereal* [7], employ a number of such protocol decoders and may be used to make or validate a protocol identification.

Specific flows will still remain unclassified even at this stage and will require inspection of their entire payload. This operation may be manual or automated for particular protocols. From our experience, focusing on the protocol semantics of FTP led to the identification of a very significant fraction of the overall traffic limiting the unknown traffic to less than 2%. At this point the classification procedure can end. However, if 100% accuracy is to be approached we envision that the last stage of the classification process may involve the manual inspection of all unidentified flows. This stage is rather important since it is likely to reveal new applications. While labour-intensive, the individual examination of the remaining, unidentified, flows caused the creation of a number of new signatures and protocol-templates that were then able to be used for identifying protocols such as PCAnywhere, the sdserver and CVS. This process also served to identify more task-specific systems. An example of this was a host offering protocol-specific database services.

On occasion flows may remain unclassified despite this process; this takes the form of small samples (e.g., 1–2 packets) of data that do not provide enough information to allow any classification process to proceed. These packets used unrecognised ports and rarely carried any payload. While such *background noise* was not zero in the context of classification for accounting, Quality-of-Service, or resource planning, these amounts could be considered insignificant. The actual amount of data in terms of either packets or bytes that remained unclassified represented less than 0.001% of the total.

3.4   Validation Process

Accurate classification is complicated by the unusual use to which some protocols are put. As noted earlier, the use of one protocol to carry another, such as the use of HTTP to carry peer-to-peer application traffic, will confuse a simple signature-based classification system. Additionally, the use of FTP to carry an HTTP transaction log will similarly confuse signature matching.

Due to these unusual cases the certainty of any classification appears to be a difficult task. Throughout the work presented in this paper validation was performed manually in order to approach 100% accuracy in our results. Our validation approach features several distinct methods.

Each flow is tested against multiple classification criteria. If this procedure leads to several criteria being satisfied simultaneously, manual intervention can allow for the identification of the true causal application. An example is the peer-to-peer situation. Identifying a flow as HTTP does not suggest anything more than that the flow contains HTTP signatures. After applying all classification methods we may conclude that the flow is HTTP alone, or additional signature-matching (e.g. identifying a peer-to-peer application) may indicate that the flow is the result of a peer-to-peer transfer.

If the flow classification results from a well-known protocol, then the validation approach tests the conformance of the flow to the actual protocol. An example of this procedure is the identification of FTP PASV flows. A PASV flow can be valid only if the FTP control-stream overlaps the duration of the PASV flow — such cursory, protocol-based, examination allows an invalid classification to be identified. Alongside this process, flows can be further validated against the perceived function of a host, e.g., an identified router would be valid to relay BGP whereas for a machine identified as (probably) a desktop Windows box behind a NAT, concluding it was transferring BGP is unlikely and this potentially invalid classification requires manual-intervention.

4   Results

Given the large number of identified applications, and for ease of presentation, we group applications into types according to their potential requirements from the network infrastructure. Table 3 indicates ten such classes of traffic. Importantly, the characteristics of the traffic within each category is not necessarily unique. For example, the BULK category which is made up of ftp traffic consists of both ftp control channel: data on both directions, and the ftp data channel which consists of a simplex flow of data for each object transferred.

Table 3. Network traffic allocated to each category

Classification	Example Application
BULK	ftp
DATABASE	postgres, sqlnet, oracle, ingres
INTERACTIVE	ssh, klogin, rlogin, telnet
MAIL	imap, pop2/3, smtp
SERVICES	X11, dns, ident, ldap, ntp
WWW	www
P2P	KaZaA, BitTorrent, GnuTella
MALICIOUS	Internet work and virus attacks
GAMES	Half-Life
MULTIMEDIA	Windows Media Player, Real

In Table 4 we compare the results of simple port-based classification with content-based classification. The technique of port-analysis, against which we compare our approach, is common industry practise (e.g., Cisco *NetFlow* or [1, 2]). UNKNOWN refers to applications which for port-based analysis are not readily identifiable. Notice that under the content-based classification approach we had nearly no UNKNOWN traffic; instead we have 5 new traffic-classes detected. The traffic we were not able to classify corresponds to a small number of flows. A limited number of flows provides a minimal sample of the application behavior and thus cannot allow for the confident identification of the causal application.

Table 4 shows that under the simple port-based classification scheme based upon the IANA port assignments 30% of the carried bytes cannot be attributed

Table 4. Contrasting port-based and Content-based classification

Classification Type	Port-Based		Content-Based	
	Packets	Bytes	Packets	Bytes
	As a percentage of total traffic			
BULK	46.97	45.00	65.06	64.54
DATABASE	0.03	0.03	0.84	0.76
GRID	0.03	0.07	0.00	0.00
INTERACTIVE	1.19	0.43	0.75	0.39
MAIL	3.37	3.62	3.37	3.62
SERVICES	0.07	0.02	0.29	0.28
WWW	19.98	20.40	26.49	27.30
UNKNOWN	28.36	30.43	<0.01	<0.01
MALICIOUS	—	—	1.10	1.17
IRC/CHAT	—	—	0.44	0.05
P2P	—	—	1.27	1.50
GAMES	—	—	0.17	0.18
MULTIMEDIA	—	—	0.22	0.21

to a particular application. Further observation reveals that the BULK traffic is underestimated by approximately 20% while we see a difference of 6% in the WWW traffic. However, the port-based approach does not only underestimate traffic but for some classes, e.g., INTERACTIVE applications, it may over-estimate it. This means that traffic flows can also be misidentified under the port-based technique. Lastly, applications such as peer-to-peer and mal-ware appear to contribute zero traffic in the port-based case. This is due to the port through which such protocols travel not providing a standard identification. Such port-based estimation errors are believed to be significant.

4.1   Examining Under and Over-Estimation

Of the results in Table 4 we will concentrate on only a few example situations. The first and most dominant difference is for BULK — traffic created as a result of FTP. The reason is that port-based classification will not be able to correctly identify a large class of (FTP) traffic transported using the PASV mechanism. Content-based classification is able to identify the causal relationship between the FTP control flow and any resulting data-transport. This means that traffic that was formerly either of unknown origin or incorrectly classified may be ascribed to FTP which is a traffic source that will be consistently underestimated by port-based classification.

A comparison of values for MAIL, a category consisting of the SMTP, IMAP, MAPI and POP protocols, reveals that it is estimated with surprising accuracy in both cases. Both the number of packets and bytes transferred is unchanged between the two classification techniques. We also did not find any other non-MAIL traffic present on MAIL ports. We would assert that the reason MAIL is found exclusively on the commonly defined ports, while no other MAIL transactions are found on other ports, is that MAIL must be exchanged with other sites and other hosts. MAIL relies on common, Internet-wide standards for port and protocol assignment. No single site could arbitrarily change the ports on which MAIL is exchanged without effectively cutting itself off from exchanges with other Internet sites. Therefore, MAIL is a traffic source that, for quantifying traffic exchanged with other sites at least, may be accurately estimated by port-based classification.

Despite the fact that such an effect was not pronounced in the analysed data set, port-based classification can also lead to over-estimation of the amount of traffic carried by a particular application. One reason is that mal-ware or attack traffic may use the well-known ports of a particular service, thus inflating the amount of traffic attributed to that application. In addition, if a particular application uses another application as a relay, then the traffic attributed to the latter will be inflated by the amount of traffic of the former. An example of such a case is peer-to-peer traffic using HTTP to avoid blocking by firewalls, an effect that was not present in our data. In fact, we notice that under the content-based approach we can attribute more traffic to WWW since our data included web servers operating on non-standard ports that could not be detected under the port-based approach.

Table 5. Analysis method compared against percentage of UNKNOWN and correctly identified data

Method									UNKNOWN Data %		Correctly Identified	
I	II	III	IV	V	VI	VII	VIII	IX	Packets	Bytes	Packets	Bytes
•									28.36	30.44	71.03	69.27
•	•							•	27.35	30.33	72.05	69.38
•	•	•						•	27.35	30.32	72.05	69.39
•	•	•	•					•	27.12	30.09	72.29	69.62
•	•	•	•	•				•	25.72	28.43	74.23	71.48
•	•	•	•	•	•			•	19.11	21.07	80.84	78.84
•	•	•	•	•	•	•		•	1.07	1.22	98.94	98.78
•	•	•	•	•	•	•	•	•	<0.01	<0.01	>99.99	>99.99

Clearly this work leads to an obvious question of how we *know* that our content-based method is correct. We would emphasise that it was only through the labour-intensive examining of all data-flows along with numerous exchanges with system administrators and users of the examined site that we were able to arrive at a system of sufficient accuracy. We do not consider that such a laborious process would need to be repeated for the analysis of similar traffic profiles. However, the identification of new types of applications will require a more limited examination of a future, unclassifiable anomaly.

4.2   Overheads of *Content-Based* Analysis

Alongside a presentation of the effectiveness of the content-based method we present the overheads this method incurs. For our study we were able to iterate through traffic multiple times, studying data for many months after its collection. Clearly, such a labour-intensive approach would not be suitable if it were to be used as part of real-time operator feedback.

We emphasise that while performing this work, we built a considerable body of knowledge applicable to future studies. The data collected for one monitor can be reapplied for future collections made at that location. Additionally, while specific host information may quickly become out-of-date, the techniques for identifying applications through signatures and protocol-fitting continue to be applicable. In this way historical data becomes an a-priori that can assist in the decision-making process of the characterisation for each analysis of the future.

Table 5 indicates the relationship between the complexity of analysis and the quantity of data we could positively identify — items are ordered in the table as increasing levels of complexity. The Method column refers to methods listed in Table 2 in Section 3.

Currently our method employs packet-header analysis and host-profile construction for all levels of complexity. Signature matching is easier to implement and perform than protocol matching due to its application of static string matching. Analysis that is based upon a single packet (the first packet) is inherently less complex than analysis based upon (up to) the first KByte. The first KByte may require reassembly from the payload of multiple packets. Finally, any form of flow-analysis is complicated although this will clearly reduce the overheads of analysis if the number of flows that require parsing is limited.

Table 5 clearly illustrates the accuracy achieved by applying successively-more-complicated characterisation techniques. The correctness of classification reported in Table 5 is computed by comparing the results using that method and the results using the content-based methodology. Importantly, the quantity of UNKNOWN traffic is not simply the difference between total and identified traffic. Traffic quantified as UNKNOWN has no category and does not account for traffic that is mis-classified. It may be considered the residual following each classification attempt.

Table 5 shows that port-based classification is actually capable of correctly classifying 69% of the bytes. Contrasting this value with the known traffic in Table 4 further demonstrates that the mis-identified amount of traffic is rather limited. Nonetheless, 31% of the traffic is unknown. Applying host-specific knowledge is capable of limiting the unknown traffic by less than 1% and signature and application semantics analysis based on the first packet of the flow provides an additional benefit of less than 1%. It's only after we observe up to 1 KByte of the flow that we can increase the correctly-identified traffic from approximately 70% to almost 79%. Application of mechanism **VII** can further increase this percentage to 98%. In Table 2 we have listed example applications that are correctly identified when the particular mechanism is applied.

In summary, we notice that port-based classification can lead to the positive identification of a significant amount of the carried traffic. Nonetheless, it contains errors that can be detected only through the application of a content-based technique. Our analysis shows that typically the greatest benefit of applying such a technique, unfortunately, comes from the most complicated mechanisms. If a site contains a traffic mix biased toward the harder-to-detect applications, then these inaccuracies may have even more adverse consequences.

5   Summary and Future Work

Motivated by the need for more accurate identification techniques for network applications, we presented a framework for traffic characterisation in the presence of packet payload. We laid out the principles for the correct classification of network traffic. Such principles are captured by several individual building blocks that, if applied iteratively, can provide sufficient confidence in the identity of the causal application. Our technique is not automated due to the fact that a particular Internet flow could satisfy more than one classification criterion or it could belong to an emerging application having behaviour that is not yet common knowledge.

We collected a full payload packet traces from an Internet site and compared the results of our content-based scheme against the current state of the art —

the port-based classification technique. We showed that classifying traffic based on the usage of well-known ports leads to a high amount of the overall traffic being unknown and a small amount of traffic being misclassified. We quantified these inaccuracies for the analysed packet trace.

We then presented an analysis of the accuracy-gain as a function of the complexity introduced by the different classification sub-methods. Our results show that simple port-based classification can correctly identify approximately 70% of the overall traffic. Application of increasingly complex mechanisms can approach 100% accuracy with great benefits gained even through the analysis of up to 1 KByte of a traffic flow.

Our work should be viewed as being at an early stage and the avenues for future research are multiple. One of the fundamental questions that need investigation is how such a system could be implemented for real-time operation. We would argue that an adapted version of the architecture described in [5], which currently performs on-line flow analysis as part of its protocol-parsing and feature-compression, would be a suitable system. Such an architecture overcomes the (potential) over-load of a single monitor by employing a method work-load sharing among multiple nodes. This technique incorporates dynamic load-distribution and assumes that a single flow will not overwhelm a single monitoring node. In our experience such a limitation is sufficiently flexible as to not be concerning.

We clearly need to apply our technique to other Internet locations. We need to identify how applicable our techniques are for other mixes of user traffic and when our monitoring is subject to other limitations. Examples of such limitations include having access to only unidirectional traffic or to a sample of the data. Both these situations are common for ISP core networks and for multi-homed sites. We already identify that the first phase of identification and *culling* of simplex flows would not be possible if the only data available corresponded to a single link direction.

We emphasise that application identification from traffic data is not an easy task. Simple signature matching may not prove adequate in cases where multiple classification criteria seem to be satisfied simultaneously. Validation of the candidate application for a traffic flow in an automated fashion is an open issue. Further research needs to be carried out in this direction. Moreover, we envision that as new applications appear in the Internet there will always be cases when manual intervention will be required in order to gain understanding of its nature.

Lastly, in future work we intend to address the issue of how much information needs to be accessible by a traffic classifier for the identification of different network applications. Our study has shown that in certain cases one may need access to the entire flow payload in order to arrive to the correct causal application. Nonetheless, if system limitations dictate an upper bound on the captured information, then the knowledge of the application(s) that will evade identification is essential.

A technical report describing the (manual) process we used is provided in [8].

Acknowledgments

We gratefully acknowledge the assistance of Geoff Gibbs, Tim Granger, and Ian Pratt during the course of this work. We also thank Michael Dales, Jon Crowcroft, Tim Griffin and Ralphe Neill for their feedback.

References

1. Moore, D., Keys, K., Koga, R., Lagache, E., kc Claffy: CoralReef software suite as a tool for system and network administrators. In: Proceedings of the LISA 2001 15th Systems Administration Conference. (2001)
2. Connie Logg and Les Cottrell: Characterization of the Traffic between SLAC and the Internet (2003) http://www.slac.stanford.edu/comp/net/slac-netflow/html/SLAC-netflow.html.
3. Fraleigh, C., Moon, S., Lyles, B., Cotton, C., Khan, M., Moll, D., Rockell, R., Seely, T., Diot, C.: Packet-level traffic measurements from the sprint IP backbone. IEEE Network (2003) 6–16
4. Choi, T., Kim, C., Yoon, S., Park, J., Lee, B., Kim, H., Chung, H., Jeong, T.: Content-aware Internet Application Traffic Measurement and Analysis. In: IEEE/IFIP Network Operations & Management Symposium (NOMS) 2004. (2004)
5. Moore, A., Hall, J., Kreibich, C., Harris, E., Pratt, I.: Architecture of a Network Monitor. In: Passive & Active Measurement Workshop 2003 (PAM2003). (2003)
6. Roesch, M.: Snort - Lightweight Intrusion Detection for Networks. In: USENIX 13th Systems Administration Conference — LISA '99, Seattle, WA (1999)
7. Orebaugh, A., Morris, G., Warnicke, E., Ramirez, G.: Ethereal Packet Sniffing. Syngress Publishing, Rockland, MA (2004)
8. Moore, A.: Discrete content-based classification — a data set. Technical Report, Intel Research, Cambridge (2005)

A Traffic Identification Method and Evaluations for a Pure P2P Application

Satoshi Ohzahata[1], Yoichi Hagiwara[1], Matsuaki Terada[1], and Konosuke Kawashima[1]

Tokyo University of Agriculture and Technology, 2-24-16 Nakacho Koganei City
Tokyo 184–8588, Japan
{ohzahata, hagi, m-tera, k-kawa}@cc.tuat.ac.jp

Abstract. Pure P2P applications are widely used nowadays as a file sharing system. In the overlay networks, music and video files are the main items exchanged, and it is known that the traffic volume is much larger than that of classical client/server applications. However, the current status of the P2P application traffic is not well known because of their anonymous communication architectures. In particular, in cases where the application does not use the default service port, and the communication route and the shared file are also encrypted, the identification traffic has not been feasible. To solve this problem, we have developed an identification method for pure Peer-to-Peer communication applications, especially for traffic for Winny, the most popular Peer-to-Peer application in Japan, by using server/client relationships among the peers. We will give some evaluation results for our proposed identification method.

1 Introduction

The Internet applications of end users are changing with the spread of high-performance PCs connected, with broadband links, through the Internet. The traffic volume is also increasing drastically increasing with the change in applications. In particular, the number of users of Peer-to-Peer (P2P) network applications is increasing rapidly since the users are easily able to use network resources over the overlay networks.

The characteristic feature of a pure P2P network is that it is a distributed autonomous system which does not rely on a specific server for communications.

Because of this fact, such systems are expected to exhibit scalability in processing power and load balancing at the end computers. However, the traffic volume is becoming much larger than that of the previous Internet applications and the bottlenecks in processing power are shifting from the end computers to the network. In addition, traffic control is very difficult because there is no administrator in the overlay networks and on account of the anonymous nature of the traffic.

Consequently, we need to estimate the effect of P2P traffic to on other forms of traffic in order to construct networks and manage them appropriately. When

C. Dovrolis (Ed.): PAM 2005, LNCS 3431, pp. 55–68, 2005.

we start evaluating the P2P traffic, we need first to identify the P2P traffic in the total Internet traffic. Much researches has been done to identify the application traffic and evaluate its characteristics.

The service port number in TCP or UDP is often used as a method of identifying the application traffic, since major Internet applications have use their well known service ports (0–1023) and the server has to use the TCP or UDP port number as the identification number [1]. If the identification number is used correctly by all applications, we can easily identify the application traffic.

Many P2P applications also have their default service port number, Gnutella [2] (6346, 6347), Kazaa [3] (1214), BitTorrent [4] (6881–6889) and so on. In consequence, many research studies for P2P traffic use the default service port number identification methods in [5], [6] and [7]. However, some recent P2P applications, WinMX [8] and Winny [9], do not use a default service port number, which would allow their services to be identified. For these applications, this identification method does not work well.

Signature matching identification methods [10], [11] are effective when the applications exchange the specific characters in the payload of packets. This traffic identification method is widely applied for Intrusion Detection Systems (IDS) [12], [13] to manage traffic. In this method, every packet needs to be analyzed and it requires huge computation power. In [14], the authors propose a scalable signature matching identification system for P2P traffic, and compare identification methods their application level signature matching method with the default service port number identification method. In these signature matching methods, the application signatures need to be updated with changing the application protocols.

There is further difficult problem in the case of Winny, which is one of the most popular pure P2P file sharing application in Japan. The payloads of the packets are encrypted and the protocol details are also not disclosed. These facts make it difficult to identify the Winny traffic since the signature matching is not also useful for the an encrypted payload.

This paper proposes an improved default service port number identification method specifically designed for pure P2P application traffic, to address the above problems. In our method, the service port number may be identified even in cases where the pure P2P applications do not use its their default service port numbers. In the Internet communication, each connection is identified by a tuple of the IP addresses, port numbers and a protocol number (TCP or UDP), and many classical Internet applications play function only as a server or client in the communications. In the classical client/server application, only one connection or relationship is used between the entities involved in the communication. Pure P2P peers, however, play function as both server and client between the peers, and two kinds of connection need to be established. In our method, the pure P2P traffic is identified by the patterns of connection to the server/client ports among the communicating entities. To realize our proposed method, we adopt active measurement and passive measurement for the pure P2P traffic. With the combination of the measurement logs, the service port of a peer is identified

through a series of steps. We adopt have apply the proposed method to the
Winny network and evaluate our proposed identification method.

The rest of this paper is structured as follows. In section 2 we briefly describe
the P2P application of Winny. Section 3 we describes our traffic measurement
method. In Section 4, describes our proposed traffic identification method, and
section 5 provides conclusions.

2 About Winny

In Japan, the famous well-known P2P file exchange applications, KaZaa [3],
emule [15] and BitTorrent [4], are rarely used since they cannot deal with
Japanese language characters in the key words for file searching or the file name.
WinMX [8] is the first P2P application for which is available Japanese language
characters may be used, though the application of a by applying the patch, and
so WinMX was the main file sharing application used until the appearance of
Winny.

Currently, Winny is one of the most popular P2P file sharing application in
Japan, since it was developed in Japan and the has a freenet-like [16] anonymous
architecture. About 200,000 peers always compose the Winny network, and between June 13 and October 23, 2004, we measured over 4,000,000 unique joined
peers which had a unique tuple of IP address and the service port from June 13
to October 23, 2004. (The measurement point is at Point A in Figure 1.)

Winny is a pure P2P application and does not depend on any central server
or super peer for file searching and sharing. The communication is encrypted,
the service port is different in each peer and the protocol is also not an open
one. This architecture makes it difficult for a network administrator to identify
the traffic. Winny is not a file exchange application but rather a file sharing
application. By setting some keywords for the file name, the peer always collects
the matching files. The file transfer technique also strengthens the anonymous
communication architecture. When a file is shared between two peers, the file
holder should transfer the file to the file receiver via an intermediate peer, so that
the two exchanging entities never know each other directly. In the intermediate
peer, the transferred file is locally cached as an encrypted one. A duplicate of the
file is also distributed to the receiving peer as the duplicate one. This architecture
results in an enormous volume of traffic as the size of shared files increases size.
In the Winny network, the shared files are mainly video files (mpeg, avi and
DVD ISO image) and the average shared file size is around 1GB. The Winny
network is composed of three kinds of networks/links, as described below.

1. Adjacent peer check/search network.
2. File search network.
3. File exchange network.

1. When a new Winny peer joins the Winny network, the peer needs to obtain
a pairs of data items, that is the IP address and service port number, of the other
peers which haves already joined the network. Then the new peer establishes and

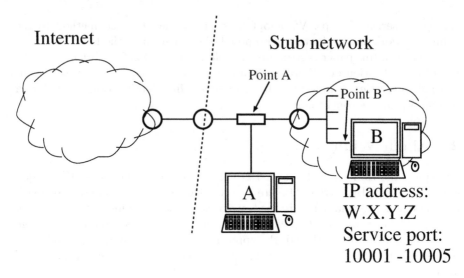

Fig. 1. Traffic measurement points

keeps maintains the links up to several hundred peers. The network comprises a nearly random network.

2. Some of the links are selected as the file search network. In this network, information is exchanged regarding who has a specified file and who wants it.

3. When file exchange conditions are satisfied, a separate route is established via the other file sharing entity.

In these networks there is a server/client relationship between two peers and many such networks are characterized by many access requests between peersaccesses are always required among the peers to keep these networks. These characteristics of networks give us hints as to how to identify Winny traffic.

3   Measurement Methods

We adopt a combination of active and passive measurements to identify Winny traffic, and have two measurement points, as shown in Figure 1. At Point A, the back-bone traffic is measured with by means of a passive measurement, while Point B is placed inside the stub network and measures Winny traffic by acting as a decoy peer with an active measurement.

The Point A (which collects data in log A) is in a switching hub which is placed between an edge router of the Internet and an edge router of the stub network. The link speed is the 100Mbps of full duplex Ethernet but the transfer speed is restricted to 10Mbps at the edge router of the Internet for both directions. We can measure the traffic in the switching hub without affecting the backbone traffic itself by port mirroring. We measured the traffic for 24 hours, from 0:00–24:00 on January 11, 2005 and found 2461 unique IP addresses of the stub network in the traffic log. The combined total traffic volume for both directions was 166.1 GB.

We only logged information of the IP and TCP headers of all these packets. We obtained limited information from the log, but we can reduce the log size and still obtain enough information from it. We define a flow, in the following, as a connection which has the same tuple of IP addresses, port numbers and a protocol number (TCP) between the packet containing the SYN segment flag and the that containing the FIN segment flag. In the measurement, some flows were not evaluated since these flows had no SYN or FYN packet flag of packet in the log. These flows are ignored in the evaluations.

At the point B (log B), the network speed is the 100Mbps of full duplex Ethernet. We measured the traffic log for 13 days, from 0:00 January 5 to 0:00 January 17, 2005. We were able to directly measure the access log from/to the peers in the Winny network at the Point B because the PC B belongs to the Winny network, acting as a decoy peer. By repeatedly changing the point of connection to the Winny network at short intervals, we were able to collect about 40,000 of unique pairs of IP address and service port of the Winny peers per one day by using 5 decoy peers. We used a different measuring period of the log for each analysis.

Both traffic logs are necessary for our traffic identification method, and the log A is used for the back-bone traffic evaluations. The detailed specifications of the PCs are below.

[Point A: for the Back-bone traffic]

- The PC is a Dell PRECISION 450 with dual Xeon 3.2Ghz CPUs and the main memory size is 2GB. The OS is FreeBSD.
- The traffic is measured by Snort version 2.0.

[Point B: for the Decoy peer traffic]

- The PC is a Dell PRECISION 450 with dual Xeon 3.2Ghz CPUs and the main memory size is 2GB. The OS is Windows XP professional.
- The version of the Winny is Winny2β6.6.
- We run 5 Winny programs in parallel in the PC in each user session , and the service port numbers is assigned are 10001–10005, respectively.
- Safeny [17] is used to disconnect all connections to/from the decoy peer after 10 seconds.
- All connections to the service port numbers are disconnected by a firewall in the PC B so as not to transfer any files to the Winny network.
- The traffic is measured by Snort version 2.0.

4   Proposed Identification Method for Pure P2P Traffic

Traditional Internet applications, WWW, FTP, E-mail, etc, are based on the client/server computing model. In thise computing model, each of the communication entitiesy is categorized by only one of the two roles, a server or a client.

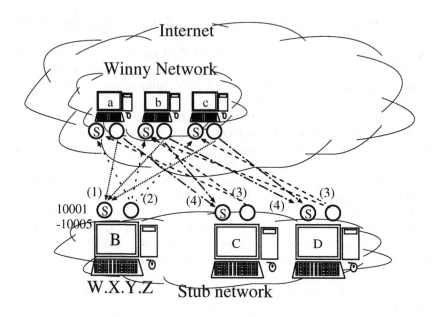

Fig. 2. Procedures of our proposed identification method 1

The server computer only supplies its service and the client computer only receives the service. When the communications start, the client computer accesses the service port of server computer with using its client port, and the server serves provides its service over the connection. Thus, only one identification of the connection between them identification (a tuple of source/destination IP addresses, source/destination port numbers and protocol number) between them is used in the communications. In pure P2P communications, however, one peer plays acts as both a server and a client simultaneously in the communications simultaneously, and there are two kinds of connections between the peers during their communications. Our traffic identification method focuses on the access relations to the ports among the peers.

4.1   Proposed Identification Method 1

The basic idea of our proposed identification method is a decoy peer which collects all pairs of IP addresses and service ports of the Winny peers. However, collecting all of these is difficult because of the restricted search capacity of the decoy peers. Therefore, we need to find the missing Winny peers by using a server/client relationship between the peers.

Figure 2 shows the procedures for our proposed identification method. We place the Peer B as a decoy peer in the stub network whose IP address is W.X.Y.Z and the service port numbers are assigned as 10001–10005 for each decoy peer. The service ports and client ports are depicted by the circles beside the PCs. As soon as the decoy peer joins the Winny network, the peers in the Winny network access the decoy peer and the decoy peer continuously accesses the peers in the Internet to configure the overlay networks. Each arrow corresponds to a connection made by one peer to another peers service port. Thus, these accesses are measured in the PCs A and B, as shown in Figure 1. The procedures are described below.

First we identify the service port number and IP address of the Winny peers connected to the Internet. In the procedures (1) and (2), only log B is used.

(1) When the decoy peer B joins the Winny network, some of the Winny peers in the Internet access the service port of the decoy peer B. The accesses come from the client port, and we can only identify the IP address of the Winny peers. In this connection, the decoy peer B functions as the server. We add the IP addresses to database α (this applies peer a IP, peer b IP and peer c IP).

(2) Using its client port, the decoy peer B accesses the service ports of the Winny peers in the Internet. If the decoy peer B access the peers in database α, we can identify the service port and IP address of the Winny peers (including peers in the stub network). We add the IP addresses and service ports to database β (this applies peer a IP:service port number, peer b IP:service port number and peer c IP:service port number). In this connection, the peer B functions as a client then the two relations are established between the two peers.

Next, we identify the IP address and the service port number of the Winny users in the stub network, and define "Winny" and "Port 0" peers in the following procedures. Port 0 setting is originally prepared for the peers which are behind the firewall or NAT, but many of the Port 0 users use the setting not to upload any files to the other peers. This is because many shared files in the Winny network are illegal and these files are also automatically shared. In addition, in most cases a file is transferred via a "Winny" peer, and then such "Winny" peers will unintentionally upload and cache these illegal files.

The procedures (3) and (4) use log A and database β.

(3) In the case of a node inside the stub network which accesses a service port of a peer in database β, the node access has the capability of a Winny peer. However, we define a Winny peer in the stub network as a peer which accesses more than two peers in database β, to improve the identification probability. In addition, in this case, the access is initiated by a node inside the stub network, and the accessed port is a service port of the peer. Then, we find that the source IP address node is a Winny peer and add its IP address to database γ (this applies to peer C IP and peer D IP).

(4) The Winny peers in database β access the service ports of peers in database γ. If more than two peers in database β access an identical IP address (database γ) and port number in the stub network of IP and service port number may be identified. We define the peer as "Winny" in the stub network, for the following description, and add the peers to database δ (peer C IP:service

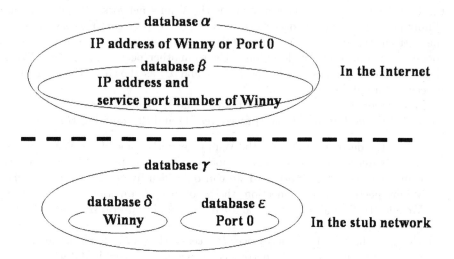

Fig. 3. Relationship between databases of Winny peers

port number and peer D IP:service port number). However, some peers in database β do not return to the peers in database γ by using their client ports.

This is because some Winny peers do not prepare their service port in their setting. We call these peers in the stub network, which do not open their service port to the Winny networks, "Port 0" in the following, and add their IP addresses to the database ϵ.

In the identification procedures (3) and (4), the peers may not be Winny peers, but the probability of this is very low. This is because the value is factorial of the number of port in the TCP or UDP header (less than $1/65536^2$) and our method also considers the port access direction.

From these procedures, (1)–(4), we can find the IP addresses and service ports of Winny peers in the Internet and the stub network. Figure 3 shows the relationship of the various databases of Winny peers. Using databases β, γ and ϵ, we can select the Winny and Port 0 traffic from the log A with this improved port number based application traffic identification method.

4.2   Proposed Identification Method 2

By extending the identification method proposed in the previous subsection, we can find new Winny peers one after another (Figure 4). In the following procedures, the service ports of Winny peers in the stub network play the same role as the decoy peer in the previous subsection. These procedures are described in below.

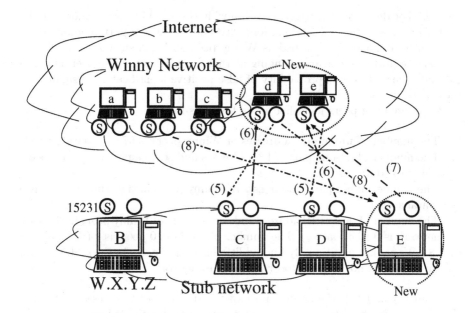

Fig. 4. Procedures of our proposed identification method 2

(5) When the peers d accesses to the service ports of peers C and D, which are not in database α, the peer d becomes newly found Winny peers. We add its IP addresses to database α.

(6) When peers C or D uses their client ports to access peer d, the service port of peer d is identified and the information is added to the database β . (Peer e is also found by the same procedures.)

(7) From inside the stub network, peer E, which is not in databases δ and ϵ, accesses the newly found peers of these service ports (peers d and e), and so we identify their IP address and add them to database γ.

(8) If more than two peers in database β use their client ports to access peer E, the IP address and service port of the peer in the stub network are identified and the peers added to database δ. The "Port 0" peers are also found in database γ and we add them to database ϵ.

By repeating above procedures, we can eventually find new Winny peers even if these peers are not found in the first few implementations of the procedures.

The next section shows the results of analysis of our proposed methods.

5   Analysis Results

5.1   Analysis Results 1

To identify Winny peers in the stub network, we used two traffic logs in section 3. First, we determined the same 24-hour measurement period for the point A and point B. The decoy peer is logged as "Winny" in the log A but we exclude it in the analysis results of this subsection. Some "IP address:service port" combinations in log B have not been identified as Winny peers in log A since a node address and service port number are changing at that time. However, our identification method ensures that the probability of false positive identification is small with these procedures.

The number of peers identified in each step is follows.

(1) The number of unique IP addresses of Winny peers is 67,984 (database α).
(2) The number of unique pair of IP addresses and service port of Winny peers is 45,873 (database β).
(3) The number of unique IP addresses of Winny peers in the stub network is 9 (database γ).
(4) The number of unique IP address and service port of Winny peers in the stub network is 0 (database δ). The number of the Port 0 peers is 9 (database ϵ).
(5)–(8) We cannot additionally find additional Winny peers in the stub network since there is no "Winny" peer in the stub network.

From (1) and (2), the service port of the decoy peer is accessed by many Winny peers in the Internet, when the decoy peer joins the Winny network. In the default setting of Winny, each peer has a few active file search connections to the other peers, but each the peer previously searches for further connectable peers to maintain the file search network. With these procedures, several hundred peer search connections are always maintained by keeping, in each Winny peer, the IP:service port number of other Winny peers (the default upper limit is 600.).

The number of IP addresses in (2) is lower than that in (1) since the information of (1) comes from the connection of the service port of the decoy peer but that in (2) is from the connection of the client port of the decoy peer into the result of (1). This fact depends on the search capacity of the decoy peers and the number of Port 0 peers, which do not have their own service port.

In (3), the Winny peers in the stub network (database γ) access 1–3216 peers of the service port in database β in Table 1. In our procedures, peers D and F are not identified as Winny peers since there are only one access to these peers. Since a Winny peer regularly accesses to the service port of the other peers to maintain the peer search connections, this identification method works well. However, we cannot find "Winny" in the definition (4). No connection between the service port of a Winny peer in the stub network and the client port of a Winny peer in the Internet is ever established. This is because that Winny users in the stub network will not upload any file to the other Winny peers.

Next we investigate the effect of the measurement period for log B. The number of nodes identified does not vary but the number of identified flows is different. A longer measurement period finds many IP addresses and service ports of Winny peers in the Internet (database β), and many flows are also identified. Comparing (e) with (h), (h) gave better results in spite of the fact that the number of peers in database β is almost the same because earlyer logged peers

Table 1. Number of accesses to database β peers from the stub network per day

Suspected peer	A	B	C	D	E	F	G	H	I	J	K
Number of accesses	2446	623	166	1	1626	1	2753	2122	3216	3027	2899

Table 2. Relationship between log period at measurement point B and identified flows

Measurement period	Databaseβ	Identified peers	Identified flows	Av. flow size
(a) 0:00 Jan. 11 – 0:00 Jan. 12	45873	9	111064	32.4KB
(b) 0:00 Jan. 10 – 0:00 Jan. 11	48434	9	57872	78.1KB
(c) 0:00 Jan. 9 – 0:00 Jan. 11	84129	9	71525	67.5KB
(d) 0:00 Jan. 8 – 0:00 Jan. 11	114715	9	78074	64.1KB
(e) 0:00 Jan. 4 – 0:00 Jan. 11	215557	10	87450	61.5KB
(f) 12:00 Jan. 10 – 12:00 Jan. 12	84129	9	120042	46.7KB
(g) 0:00 Jan. 10 – 0:00 Jan. 12	110097	9	123589	48.2KB
(h) 0:00 Jan. 8 – 0:00 Jan. 14	213968	9	128486	47.7KB

were not joining the Winny network during the measurement period of the log A. However, the differences in the number of identified flows between (g) and (h) is small. This fact will depend on the connection period of each peer. In (a), (g) and (h), the average flow size becomes small, because the additionally identified flows are used for composing the Adjacent peer check/search network.

5.2   Analysis Results 2

As shown in the previous subsection, we found that there are no "Winny" peers in the stub network except for the 5 decoy peers. We do have not shown whether the proposed identification method 2 works effectively and need to evaluate whether it does. We ran 5 Winny applications in the PC in parallel and treated all the 5 peers as decoy peers in the previous subsection. However, for the next set of evaluations we treated one of the decoy peers as the decoy peer, and the others 4 peers is as general "Winny" peers in the following evaluations. By With these analyses, we can find one decoy peer and 4 "Winny" peers in the stub network. These 5 "Winny" peers are all treated as "Winny" in this subsection. Note that the 5 decoy peers run on one PC and the relationships in Figure 3 are different for the following results.

First, we determined the same 24-hour measurement period for point A and point B. The number of identified peers in each step is as follows.

(1) The number of unique IP addresses of Winny peers is 19136 (database α).
(2) The number of unique pairs of IP addresses and service ports of Winny peers is 13791 (database β).
(3) The number of unique IP addresses of Winny peers in the stub network is 11 (database γ).

Table 3. Relation between the number of peers in databaseβ and the identified peers

No.of peer in databaseβ	10	100	1000	one docoy peer
Database α (old→updated)	$- \to 39133$	$- \to 46264$	$- \to 46264$	$19136 \to 51365$
Database β (old→updated)	$10 \to 26740$	$100 \to 31211$	$1000 \to 31263$	$13791 \to 34868$
Database γ	5	8	10	10
Database δ	4	5	5	5
Database ϵ	4	7	9	9

(4) The number of unique IP address and service port of Winny peers in the stub network is 5 (database δ). The number of the Port 0 peers is 9 (database ϵ).

(5) The IP addresses of an additional 32229 peers are found and added to database α by the Winny peer.

(6) The database β is also updated and the number of peers is 34868.

(7)–(8) All the "Winny" and "Port 0" peers in the stub network were identified in the step (4): no additional peers are found in this step.

These "Winny" peers accesses the other peers much more than normal Winny peers but we can show our procedures work well from this result.

Next, we investigate the effect of the size of database β. We change the size from 10, 100, 1000 and 13791. These peers are the first accessed peers by the decoy peer (one of the five decoy peers) from 0:00 Jan. 11. Table3 shows that the relation between the numbers of peers in database β and the identified peers in each database. In the case of 10 or 100 peers in database β, some Winny peers are not identified. But, when 1000 peers were used, the results are almost same as the "one decoy peer" case. This means that if there are many "Winny" peers in the stub network, our identification performance will be improved.

6   Conclusion

We have proposed an identification method for pure P2P traffic, Winny, and evaluated its the basic characteristics of it. Using the a decoy node, we identified the IP address and service port of Winny peers and can select the identified IP and service port number in the traffic log of the back-bone. Our identification method will be effective for pure P2P applications which will appears in the future since our methodits depends on the basic relationships among in client/server computing in the Internet applications.

In the a stub network, the number of Winny users is small. We may not find "Winny" traffic since the Winny users in the stub network are use Port 0. We only a collect traffic log from the other stub networks which haves many Winny users, even if search capacity of the decoy peer is current one, characteristics of the traffic will be much clearly analyzed. The introduced identification method is one of thean example, and we should improve the method with by analyzing the access patterns among the peers. Our identification method depends on the access number of accesses of the decoy peers from by peers in the Winny networks and the number of users in the stub network. ThenAs a result, some flows may not be identified by our method. If we prepare many decoy peers or there are many users in the stub network, our method improves the identification performance of our method improves.

When we control traffic, we should need know the status and deal themmanage it in real time. Our proposed procedure will require this improvement for the usageapplication.

Acknowledgments

The authors wish to express their gratitude to Mr. Hideaki Suzuki of the Tokyo University of Agriculture and Technology for their support on traffic analysis support. The authors also thank the anonymous reviewers for their useful comments and advice to improve this paper. This research is partly supported by Grants-in-Aid for Scientific Research (KAKENHI), No. 15500032.

References

1. M. St. Johns and G. Huston, "Considerations on the use of a Service Identifier in Packet Headers," RFC 3639, 2003.
2. Gnutella, "http://www.gnutella.com/"
3. Kazaa, "http://www.kazaa.com/"
4. BitTorrent Protocol, "http://bitconjurer.org/BitTorrent"
5. S. Saroiu, P. Gummadi and S. D. Gribble, "Measurement study of peer-to-peer file sharing systems," *Multimedia Computing and Networking 2002,* 2002.
6. S. Sen and J. Wang, "Analyzing Peer-To-Peer Traffic Across Large Networks," *IEEE/ACM Trans. on Networking,* Vol. 12, No. 2, pp. 219–232, 2004.
7. M. Kim, H. Kang and J. W. Hong, "Towards Peer-to-Peer Traffic Analysis Using Flows," *Proc. of 14h IFIP/IEEE Workshop Distributed Systems: Operations and Management,* 2003.
8. WinMX, "http://www.winmx.com/"
9. Winny, "http://www.nynode.info/"
10. C. Dewes, A. Wichmann and A. Feldmann, "An Analysis of Internet Chat Systems," *Proc. of ACM SIGCOMM Internet Measurement Workshop 2003,* pp. 51–64, 2003.
11. K. P. Gummadi, R. J. Dunn and S. Saroiu, "Measurement, Modeling and Analysis of a Peer-to-Peer File-Sharing Workload," *Proc. of ACM SOSP'03 2003,* pp. 314–329, 2003.
12. Snort, "http://www.snort.org/"
13. P. Barford, J. Kline, D. Plonka and A. Ron, "A Signal Analysis of Network Traffic Anomalies," *Proc. of ACM IMW'02,* pp. 71–82, 2002.
14. S. Sen O. Spatscheck and D. Wang, "Accurate, Scalable In-Network Identification of P2P Traffic Using Application Signatures," *Proc. of ACM WWW'04,* 2004.

15. K. Tutscheku, "A Measurement-based Traffic Profile of the eDonkey Filesharing Service," Proc. of PAM'04, 2004.
16. I. Clarke et al, "Freenet: A Distributied Anonymous Information Strage and Retrieval Systems," Proc of ICSI Workshop on Design Issues in Anonymity and Unobsenvability, Springer-Verlag, LNCS 2009, pp. 46–66, 2001.
17. Safeny, "http://www.geocities.co.jp/SiliconValley-SanJose/7063/"

Analysis of Peer-to-Peer Traffic on ADSL[*]

Louis Plissonneau, Jean-Laurent Costeux, and Patrick Brown

France Telecom R&D,
905, rue Albert Einstein,
06921 Sophia-Antipolis Cedex - France
{louis.plissonneau, jeanlaurent.costeux}@francetelecom.com

Abstract. Peer-to-Peer (P2P) applications now generate the majority
of Internet traffic, particularly for users on ADSL because of flatrate
tarification. In this study, we focus on four popular P2P systems to
characterize the utilization, the performance and the evolution of P2P
traffic in general. We observe and compare the influence of each P2P
application over the traffic, and we evaluate the evolution of these P2P
systems over a year. Our analysis is based on ADSL traffic captured at
TCP level on a Broadband Access Server comprising thousands of users.
Thus, we characterize the P2P traffic and users, and we draw interesting
results on connectivity and cooperation between peers, localization of
sources, termination of connections and performance limitations. The
evolution of the traffic over the year allows us to see the dynamics of
the use of P2P systems. The difference between week days and week-end
days informs us about the behavior of P2P users.

1 Introduction

This study is based on TCP captures on ADSL, which are used to establish
general characteristics of P2P systems. The fact that we take into account only
ADSL traffic is important, because these users are predominantly present in P2P
traffic. Indeed 24 hours per day, unlimited connection is proposed by ISPs to the
ADSL customers. And as we shall see, P2P file sharing systems thus account for
more than 60% of the total ADSL traffic.

The originality of our measures lies in the fact that, firstly, we analyze all
the TCP flows of a regional ADSL concentrating point, secondly, we observe
only ADSL traffic (excluding modems 56k) which is more representative of P2P
utilization, and thirdly, the data collected is representative of general ADSL
users and not restricted to a specific class of users or hosts (*e.g.* a University or
a private network). Furthermore, our data include several thousands of users.

We shall differentiate systematically between the P2P users with the help of
a unique ADSL user identification. As noticed in [7], an analysis based on IP addresses can have a negative influence on the interpretation of the traces because

---

[*] This work is partly supported by project Métropolis of RNRT (French Network for
Research in Telecommunications).

C. Dovrolis (Ed.): PAM 2005, LNCS 3431, pp. 69–82, 2005.

of NATs (Network Address Translator) and dynamic IP addresses. Indeed, we noticed a significant qualitative difference between graphs based on IP addresses and those based on ADSL users.

In this study, we compare four popular P2P networks: eDonkey [2], BitTorrent [1], FastTrack and WinMX [3].

The evolution over a year shows that the popularity of P2P networks is very volatile, this popularity is also very dependent on the country.

Our flow level analysis enables us to describe volumetric properties, connection duration, traffic pattern over time, host connectivity and geographical location of peers. Then we map some of these experimental distributions into classical statistical laws. Our packet level analysis allows us to clearly identify beginning and termination of connections leading to some findings on performance limitations. We mention here two interesting results:

- about 40% of connections are only connection reattempts, and it concerns about 30% of peers;
- there are two main classes of peers: those contributing to most of the traffic volumes, and the other. The first class affects strongly the main characteristics of the P2P system, while the second one softly influences these characteristics.

The remainder of the paper is organized as follows: Section 2 details the methodology for our measurements. In Section 3, we elaborate some relevant characteristics of P2P traffic, such as proportion of signaling traffic, comparison of upstream and downstream volumes, connection duration, traffic pattern over time, geographical distribution of peers and termination of connections. Section 4 deals with the number of connections a peer establishes. We summarize the main results and conclude the paper in Section 5.

2   Capture Methodology and P2P Overview

2.1   Measurement Details

First of all, we detail our experimentation protocol. As shown in Figure 1, the BAS[1] collects the traffic issued from the DSLAM[2] before forwarding it through the POP[3] to the France Telecom IP backbone. Our probe is located between a BAS and the IP backbone. We draw attention to the fact that we capture all TCP packets without any sampling or loss. We perform an analysis of the traffic over week days and week-end days of September 2004 and we compare these results with those computed over data recorded one year ago (in June 2003).

The identification of P2P protocols is done through a port analysis: a connection is classified as a P2P protocol if one of its TCP ports is a standard port of this protocol. We shall discuss the accuracy of this method in Section 2.2.

---

[1] Broadband Access Server.
[2] Digital Subscriber Line Access Multiplexer.
[3] Point-Of-Presence.

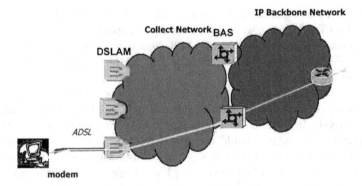

Fig. 1. ADSL architecture

Table 1. Distribution of protocol traffic over P2P traffic

Protocol	June 2003		September 2004	
	Volume	# Connections	Volume	# Connections
eDonkey	84%	96%	91%	93%
BitTorrent	0.8%	0.009%	6%	2.7%
Gnutella	0.8%	0.9%	1%	3.6%
WinMX	1.3%	0.06%	1%	0.08%
FastTrack	12%	1.8%	1%	0.01%
other protocols	1.1%	1.2%	0%	0.6%

We shall denote as *local* peers or users the ADSL hosts connected to the observed BAS, and as *non-local* or *distant* peers the remainder of the hosts. The *upstream traffic* will represent the packets transferred from local peers to the backbone, and the *downstream traffic* those transferred from the backbone to the local peers.

2.2   Overall P2P Overview

In our data, about 60% of the traffic lies on P2P ports in September 2004. It represents a small drop compared to the proportion of traffic on P2P ports in June 2003 which was about 65%.

In Table 1, we reported the distribution of the main P2P protocols over P2P traffic. In September 2004, eDonkey is by far the most popular protocol in terms of volume, BitTorrent is the second most popular and all the other protocols are almost negligible in volume as compared to eDonkey.

The popularity of each P2P file sharing system is very variable among location and time. According to [11] in October 2003, in Europe, eDonkey is overwhelmingly popular whereas in U.S. FastTrack is the most popular followed by WinMX. The evolution over time on our data shows that FastTrack lost its popularity in France (more than a year back, in June 2003, the proportion of volume of FastTrack traffic was the second most important).

In the remainder of the paper, we shall discuss only the protocols eDonkey, BitTorrent, FastTrack and WinMX, because of their popularity and the diversity of their working processes.

As reported by Karagiannis *et al.* in [9] and [10], some of P2P traffic might use non-standard port numbers so that we miss some traffic by restricting ourselves to a port analysis. In [12], Sen *et al.* reported that an identification of P2P traffic using application signatures could increase threefold the volume compared to a port based identification. But in this study, we remark that on the one hand, only Kazaa (using the FastTrack network) has a huge hidden traffic, and on the other hand, eDonkey and BitTorrent peers use mainly standard ports. Indeed, on the FastTrack network there is no limitation based on the port used by the P2P application, and some users (in fact many users) might change it. But on eDonkey network, the peers running their application on non-standard port receive a *Low ID* when they connect to an eDonkey server while other peers get a *High ID*. The High ID peers have no restrictions while Low ID peers can only download from High ID peers, so that eDonkey peers are strongly encouraged not to change the port number of their application. As we shall see, the main part of P2P traffic in France is on eDonkey network, and the port based identification of P2P protocols is relevant in this situation.

3   Characteristics of P2P Traffic

3.1   Signaling Traffic

P2P traffic can be split into two parts:

- the traffic generated strictly for downloading data,
- the traffic generated for maintaining the network and performing queries, that we shall denote as *signaling traffic*.

We separate these two kinds of traffic according to a threshold of the volume transmitted by each connection. In Figure 2, we plot the cumulative distribution function of the volume of connections for each P2P protocol. Note that Figure 2 *(a)* informs us on the frequency of a connection size, whereas Fig 2 *(b)* indicates the percentage of volume generated by the connections.

We choose a threshold of 20 kbytes for signalling traffic according to Figure 2 *(a)*. A direct identification of signalling connections, as in [14], leads to an average size of non-download streams of 16.7 kbytes, which is coherent with our data.

As also observed in [13] and [6], the overwhelming part (more than 90%) of P2P connections consists of signaling ones whereas they represent only a small proportion of the volume transferred: eDonkey has the biggest proportion of volume for signaling traffic with 6% (see Figure 2 *(b)*).

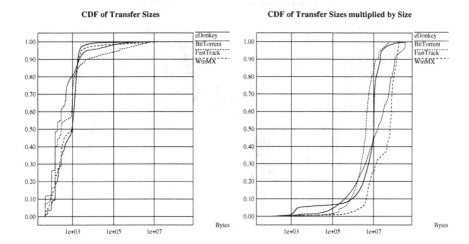

Fig. 2. Volume of P2P Connections

3.2   Upstream *Versus* Downstream Volumes

User-Based Comparison. In our data, the amount of downstream traffic is larger than the upstream traffic in terms of volume for each protocol. It means that local peers (*i.e.* several thousands of users) tend to download more than they upload on our observation point. This is a consequence of ADSL, which tends to offer much lower outbound than inbound capacity, whereas overall download and upload rates must be equal in a P2P file-sharing system .

On Figure 3, we have plotted a cloud of points representing for each eDon- key user its downstream volume versus its upstream one. On the figure we can identify the users contributing to big volumes (on the upper right corner), those generating small volumes (above the diagonal, in the middle), and a certain num- ber of peers having no upstream volume whereas they have some downstream one (to be explained in section 3.6).

We analyze the behavior of these two kinds of users:

- peers having small volume, download files but upload very few, thus they can have downstream-to-upstream ratio up to 1000;
- peers contributing to big volumes have comparable upstream and down- stream volumes with a downstream-to-upstream ratio of about 1.2.

This is the case of eDonkey peers which represents the vast majority of peers, and other P2P users have similar trends. The first class of users share few files, or disconnect themselves after download. And the second class of users have to stay connected to the P2P system for long periods to obtain high downstream volume, thus they share files (at least those being downloaded).

By recalling that less than 10% of the users contribute to the most significant part (98%) of the traffic (see also [13]), we can see that the non-cooperative behavior of small volume users doesn't disturb the balance of P2P system. The

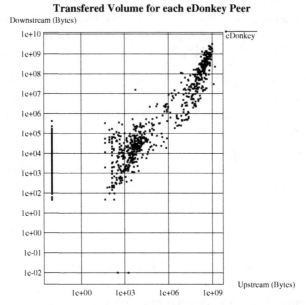

Fig. 3. Downstream volume versus upstream volume for each eDonkey peer

downstream-to-upstream ratio for overall P2P traffic is about 1.2 over our pool of users, whereas the mean ratio is of 38.

In our analysis, we identified that about 20% of the peers are probably *free-riders* (*i.e.* peers that do not share files) over the eDonkey network.

In [4] and [5], the number of *free-riders* on Gnutella network is evaluated at 70% and 42% in 2000 and 2001 respectively. Mechanisms like multi-part download (now used by most P2P applications) allow peers to share already downloaded file chunks. This explains the reduction of the number of peers who do not share data compared to previous studies.

Connection-Based Comparison. Most of the connections bring in a very small proportion of the volume of each P2P network. Indeed, the connections transferring less than 100 kbytes represent less than 8% of the traffic volume, whereas they account for more than 90% of the connections (see Figure 2).

We explain this overwhelming number of small connections as follows:

- signalization generates a lot of small transfers,
- many transfers are interrupted,
- many peers attempt to connect to offline peers (see Section 3.6).

For BitTorrent, the distribution of transfer sizes is different. BitTorrent generates a higher proportion of *big* transfers, indeed there is no search process included in this protocol, only the coordination of transfers is taken care of.

We have approximated the observed distribution of transfer sizes by classical statistical laws using Kolmogorov-Smirnov (K-S) goodness-of-fit test, and we conclude that:

- eDonkey can be approximated by a lognormal distribution;
- FastTrack by a lognormal one, but the tails of the distribution (*i.e.* big transfers) fits better a Pareto one;
- BitTorrent by a Weibull one.

To conclude this section, we mention that not only is the median volume per connection very small (less than 1 kbyte) due to numerous small connections, but so is the median volume per user (10 kbytes). The huge proportion of signaling traffic induces a mean volume per connection of 10 kbytes. On the contrary, due to some users contributing to large traffic volumes, the mean volume per user per day amounts to 70 Mbytes.

3.3   Connection Duration

We present the cumulative distribution function of the connection duration in Figure 4.

Connection durations are very long in view of their size. Indeed, more than 85% of the connections stay open for more than 10 seconds while more than 90% of the connections comprise less than 20 kbytes. This reveals long idle periods during the connections. These idle periods often encourage authors to consider

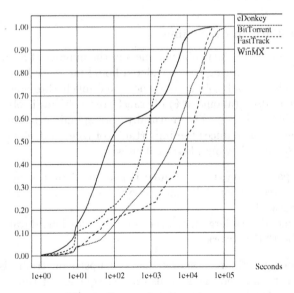

Fig. 4. Connection Duration

Table 2. Mean value of connection duration

	eDonkey	BitTorrent	FastTrack	WinMX
Signaling traffic	12s	22s	7s	13s
Non-signaling traffic	1436s	1670s	356s	2721s

as finished a connection after 5 seconds of inactivity as in [6] or filter out these periods as in [14].

eDonkey and FastTrack connections are of shorter duration than BitTorrent connections. BitTorrent encounters longer connection durations and also bigger throughputs. We give the mean value of connection durations in Table 2.

As far as BitTorrent is concerned, a peer distributing or downloading a file has a pool of 5 upload slots, *i.e.* only 5 other peers can download the file from him. To ensure a certain fairness, these slots are reallocated periodically, but the TCP connections are not closed. As a result, BitTorrent also encounters idle times during the transfers and long connections.

The case of eDonkey is different: a file is splitted into chunks. With the most popular eDonkey client (eMule), when a peer has finished to download a chunk, it should wait for another chunk in a queue, but the TCP connection between the peers stays open.

The difference between BitTorrent on the one hand and eDonkey and Fast-Track on the other hand is confirmed by the connection duration distributions. We find again that a lognormal law fits better eDonkey and FastTrack, whereas BitTorrent's connection durations fail to be well approximated by a classical law (the K-S test doesn't give a satisfactory answer).

3.4   Traffic Pattern over Time

In figure 5, we represent the traffic volume and the number of peers (transferring more than one packet during the considered hour) for every hour of a week day for eDonkey. The overall volume does not vary much throughout the day for eDonkey: a small drop (about 20%) of transferred volumes is observed between 12pm (midnight) and 9am. This naturally correspond to a diminution of the number of peers. This reduces the importance of *time-of-day* effect observed in [13], [7] and [8], the following ideas explain why:

- the main part of the traffic volume is generated by a small proportion of users (10%, see Section 3.2), and these users are permanently connected to the P2P network,
- peers download very large files which accounts for long durations of connections to the P2P network.

A closer look at Figure 5 shows that the upstream traffic is more stable than the downstream one. This is also a consequence of the nature of peers involved in the eDonkey network. The peers contributing to big volumes and permanently connected are responsible for the main part of downstream and

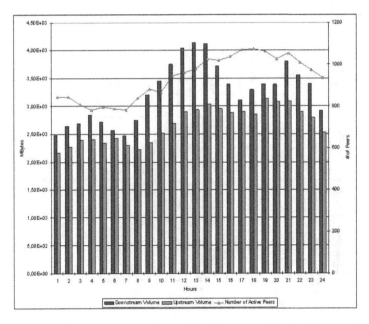

Fig. 5. eDonkey Traffic Volume and Number of Peers over Time during a Week Day

upstream volumes, which are comparable. The other peers introduce a lot of downstream traffic whereas they contribute in a smaller proportion to upstream traffic.

For BitTorrent, the behavior is completely different. The connections are actually longer and the receiver sees many *on* and *off times* during its transfer, so the traffic pattern is a bit skewed. It is interesting to note that one year ago, the time-of-day effect was clearly observable on BitTorrent's traffic, as most of the volume (85%) was usually transferred between 3p.m. and midnight in France. With the increase of BitTorrent users, the traffic pattern is more regular over the day.

In Figure 6, we expose the traffic volume and the number of active peers for every hour of a week day for BitTorrent. From midnight to 5pm, the number of peers is quite stable and so is the upstream volume. But it is surprising to see downstream volume having high fluctuations. We explain this by the functionnig system of BitTorrent: in [15], the model corresponding to a BitTorrent-like file-sharing system states that the service capacity (*e.g.* the total throughput) grows with the peers involved in the file-transfer, resulting in a very good response to flash crowds (when many peers simultaneously ask for the same file). The two peaks of downstream volume probably result from a sudden increase of the number of distant BitTorrent peers. We can add that the file(s) corresponding to this growth of peers should not be very big, indeed the peaks indicates that the download time is less than two hours.

For FastTrack and WinMX, we observe a dynamic traffic pattern over the day: users connect to these P2P networks mainly between 12am (noon) and

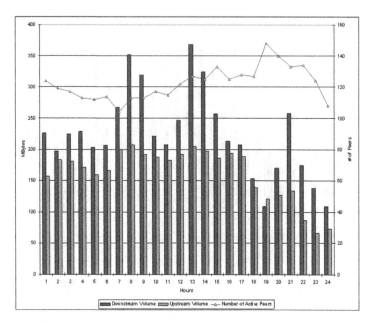

Fig. 6. BitTorrent Traffic Volume and Number of Peers over Time during a Week Day

12pm (night) for week days and all day long during week-ends. We explain this phenomenon as follows:

- the afternoons and evenings are high activity periods for ADSL activity (during week days),
- users download shorter files with Kazaa or WinMX, hence the P2P applications can be used for shorter periods of time.

For week-end days, the traffic pattern is completely different. The whole saturday and sunday morning are high activity periods for eDonkey transfers, but they end at sunday noon! BitTorrent exhibits the same pattern but the traffic distribution is more erratic.

3.5   Geographical Distribution of Peers

We locate the destinations of transfers over a week. We use a description field of IANA database to establish this geographical information.

The geographical distributions show that most of the traffic (about 30%) heads to and from France, followed by U.S. for eDonkey and BitTorrent. On the FastTrack and WinMX networks, U.S. is the primary source and destination of transfers.

We have to remark that these distributions are very sensitive to the activity of the week. Indeed in June 2003, we reported a majority of eDonkey traffic coming from Germany.

Table 3. Distribution of the TCP-Flag of the last packet of connections

Flag of last packet	eDonkey	BitTorrent	FastTrack	WinMX
NORMAL	20%	9%	15%	20%
SYN	21%	42%	39%	27%
FIN	6%	5%	4%	5%
PUSH	20%	16%	5%	20%
RESET	7%	11%	11%	9%
OTHER	23%	16%	25%	17%

The distinction of signalling traffic allows us to determine if the geographical distribution of servers follows the distribution of peers. Our first analysis indicates that the distributions are similar, but small differences can give interesting conclusions: *e.g.* Belgium encounters more small connections than big ones: indeed the most popular eDonkey indexing server (with 800,000 peers connected to *Razorback 2*) lies in Belgium.

3.6   Termination of Connections

In order to characterize the termination of connections, we expose in Table 3 the connections with a normal TCP-ending (four-way handshake, denoted as "NORMAL"), and for the other connections, we observe the TCP-Flag of the last packets.

Here all four protocols have similar trends: only few connections end normally. We have identified that a P2P client disconnecting from the P2P network results in sending a RESET packet, this accounts for many connections. We also observe a high percentage of connections that end abnormally, *e.g.* by a PUSH.

But the main remark here is that 20 to 40% of connections are only connection attempts (the last packet of these connections is a SYN). The peers involved in these connections (about 30% of peers) receive connection requests whereas they are no longer connected to the P2P network. This is observable in Figure 3 by looking at users who have no upstream volume whereas they have some downstream one (the vertical line at the left side of the figure). We explain this by the delay in forwarding the information on peer availability across the P2P network. At present, we are further investigating this issue because it can have some influence on P2P networks.

4   Connectivity of Peers

4.1   Local Peer Connectivity

Now, we study the number of distant hosts contacted by a local user during a day, *i.e.* the *connectivity* of a local peer. Figures 7 *(a)* and *(b)* show these data over a complete day and over periods of two minutes spanning the day, respectively. The results of the two minutes period graph represent the number of simultaneous connections by user.

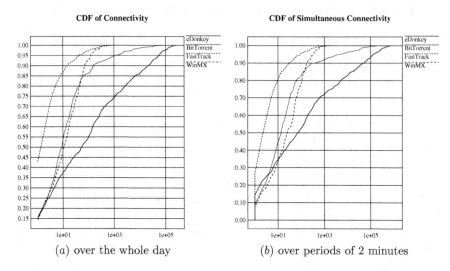

Fig. 7. Number of different IP addresses (X axis) connected to a local user

Firstly, the connectivity of eDonkey peers is greater. Indeed this protocol generates a lot of connections between peers. Only 18% of the local eDonkey peers contact a single peer, almost 50% of the local peers contact more than 10 other peers. Some eDonkey peers connect to more than 100,000 other peers, which shows that we have a small number of eDonkey indexing servers among the set of local peers.

The similarity of Figures 7 *(a)* and *(b)* tend to prove that peers establish connections with other peers within periods of 2 minutes. Indeed, when a peer begins to download a file, it tries to connect to all the peers having this file.

For BitTorrent, the downloading traffic presents a higher connectivity than the signaling traffic: this is due to the information management, which is done by a single *tracker* for each single file but the download is made from several sources.

4.2   Distant Peer Connectivity

Now we investigate the connectivity of a non-local peer, *i.e.* the number of local hosts connected to each distant peer. For this section, we can only distinguish distant peers by their IP addresses. These results are computed over a pool of 1 Million distant peers, mainly eDonkey ones.

Only few local peers are connected to the same distant peer. We don't see any accumulation of connections on a distant IP address: traffic and requests are well distributed over distant hosts.

eDonkey peers still have the densiest connectivity with 20% of the distant peers being contacted by more than three local peers. For the three other protocols, about 80% of the distant peers are contacted only once by local peers.

The previous analysis allows us to identify highly connected peers. In this paragraph, we characterize the traffic of eDonkey peers which connect to more than 10,000 other peers during the day. We report about 130 peers with this property. These peers encounter same mean of connections sizes than other peers. But the mean downstream volume transfered amounts to 500MBytes per peer and the upstream one to 340MBytes per peer. As most of the traffic is generated by these peers, their cumulative distribution function of the tranfers sizes is very similar to those exposed in Figure 2.

5   Conclusion

In this paper, we compare the performance and characteristics of four P2P applications. Our measurement methodology allows us to deeply analyze a complete set of traffic traces stemming from all the users of a regional ADSL area. We derive many results dealing with characteristics of connections (volume, duration, termination of connections), localization of peers, traffic pattern over the day and connectivity of peers.

Our study indicates firstly that, even for P2P traffic, most of the connections are very short and represent a small volume, and secondly that very few users contribute to the most significant part of the traffic volume. The two kinds of peers involved in P2P networks (*i.e.* those conributing to big volumes and the other) strongly influence the functionning system of P2P file-sharing. We reveal that even on a whole regional concentration point, users tend to download more than they upload. We also find that local peers tend to contact many different distant peers. Focusing on the packets sent at the termination of connections, we detect that unsuccessful connection attemps represent a lot of connections, and concern many users.

The persistency of our measures has allowed us to see a change of popularity in P2P applications (FastTrack being overtaken by BitTorrent) and some changes in the location of sources over a year.

Acknowledgements

We would like to thank our colleagues Anne-Marie Bustos and Denis Collange for their contributions to the geographical analyses and the utilization of measurements. A lot of improvements in the final paper is due to anonymous reviewers, we thank them for their detailed comments.

References

1. BitTorrent. www.bittorrent.com.
2. eDonkey. www.edonkey2000.com.
3. WinMX. www.winmx.com.
4. E. Adar and B. Huberman. Free riding on Gnutella. *First Monday*, 5(10), October 2000.

5. A. Asvanund, K. Clay, R. Krishnan, and M. Smith. An Empirical Analysis of Network Externalities in P2P Music-Sharing Networks. In *The 23rd Annual International Conference on InformationSystems (ICIS 02)*, Barcelona, Spain, December 15–19 2002.

6. N. Ben Azzouna and F. Guillemin. Analysis of ADSL traffic on an IP backbone link. In *Proceedings of Globecom 2003*, San Fransisco, CA, USA, December 2003.

7. R. Bhagwan, S. Savage, and G. Voelker. Understanding Availability. In *Proceedings of the 2nd International Workshop on Peer-to-Peer Systems*, Berkeley, CA, USA, February 2003.

8. J. Chu, K. Labonte, and B. Levine. Availability and Locality Measurements of Peer-to-Peer File Systems. In *Proceedings of ITCom: Scalability and Traffic Control in IP Networks*, July 2002.

9. Thomas Karagiannis, Andre Broido, Nevil Brownlee, K. C. Claffy, and Michalis Faloutsos. Is P2P dying or just hiding? In *IEEE Globecom 2004 - Global Internet and Next Generation Networks*, Dallas, Texas, USA, December 2004.

10. Thomas Karagiannis, Andre Broido, Michalis Faloutsos, and K. C. Claffy. Transport Layer Identification of P2P Traffic. In *Internet Measurement Conference (IMC)*, Taormina, Sicily, Italy, October 2004.

11. Sandvine Corporation. Regional characteristics of P2P, October 2003. www.sandvine.com.

12. Subhabrata Sen, Oliver Spatscheck, and Dongmei Wang. Accurate, Scalable In-Network Identification of P2P Traffic Using Application Signatures. In *13th International World Wide Web Conference*, New York City, 17–22 May 2004.

13. Subhabrata Sen and Jia Wang. Analyzing Peer-to-Peer Traffic Across Large Networks. *IEEE/ACM Transactions on Networking*, 2004.

14. K. Tutschku. A Measurement-based Traffic Profile of the eDonkey Filesharing Service. In *Proceedings of Passive and Active Network Measurement (PAM 2004)*, Antibes Juan-les-Pins, France, April 2004.

15. Xiangying Yang and Gustavo de Veciana. Service capacity in peer-to-peer networks. In *Procedings of IEEE INFOCOM 2004*, Hong Kong, March 2004.

Analysis of Communities of Interest in Data Networks*

William Aiello[1], Charles Kalmanek[2], Patrick McDaniel[3],
Subhabrata Sen[2], Oliver Spatscheck[2], and Jacobus Van der Merwe[2]

[1] Department of Computer Science, University of British Columbia,
Vancouver, B.C. V6T 1Z4, Canada
aiello@cs.ubc.ca
[2] AT&T Labs – Research, Florham Park, NJ 07932, U.S.A
{crk,sen,spatsch,kobus}@research.att.com
[3] Department of Computer Science and Engineering, Penn State University,
University Park, PA 16802, U.S.A
mcdaniel@cse.psu.edu

Abstract. *Communities of interest* (COI) have been applied in a variety of environments ranging from characterizing the online buying behavior of individuals to detecting fraud in telephone networks. The common thread among these applications is that the historical COI of an individual can be used to predict future behavior as well as the behavior of other members of the COI. It would clearly be beneficial if COIs can be used in the same manner to characterize and predict the behavior of hosts within a data network. In this paper, we introduce a methodology for evaluating various aspects of COIs of hosts within an IP network. In the context of this study, we broadly define a COI as a collection of interacting hosts. We apply our methodology using data collected from a large enterprise network over a eleven week period. First, we study the distributions and stability of the size of COIs. Second, we evaluate multiple heuristics to determine a stable core set of COIs and determine the stability of these sets over time. Third, we evaluate how much of the communication is not captured by these core COI sets.

1   Introduction

Data networks are growing in size and complexity. A myriad of new services, mobility, and wireless communication make managing, securing, or even understanding these networks significantly more difficult. Network management platforms and monitoring infrastructures often provide little relief in untangling the *Gordian knot* that many environments represent.

In this paper, we aim to understand how hosts communicate in data networks by studying host level *communities of interest* (COIs). A community of interest is a collection of entities that share a common goal or environment. In the context of this study, we broadly define a community of interest as a collection of interacting hosts. Using data collected from a large enterprise network, we construct community graphs representing the existence and density of host communications. Our hypothesis is that the behavior of a collection of hosts has a great deal of regularity and structure. Once such structure is illuminated, it can be used to form parsimonious models that can become

---

* This research was conducted when the authors were with AT&T Labs – Research.

C. Dovrolis (Ed.): PAM 2005, LNCS 3431, pp. 83–96, 2005.

the basis of management policy. This study seeks to understand the structure and nature of communities of interest ultimately to determine if communities of interest are a good approximation of these models. If true, communities of interest will be useful for many purposes, including:

- *network management* - because of similar goals and behavior, communities will serve as natural aggregates for management.
- *resource allocation* - allocating resources (e.g., printers, disk arrays, etc.) by community will increase availability and ensure inter-community fairness.
- *traffic engineering* - profiles of communal behavior will aid capacity planning and inform prioritization of network resource use.
- *security* - because communities behave in a consistent manner, departure from the norm may indicate malicious activity.

Interactions between social communities and the Web have been widely studied [1, 2]. These works have shown that the web exhibits the *small world phenomena* [3,4], i.e., any two points in the web are only separated by a few links. These results indicate that digital domains are often rationally structured and may be a reflection of the physical world. We hypothesize that host communication reflects similar structure and rationality, and hence can be used to inform host management. In their work in network management, Tan et. al. assumed that hosts with similar connection habits play similar roles within the network [5]. They focused on behavior within local networks by estimating host *roles*, and describe algorithms that segment a network into host role groups. The authors suggest that such groups are natural targets of aggregated management. However, these algorithms are targeted to partitioning hosts based on some *a priori* characteristic. This differs from the present work in that we seek to identify those characteristics that are relevant. Communities of interest can also expose aberrant behavior. Cortes et. al. illustrated this ability in a study of fraud in the telecommunications industry [6]. They found that people who re-subscribed under a different identity after defaulting on an account could be identified by looking at the similarity of the new account's community.

This paper extends these and many other works in social and digital communities of interest by considering their application to data networks. We begin this investigation in the following section by outlining our methodology. We develop the meaning of communities of interest in data networks and then explain how our data was collected and pre-processed. While the data set that we analyze is limited to traffic from an enterprise network, we believe that the methodology is more broadly applicable to data networks in general. In Section 3 we present the results of our analysis and conclude the paper in Section 4 with a summary and indication of future work.

2   Methodology

In this section we consider the methodology we applied to the COI study. First we develop an understanding of what COI means in the context of a data network. Then we explain how we collected the data from an enterprise network and what pre-processing we had to perform on the data before starting our analysis.

2.1   Communities of Interest

We have informally defined COI for a data network as a collection of interacting hosts. In the broadest sense this would imply that the COI of a particular host consists of *all* hosts that it interacts with. We call the host for which we are trying to find a COI the **target-host**. We begin our analysis by exploring this broad COI definition, by looking at the total *number* of hosts that target-hosts from our data set interact with. Thus in this first step we only look at the COI set size and its stability over time.

Considering all other hosts that a target-host ever communicates with to be part of its COI might be too inclusive. For example, this would include one-time-only exchanges which should arguably not be considered part of a host's COI. Intuitively we want to consider as part of the COI the set of hosts that a target-host interact with *on a regular basis*. We call this narrower COI definition the *core* COI.

In this work it is not our goal to come up with a single core COI definition. Instead, it is our expectation that depending on the intended *application* of COI, different definitions might be relevant. For example, in a resource allocation application the relevant COI might be centered around specific protocols or applications to ensure that the COI for those applications receive adequate resources. On the other hand an intrusion detection application might be concerned about deviations from some "normal" COI. However, in order to evaluate our methodology, we do suggest and apply to our data two example definitions of a core COI:

- **Popularity.** We determine the COI for a *group* of target-hosts by considering a host to be part of the COI if the percentage of target-hosts interacting with it exceeds a threshold T, over some time period of interest Y.
- **Frequency.** A host is considered to be part of the COI of a target-host, if the target-host interacts with it at least once *every* small time-period Z (the bin-size) within some larger time period of interest Y.

Intuitively these two definitions attempt to capture two different constituents of a core COI. The most obvious is the *Frequency* COI which captures any interaction that happens frequently, for example access to a Web site containing news that gets updated frequently. The *Popularity* COI attempts to capture interactions that might happen either frequently or infrequently but is performed by a large part of the user population. An example would be access to a time-reporting server or a Web site providing travel related services.

From the COI definitions it is clear that the *Popularity* COI becomes more inclusive in terms of allowing hosts into the COI as the threshold (T) decreases. Similarly the *Frequency* COI becomes more inclusive as the bin-size increase. For the *Popularity* case where the threshold is zero, *all* hosts active in the period-of-interest are considered to be part of the COI. Similarly, for the *Frequency* case where the bin-size is equal to the period-of interest, all hosts in that period are included in the COI. When the period-of-interest, Y, is the same for the two core COI definitions, these two special cases (i.e., $T = 0$ for the *Popularity* COI and $Z = Y$ for the *Frequency* COI), therefore produce the same COI set.

Notice that the *Popularity* COI defines a core COI set for a "group" of hosts, whereas the *Frequency* COI defines a per-host COI. We have made our core COI definitions in the most general way by applying it to "hosts", i.e., not considering whether the host was the initiator (or client) or responder (or server) in the interaction[1]. While these general definitions hold, in practice it might be useful to take directionality into account. For example, the major servers in a network can be identified by applying the *Popularity* definition to the percentage of clients initiating connections to servers. Similarly, the *Frequency* definition can be limited to clients connecting to servers at least once in every bin-size interval to establish a per-client COI.

In the second step of our analysis we drill deeper into the per-host interactions of hosts in our data set to determine the different core COI sets. Specifically, we determine the *Popular* COI and the *Frequency* COI from a client perspective and consider their stability over time.

Ultimately we hope to be able to predict future behavior of hosts based on their COIs. We perform an initial evaluation of how well core COIs capture the future behavior of hosts. Specifically, we combine all the per-host *Client-Frequency* COIs with the shared *Popularity* COI to create an **Overall** COI. We construct this COI using data from a part of our measurement period and then evaluate how well it captures host behavior for the remainder of our data by determining how many host interactions are *not* captured by the *Overall* COI.

2.2 Data Collection and Pre-processing

To perform the analysis presented in this paper we collected eleven weeks worth of flow records from a single site in a large enterprise environment consisting of more than 400 distributed sites connected by a private IP backbone and serving a total user population in excess of 50000 users. The flow records were collected from a number of LAN switches using the Gigascope network monitor [7]. The LAN switches and Gigascope were configured to monitor *all* traffic for more than 300 hosts which included desktop machines, notebooks and lab servers. This set of monitored hosts for which we captured traffic in both directions are referred to as the **local hosts** and form the focal point of our analysis. In addition to some communication amongst themselves, the local hosts mostly communicated with other hosts in the enterprise network (referred to as **internal hosts**) as well as with hosts outside the enterprise environment (i.e., **external hosts**). We exclude communication with external hosts from our analysis as our initial focus is on intra-enterprise traffic. During the eleven week period we collected flow records corresponding to more than 4.5 TByte of network traffic. In our traces we only found TCP, UDP and ICMP traffic except for some small amount of RSVP traffic between two test machines which we ignored. For this initial analysis we also removed weekend data from our data set, thus ensuring a more consistent per-day traffic mix. Similarly, we also excluded from the analysis any hosts that were not active at least once a week during the measurement period.

Our measurement infrastructure generated unidirectional flow-records for monitored traffic in 5 minute intervals or bins. A flow is defined using the normal 5-tuple of IP protocol type, source/destination addresses and source/destination port numbers. We record the number of bytes and number of packets for each flow. In addition, each flow

---

[1] We provide an exact definition of client and server in the next section.

record contains the start time of the 5 minute bin and timestamps for the first packet and last packet of the flow within the bin interval. The collected "raw" flow-records need to be processed in a number of ways before being used for our analysis:

Dealing with DHCP: First, because of the use of Dynamic Host Configuration Protocol (DHCP), not all IP addresses seen in our raw data are unique host identifiers. We use IP address to MAC address mappings from DHCP logs to ensure that all the flow records of each unique host are labeled with a unique identifier.

Flow-record processing: The second pre-processing step involves combining flows in different 5 minute intervals *that belong together from an application point of view.* For example, consider a File Transfer Protocol (FTP) application which transfers a very large file between two hosts. If the transfer span several 5 minute intervals then the flow records in each interval corresponding to this transfer should clearly be combined to represent the application level interaction. However, even for this simple well-known application, correctly representing the *application* semantics would in fact involve associating the FTP-control connection with the FTP-data connection, the latter of which is typically initiated from the FTP-server back to the FTP-client.

Applying such application specific knowledge to our flow-records is not feasible in general because of the sheer number of applications involved and the often undocumented nature of their interactions. We therefore make the following simplifying definition in order to turn our flows records into a data set that captures some application specific semantics. We define a **server** as any host that listens on a socket for the purpose of other hosts talking to it. Further, we define a **client** as any host that *initiates* a connection to such a server port. Clearly this definition does not perfectly capture application level semantics. For example, applying this definition to our FTP interaction, only the control connection would be correctly identified in terms of application level semantics. This client/server definition does however provide us with a very general mechanism that can correctly classify all transport level semantics while capturing some of the application level semantics.

To summarize then, during the second pre-processing step we combine or splice flow-records in two ways: First, flow-records for the same interaction that span multiple 5 minute intervals should be combined. Second, we combine two uni-directional flow-records into a single record representing client-server interaction.

To splice flow-records that span multiple 5-minute intervals, we use the 5-tuple of protocol and source/destination addresses and ports. We deal with the potential of long time intervals between matching flows by defining an *aggregation time* such that if the time gap between two flow records using the same 5-tuple exceed the aggregation time, the new flow-record is considered the start of a new interaction. If the aggregation time is too short, later flow-records between these hosts will be incorrectly classified as a new interaction. Making the aggregation time too long can introduce erroneous classification for short lived interactions. We experimented with different values of aggregation time and found a value of 120 minutes provided a good compromise between incorrectly splitting flows that fit together and incorrectly combining separate flows.

The 5-tuple is again used to combine two unidirectional flows into a single interaction. For TCP and UDP, two flow-records are combined into a single record if the flows are between the same pair of hosts and use the same port numbers in a swapped fashion (i.e., the source port in one direction is the same as the destination port in the reverse direction). For ICMP traffic, flow-records are combined if they are between the same pair of hosts. The result of splicing two unidirectional flows together is an *edge-record* and we present the data as a directed graph in which each edge represents a communication between a client and a server and each node represents a unique host. The direction of the edge represents client/server designation and the labels on the edge indicate the number of packets and bytes flowing in each direction between the two nodes.

We evaluated the experimental error introduced by our flow-record processing as follows. We consider a 5 week subset of our total 11 week data set for this evaluation. We note that flows labeled with a client port number below 1024 and a server port number above 1024 is highly likely to be incorrect for all but a few services (as it is not consistent with the normal use of reserved ports), and the reverse (server port < 1024, and client port > 1024) are likely to be correct. We bound experimental error by calculating the ratio of incorrect to correct labeled flows based on this heuristic (after removing known services that violate this property, e.g., `ftp-data`, NFS traffic through `sunrpc`). This approximation yields a 2.187% role assignment error for all traffic, while the numbers for TCP and UDP are 2.193% and 2.181%, respectively. Each instance of mis-interpreted directionality introduces an additional flow into the data set. Hence, such errors do not change the structure of the community, but slightly amplify a host's role as a client or server.

Removing unwanted traffic: Since we are interested in characterizing the "useful" traffic in the enterprise network the third pre-processing step involves removing all graph edges for suspected unwanted traffic, such as network scans or worm activity. Doing such cleaning with 100% accuracy is infeasible because unwanted traffic is often indistinguishable from useful traffic. We use the following heuristics:

- *TCP:* We clean the data by removing all edges which do not have more than 3 packets in each direction. We chose the number three since a legitimate application layer data transfer needs more than three packets to open, transfer and close the TCP connection. This cleaning removes 16% of all edges indicating that a large fraction of traffic in the monitored network does not complete an application-level data transfer.
- *UDP:* We observe that there are two types of legitimate UDP uses. One is request/response type interaction such as performed by DNS and RPC. The other is a long lived UDP flow as used by many streaming applications. In both cases we expect an edge which performs a useful task to be associated with at least two packets, either in the same direction or in opposing directions. Therefore, we remove all edges for which the sum of packets in both directions is smaller than 2.
- *ICMP:* We do not perform any cleaning on the ICMP data since a single ICMP datagram is a legitimate use of ICMP.

3   Results

In this section we present the COI analysis as applied to the enterprise data we collected. After pre-processing, the final data set we used for the analysis consisted of 6.1 million edge-records representing 151 *local* hosts and 3823 *internal* hosts and corresponding

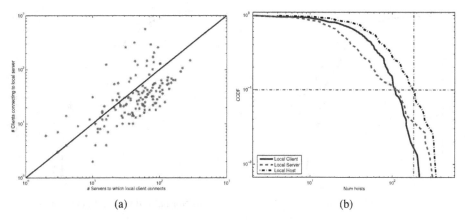

(a)   (b)

Fig. 1. (a) Scatterplot of 151 *local* hosts: Clients using the local host as a server and the local host talking to servers as a client. (b) CCDF:local host communication for total 11 week period

to 2.6 TBytes worth of network traffic. We will characterize only the set of 151 local hosts, but consider *all* their interactions, with both other *local* and *internal* hosts.

3.1 Community of Interest Set Size

First we evaluate the COI of the set of local hosts in our data set based on the broadest definition of COI. Specifically we consider the *number* of other hosts that each local host interact with. We look at the total number of such hosts and then do a breakdown based on whether the target local host was acting as a client or a server.

We first perform this analysis for all hosts over the *entire* measurement period. Figure 1(a) shows a scatter plot of the in/out-degree of the set of 151 local hosts considering all observed traffic. The Y-axis shows the number of clients connecting to the local host acting as a server (i.e., in-degree). The X-axis shows the number of servers that the local host connects to acting as a client (i.e., out-degree). Observe from Figure 1(a) that most hosts act as both client and server over the observation period. Indeed for the total traffic breakdown shown, *all* hosts act as both client and server during the measurement period. The general observation that most hosts act as both client and server, hold when data is analyzed on a per-protocol basis. Specifically, counting the number of hosts that acted purely as clients on a per protocol basis we get only 3 for TCP, 2 for UDP and 1 for ICMP. Similarly, counting the number of hosts acting purely as servers on a per protocol basis we get none for TCP, 2 for UDP and 5 for ICMP. Further, as indicated by the density below the diagonal line, the majority of local hosts are mostly acting as clients. For the plot shown, 111 hosts are below and 35 hosts above the diagonal line. The implication of the simple observation that most hosts act as both clients and servers, is that security schemes that rely on hosts acting exclusively as clients or servers, are likely to be infeasible in current enterprise networks.

Figure 1(b) shows the empirical Complementary Cumulative Distribution Function (CCDF) of the number of machines that our local hosts communicate with for all traffic over the entire 11 week measurement period. The "Local Host" curve corresponds to the total number of hosts (either *local* or *internal*) that a particular local host interacts

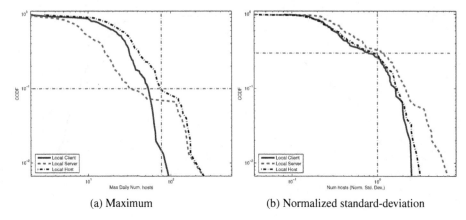

(a) Maximum   (b) Normalized standard-deviation

Fig. 2. CCDFs for the number of hosts communicated with on daily basis

with, whether as a client or as a server. The plot shows that each of the local hosts communicates with a fairly small community of other hosts even over a period of several weeks. For example, 90% of the local hosts talks to fewer than 186 other hosts. Considering the client/server breakdown, the same holds true with local hosts interacting with a fairly small number of servers and clients. The final 10% of the "Local Server" curve shows that a small number of local machines acting as servers have higher numbers of clients talking to them than the other 90% of the local servers. These machines most likely correspond to "real" servers that serve a significant client population as opposed to hosts that are servers on the basis of the protocol interaction only.

We next look at the COI of each host on a *daily* basis and examine the statistical properties of these daily values over the complete observation period. First, Figure 2(a) shows the CCDF of the maximum daily number of hosts that each local host communicates with over the entire eleven weeks. These maximum number per day CCDFs are similar to those for the maximum over the entire measurement period, Figure 1(b), but the numbers are lower (i.e., the curves are "shifted" to the left). For example, the 90th percentile number for the "Local Host" curve in Figure 2(a) is only 77 compared with 186 for the same percentile in Figure 1(b). Also similar to Figure 1(b), there is an inflection at the 10% point in Figure 2(a) for the "Local Server" (and "Local Host") curves which is likely caused by "real" servers.

The relatively small sizes of the total number of hosts communicated with over the entire period as well as the small per-day maximums for the vast majority of hosts, suggest that a simple anomaly detection approach based on monitoring the normal COI size, has the potential to detect abnormal activities like port scans and worm spreads. These anomalies are often marked by a host communicating with a large number of other machines within a very short time span.

Next we consider the variability of the per-day COI size for each local host over the entire measurement period. Figure 2(b) shows the resulting CCDF of the normalized standard deviation (normalized by the mean for each local host). Note that some of the variability is a result of hosts being inactive on some days, one contributing reason being telecommuting users. Hosts for these users might either be inactive because

they are not being used, or in the case of notebooks, might not be visible to our monitoring infrastructure. The graph shows that approximately 70% of the local hosts have
normalized standard-deviations in their per-day COI size that is less than 1. Assuming
that all of the traffic in our data set was indeed legitimate, this would mean a simplistic
approach to detect abnormal behavior for these hosts, based on a policy that restricts
"normal" per-day COI size to 3 times the respective per-day means, would result in
false alarms being generated only 5% of the time. Note also from Figure 2(b) that the
standard deviation for the "Local Client" curve is less skewed than the "Local Server"
curve. This suggests that on a daily basis the number of servers which a local client
talks to, is more stable than the number of clients that talk to a local server. The implication of this is that network management policies derived from observations close to
the initiator of communication (client) is likely to be more stable than policies derived
from traffic close to the communication responder (server).

3.2   Core Communities of Interests

We next explore our two example core COI definitions *Popularity* and *Frequency* core
COIs and their interactions.

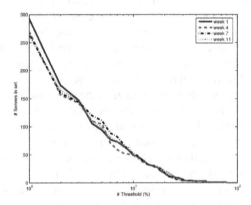

Fig. 3. Size of *Popularity* COI set for all traffic

Popularity **COI:** Recall that for the *Popularity* COI we consider a host to be part of
the COI for a group of target-hosts if the percentage of target-hosts interacting with it
exceeds a threshold T over some period of interest Y. Here we identify the *Popularity*
COI of the local hosts from a client view point, for each of the 11 weeks in our data
set (i.e., Y is one week). Figure 3 shows the size of the *Popularity* core COI set as a
function of the threshold T for 4 equally spaced weeks out of the total 11 weeks, for
traffic across all protocols. The graphs shows the expected decline of the set size as one
progresses from a threshold of 0% (which would include all hosts) to a threshold of
100% at which point the size is expected to be very small as it would require all target-
hosts to communicate with each member of the set. We observe that the size of the core
COI set as a function of the threshold is very similar across the different weeks. This

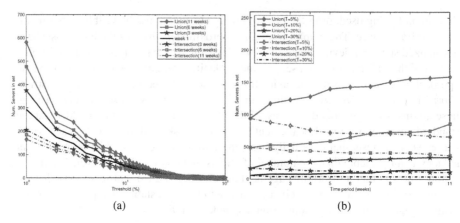

Fig. 4. *Popularity* COI: Union and intersection set size (a) As a function of threshold T (b) As a function of length of time window N weeks ($1 \leq N \leq 11$)

suggests that, deviations from the *Popularity* COI size distribution, for a set of hosts monitored over time, would be a strong indication of a network anomaly.

While the stability in the core COI set size is encouraging, we are also interested in the stability and predictability of the core COI set *membership*. To evaluate this, we determine the core COI set for each week in our data and then explore how the membership of these sets change over the measurement period. We do this by calculating the union (the set of servers that belong to the core set in at least one week) and the intersection (the set of servers that belong to the core set in every week) of the COI sets. For any two sets the difference between the size of the union and intersection represents a measure of the "churn" between the two sets - that is the total number of elements that needs to be added or removed from one set to transform it to the other set. Therefore, for a window of N COI sets, the difference between the union and intersection of **all** the sets, represents an upper bound on the churn between any two pairs in N. By looking at this bound we get a worst case estimate of how much the COI membership changes over the time window (N). By progressively increasing the length of the time window, we determine how this worst case estimate changes over time.

Figure 4(a) depicts the sizes of the union and intersection of core COI sets for weeks 1 to 3, 1 to 6 and 1 to 11, as a function of the threshold T for all traffic. For comparison the core COI set size for week 1 is also shown. For all curves (i.e., for all time periods considered), the difference between the union and intersection set sizes, i.e., the churn, tends to decrease as the threshold increases. Figure 4(b) shows the same data, but in this case we show the union and intersection set sizes for selected thresholds for increasing time windows N of interest (1 to 11 weeks), starting from week 1, i.e., 1 to 2 weeks, 1 to 3 weeks, etc. As expected, the union set size size increases and the intersection set size decreases for a given threshold as the time window increases. Notice though from Figure 4(b) that for any threshold, the union and intersection set sizes change in a sub-linear fashion with increasing N. In fact the intersection seems to flatten within 6 to 8 weeks. While the union set size shows a continued small growth, the maximum union set size, for the thresholds considered, did not increase beyond a factor of 2.5 over the

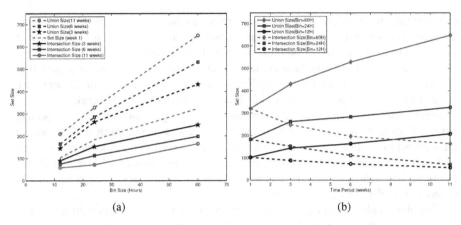

Fig. 5. *Overall Frequency* COI: Union and intersection set size (a) As a function of bin size Z (b) As a function of length of time window N weeks ($1 \leq N \leq 11$)

entire time window of interest. The observed intersection behavior implies that servers present in the *Popularity* COI in the first week, have a high probability of remaining in the COI for the entire period. This holds true independent of the threshold. For example for the 5% threshold, after the 11 weeks 66 of the initial 94 servers are still in the intersection set. The relatively small growth of the union set implies that even though servers are constantly added to the set, the number of additional servers added in a week is low. Even for the 5% threshold, the number is low enough that applications requiring "manual" verification of the status of new servers, i.e., whether they are legitimate servers, would be feasible.

The above results indicate that rapid changes in the *Popularity* COI membership would be an additional indication of anomalous network behavior. Note that this holds true if a large number of popular servers are either rapidly added to **or** removed from the COI set.

Frequency COI: We have also defined a core COI that captures the frequency of interaction. Recall that our *Frequency* Core COI considers a host to be part of the COI if a target-host interacts with it at least once in every time bin Z over some larger time period of interest Y. To evaluate this core COI definition we calculated the *Frequency* COI (client perspective) for each host for each week of our data (i.e., Y is one week) for bin-size (Z) values 12, 24, 60 and 120 hours. For each week, we define the *Overall Frequency* COI to be the union of all per-host *Frequency* COIs for a given bin-size. We explore how the membership of this set changes over time.

Similar to the approach above for *Popularity* COI, we determined the union and intersection of all the *Overall Frequency* COI sets for a specific time window of interest N. Figure 5(a) shows the size of the union and intersection sets for N equal to 3 weeks, 6 weeks and 11 weeks, for different bin sizes. (Note that we did not include a bin size of 120 hours for this plot as that would include all hosts that communicated in a week as part of the core COI for the week, which would be too inclusive for a core COI.) As a reference point, Figure 5(a) also shows the size of the *Overall Frequency* COI for the first week. As the bin size increases, the COI set becomes more inclusive and as expected the set size (shown for week 1) increases as the bin size increase to 60 hours. The same holds true for the union and intersection set sizes, i.e., for a particular size of N (e.g., 3 weeks), the size of both the union and intersection sets increase as the bin size increases. Next consider how the union and intersection set sizes for a particular bin size change for different values of N. For example, for a bin size of 24 hours, we see that the union set size increases as N increases from 3 to 6 to 11 weeks, while the intersection set size decreases for the same values of N. Again this behavior is expected, but it is interesting to note that this increase and decrease is not linear with respect to the increase in N. For example, doubling N from 3 weeks to 6 weeks does not result in doubling the union set size or halving the intersection set size. This is best shown in Figure 5(b), which depicts the union and intersection set sizes for each value of N (1 week, 3 weeks, 6 weeks and 11 weeks).

The above behavior of the *Overall Frequency* COI as a function of increasing N is similar to the behavior of the *Popularity* COI as a function of increasing N as shown in Figure 4(b). As for the *Popularity* COI, the results for the *Overall Frequency* COI indicate that rapid changes in the COI membership would be an indication of anomalous network behavior.

Overall COI: Recall that the *Popularity* and *Frequency* COI definitions attempt to capture different types of interactions that should be considered part of a core COI. Above we explored the churn in the *Popularity* and *Frequency* COIs separately, and focused on the churn in the membership of these sets. In contrast to this aggregate view, another way to explore variability is to inspect how the ability to capture the communication behavior for individual hosts is impacted by the churn in these COI sets. This is the goal of the study described next.

The *Popularity* COI is a function of the threshold parameter (T), while the *Frequency* COI is a function of the bin-size (Z) as defined earlier. For this part of the study, we computed, for a range of {threshold,bin-size} pairs, the *Overall* COI set for the **first week** of our data by combining (using set union) the *Overall Frequency* COI with the *Popularity* COI of the total local host set. Should this *Overall* COI accurately capture the core interactions of the target hosts in subsequent weeks, then one would expect that few of the target-hosts' interactions would be with hosts not in this set. We define interactions with hosts outside of the *Overall* COI to be *out-of-profile*.

For each local host we determined the number of out-of-profile interactions for subsequent weeks of our data. We calculate a distribution of the out-of-profile interactions across all hosts for each of the {threshold,bin-size} pairs. The results for this analysis are shown in Figures 6(a) and (b) for 6 and 11 weeks respectively. The figures depict the 90th and 50th percentiles for these distributions for a number of threshold values and as a function of bin-size.

We had discussed in Section 2.1 the situations under which the *Popularity* and *Frequency* COIs are identical. This explains why each set of curves (50th, 90th percentiles) converge in the 120 hour bin-size value in Figures 6(a) and (b). It also explains the horizontal lines (i.e., the cases where threshold is zero): in these latter cases the *Popularity* COI already includes all servers, so the union with the *Frequency* COI does not add any members to the *Overall* COI, and the number of out-of-profile interactions is therefore independent of the bin-size.

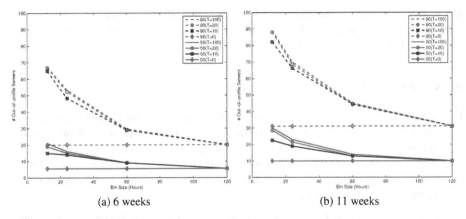

(a) 6 weeks   (b) 11 weeks

Fig. 6. Out-of-profile interactions for *Overall* COI for different threshold values and as a function of bin-size

The two horizontal lines in each figure correspond to the case where the threshold is zero, i.e., where the *Popularity* COI includes all hosts that acted as servers in week 1. From Figure 6(a), (the six week period), 50% of the local hosts (50th percentile line), had less than 6 out-of-profile interactions, while 90% of the local hosts (90th percentile line) did not exhibit more than 20 out-of-profile interactions over the entire period. The corresponding numbers for the full 11 week period shown in Figure 6(b) are 10 and 31 out-of-profile interactions for 50% and 90% of the local hosts respectively.

Now we consider the impact of the *Popularity* COI set by looking at the graphs in both figures for a fixed bin-size, (e.g., 60 hours). For this bin size, consider the 4 values of the six week distributions (Figure 6(a)). Notice that there is a significant difference between the case where the threshold is zero and the case where the threshold is 10%, when compared to the difference between the 10% and 20% (or even 100%) points. This also holds true for the 11 week distributions. This suggests that as the *Popularity* set becomes more inclusive (i.e., as the threshold gets closer to zero), it contributes *more significantly* to the *Overall* COI set. This seems to suggest that in this region the *Popularity* set indeed captures the important infrequent interactions that should be part of a host's COI. The relatively small difference between the values of the out-of-profile interactions for each set for thresholds between 10% and 100% seem to suggest that the *Overall Frequency* COI already captures most of the servers that the *Popularity* set would capture at such thresholds. This in turn suggests there may be significant overlap between the servers that communicate with a larger fraction of clients and those that interact frequently with clients, in our data set. Note that the increase in violations going from 6 to 11 weeks, does not increase proportionally with the increase in time.

In summary the graphs show that an *Overall* COI derived from one week's worth of data is not sufficient to fully capture host interactions for subsequent weeks. However, using the most inclusive variant of this COI definition (i.e., where threshold T is zero or where the bin size Z is 120), 90% of the hosts experienced on the average less than 3 violations per week (for the 11 week graph). Similarly, for the most restrictive COI we considered (i.e., T equal to 100 and Z equal to 12), 90% of the hosts experienced less than 9 violations per week over the 11 week period. Both rates are low enough that they would not preclude COI derived application that require human involvement. In fact, 50% of the hosts would only experience one third of these violations.

4   Conclusions

In this paper we presented our methodology and initial results for characterizing communities of interest (COIs) for hosts communication in data networks. We presented example definitions for COI that attempt to capture different characteristics of the underlying communities. We explained how we collected our measurement data and the pre-processing steps that were required before analysis. While this work is still maturing our initial results indicate that:

- Hosts typically act as both clients and servers which implies that any management applications or policies will have to explicitly deal with this.
- Using a very broad COI definition we saw similar distributions for the COI size over daily and monthly timescales, suggesting some stability in the COI for the community as a whole.
- COI definitions that represent core host interactions, showed significant stability of the COI over timescales of several weeks.
- Core COIs calculated over a part of our measurement period were also able to capture the actual host interaction in the remainder of the data fairly well.

We are continuing the presented work by moving from the presented aggregate COI characterization to finer grained per-host characterization. Our ongoing work aims to provide models that accurately capture host behavior. And our ultimate goal is to be able to apply such models to the many challenging network management tasks presented in the introduction.

References

1. Emily M. Jin et. al., "The structure of growing social networks," *Physics Review E*, vol. 64, pp. 845, 2001.
2. Ravi Kumar et.al., "The web and social networks," *IEEE Computer*, vol. 25, no. 11, pp. 32–36, 2002.
3. J. Kleinberg, "The Small-World Phenomenon: An Algorithmic Perspective," in *Proceedings 32nd ACM Symposium on Theory of Computing*, 2000, pp. 163–170.
4. J. Kleinberg, "Navigation in a small world," *Nature*, vol. 405, pp. 845, 2000.
5. Godfrey Tan et. al., "Role Classification of Hosts within Enterprise Networks Based on Connection Patterns," in *Proceedings of 2003 USENIX Annual Technical Conference*, June 2003, pp. 15–28, San Antonio, TX.
6. Corinna Cortes, Daryl Pregibon, and Chris T. Volinsky, "Communities of interest," *Intelligent Data Analysis*, vol. 6, no. 3, pp. 211–219, 2002.
7. Chuck Cranor et. al., "Gigascope: a stream database for network applications," in *Proceedings of ACM SIGMOD*, June 2003.

Binary Versus Analogue Path Monitoring in IP Networks

Hung X. Nguyen and Patrick Thiran

School of Computer and Communication Sciences, EPFL,
CH-1015 Lausanne, Switzerland
{hung.nguyen, patrick.thiran}@epfl.ch

Abstract. Monitoring systems that can detect path outages and periods of degraded performance are important for many distributed applications. Trivial pair-wise probing systems do not scale well and cannot be employed in large networks. To build scalable path monitoring systems, two different approaches have been proposed in the literature. The first approach [1], which we call the continuous or analogue model, takes real measurement values and infers the performance metrics of unmeasured paths using traditional $(+, \times)$ algebra. The second approach [2], which we call the Boolean model, takes binary values from measurements (e.g., whether the delay/loss of an end-to-end path is above a given threshold) and infers the performance quality of unmeasured paths using Boolean algebra. Both approaches exploit the fact that end-to-end paths share network links and hence that the measurements of some paths can be used to infer the performance on others. In this work, we are only interested in detecting whether the performance of a path is below an acceptable level or not. We show that when the number of *beacons* (nodes that can send probes and collect monitoring information) is small, the Boolean model requires fewer direct measurements; whereas for a large number of beacons the continuous model requires fewer direct measurements. When the number of beacons is significantly large, however, there is no difference in terms of the number of paths that we need to measure directly in both models. We verify the results by simulations on inferred network topologies and on real measurement data.

1 Introduction

Network dynamics may significantly affect the performance of distributed applications such as distributed system management, replicated services, and application layer multicast [3]. Robust and efficient distributed systems therefore need to adapt their behavior to environment changes. Loss rate monitoring systems that can detect path outages and periods of degraded performance can both facilitate distributed system management (such as virtual private network (VPN) or a content distribution network), and help build adaptive overlay applications, e.g., streaming media [4].

Monitoring systems that target small networks [5] usually employ pair-wise probing where each node probes the paths from itself to all other nodes. For

C. Dovrolis (Ed.): PAM 2005, LNCS 3431, pp. 97–107, 2005.

a network monitoring system with n_B beacons, the number of possible end-to-end measurements is $O(n_B^2)$. Therefore, active end-to-end measurement in such settings does not scale well and as a result cannot be deployed for complete network-wide measurement. Furthermore, this approach generates highly redundant measurements where many links in the network are repeatedly measured. It is therefore important to have a scalable overlay monitoring system that does not generate redundant information. Existing scalable network estimation systems determine network characteristics by measuring end-to-end paths periodically. To monitor a path, a node at one end of the path periodically sends probe packets to the node at the other end. From the delay characteristics and delivery status of these probe/acknowledgement packet pairs, the sending node can infer the quality of the path. This method is similar to network tomography approaches that infer the internal network characteristics based on end-to-end observations. Network tomography has been extensively studied in the literature([6] provides a detailed survey). However, to the best of our knowledge, none of the existing tomography works deal with minimizing the number of probes that need to be sent in a tomography system.

Chen et al. [1] have shown that it is possible to reconstruct complete end-to-end path properties exactly from the measurements of only a subset of paths. The results in [1] are based on a linear algebraic analysis of routing matrices of the monitoring systems where a routing matrix is the binary matrix that specifies the links that occur in a given path. Since a maximal set of independent paths can be used to recover any other path in the network, it is enough to monitor only this set. The number of independent paths in a monitoring system, which is the rank of the routing matrix, tends to be much smaller than the total number of paths. A similar approach is given in [7] where only bounded estimations for network paths can be achieved. Chua et al. [8] show that a significantly smaller number of direct path measurements than that are required by the monitoring systems of [1] can be used to approximate some network wide properties.

Padmanabhan et al. [9] studied the end-to-end packet loss rate experienced by clients of the Web server at *microsoft.com*. They report that the correlation between end-to-end loss rate and hop count is weak, which suggests that end-to-end paths are dominated by a few lossy links. Furthermore, the end-to-end loss rate is stable for several minutes. A notable feature of the model considered in [9] is that its parameters (the loss rates on the logical links) are not statistically identifiable from the data (the server-to-clients loss rate), meaning that there exist different sets of parameters that give rise to the same statistical distribution of data. Although the model is not statistically identifiable, some methods proposed in [9] are quite successful in identifying the lossiest links, both in simulated and real networks. The underlying reasons behind the success of the methods in [9] are the nature of link performance in the studied networks [10]. In such networks, suppose that we can classify links as "good" or "bad" with performance measures sufficiently far apart, then the performance experienced along a network path will be bad only if one of its constituent links is bad. Duffield et al. [10] calls this kind of link performance a *separable perfor-*

mance and identifies many separable performance metrics of network links such as connectivity, high-low loss model, and delay spike model.

A special case of separable performance, where links are either up or down, has been widely studied in the literature [2, 11, 12]. In our previous work [2], we showed that analogously to the work of Chen et al. [1], it is sufficient to monitor only a subset of end-to-end paths to infer the connectivity of all end-to-end paths in the network. The difference between [1] and our model is that our model relies on Boolean (max, ×) algebra instead of traditional (+, ×) algebra (which we also refer to as the "continuous model"). The end-to-end paths that need to be monitored are those that form the basis of the row space of the routing matrix in this Boolean algebra.

In this work, we compare the efficiency of the continuous and Boolean path monitoring systems for separable performance metrics. Since we are only interested in the classification of links as "good" or "bad", depending on whether the metric of interest has exceeded or not a given threshold, the continuous and Boolean models bring the same information. To obtain this same information, however, they will need different sets of end-to-end measurements. Thus we want to determine the system that uses fewer direct path measurements. Specifically, we are interested in comparing the dimension of the basis of the vector space in Boolean algebra for the Boolean model with the dimension of the basis of the same vector space in traditional (+, ×) algebra for the continuous model. Our main contributions are as follows.

- First, we show that the Boolean model in [2] can be used for other separable performance metrics.
- Second, by simulations on Rocketfuel topologies [13], we show that for a separable performance when the number of beacons is small, it is better to use the Boolean model. On the contrary, when the number of beacons is large, it is better to use the continuous model. However, when the number of beacons is significantly large, both models result in the same number of paths that need to be measured directly. We also provide some intuitive explanations for the simulation results.
- Finally, we verify our results on the data set gathered by the NLANR's AMP infrastructure [14].

The remainder of this paper is organized as follows. We introduces the network models and the basic algorithms in Section 2. We presents the numerical comparison of the two models on Rocketfuel ISP topologies in Section 3. Evaluations of the performance of the models on the NLANR's AMP active measurement infrastructure are given in Section 4. Finally, we conclude the paper in Section 5.

2   Network Model and Basic Algorithms

The network is modelled as an undirected graph $\mathcal{G}(\mathcal{V}, \mathcal{E})$, where the graph nodes, $\mathcal{V}$, denote the network components and the edges, $\mathcal{E}$, represent the communication links connecting them. The number of nodes and edges is denoted by $n = |\mathcal{V}|$ and $e = |\mathcal{E}|$, respectively. Suppose there are n_B beacons that belong to a single or confederated overlay monitoring system. They cooperate to share an overlay monitoring service, and are instrumented by a central authority. The set of all beacons is denoted by $\mathcal{V}_B$. Furthermore, we use $P_{s,t}$ to denote the path traversed by an IP packet from a source node s to a destination node t. Let $\mathcal{P}$ be the set of all paths between the beacons on the network and let $n_p = |\mathcal{P}|$.

For a known topology $\mathcal{G} = (\mathcal{V}, \mathcal{E})$ and a set of paths $\mathcal{P}$, we can compute the routing matrix D of dimension $n_p \times e$ as follows. The entry $D_{ij} = 1$ if the path $P_{s,t} \equiv P_i$, with $i = (s, t)$, contains the link e_j and $D_{ij} = 0$ otherwise. A row of D therefore corresponds to a path, whereas a column corresponds to a link. Note here that if a column contains only zero entries, the link corresponding to that column does not have any effect on the performance of the paths in $\mathcal{P}$. We drop these columns from the routing matrix to obtain a matrix of dimensions $n_p \times n_l$, where $n_l \leq e$ is the number of links that are covered by at least one path in $\mathcal{P}$.

Our performance model is as follows. During some measurement period, each beacon sends a set of packets to each destination (chosen among other beacons). When traversing link e_j, each packet is subject to a performance degradation (e.g. loss or delay) according to a distribution specified by a parameter ϕ_{e_j}. If the path P_i comprises links $e_1, ..., e_m$, the performance degradation along the path follows a composite distribution described by the parameters $\phi_i = \{\phi_{e_1}, ..., \phi_{e_m}\}$.

2.1   Continuous $(+, \times)$ Algebraic Model

In the continuous model [1], the performance parameters ϕ take values in $\mathbb{R}$. Let $y \in \mathbb{R}^{n_p}$ be a vector that represents a metric measured on all paths $P_i \in \mathcal{P}$. y is linearly related to the value $x \in \mathbb{R}^{n_l}$ of that same metric over the links $e_j \in \mathcal{E}$.

For example, letting ϕ_i denote the packet loss probability on path P_i and ϕ_{e_j}, the corresponding probability on link e_j, and assuming independence among loss-events on links, the relation between the path-wise and link-wise loss probability becomes

$$y = Dx = \left[\sum_{j=1}^{n_l} x_j D_{ij} \right]_{1 \leq i \leq n_p}, \tag{1}$$

where $y_i = \log(1 - \phi_i)$ and $x_j = \log(1 - \phi_{e_j})$.

There are $n_p = O(n_B^2)$ equations in (1). However, in general the matrix D is rank deficient, i.e., $k = rank(D) < n_p$. For the sake of simplicity, assume that the first k rows of D form a basis of the row vector space $\mathcal{R}(D)$ of D. Because every row vector d_i, $i > k$, of D can be represented as a linear combination of the first k independent row vectors, we can write that $d_i = \sum_{j=1}^{k} \alpha_j d_j$ for some α_j, $1 \leq j \leq k$. The metric y_i of the path P_i can be obtained from $y_1, ..., y_k$ as: $y_i = \sum_{j=1}^{k} \alpha_j y_j$. Therefore, only k independent equations of the n_p equations in (1) are needed to compute all elements of y, and as a result we only need to measure k paths, which form a basis of $\mathcal{R}(D)$, to obtain the loss rate on all paths [1].

2.2   Boolean (max, ×) Algebraic Model

In the Boolean model [2], the values of the performance parameters ϕ are partitioned into two subsets that we call "good" and "bad". We call the link e_j bad if and only if its parameter ϕ_{e_j} is bad and we call the path $P_i = \{e_1, ..., e_m\}$ bad if and only if ϕ_i is bad. The partitions are called separable when a path is bad if and only if at least one of its constituent links is bad. For example, in the loss model LM_1 of [9], good links have loss rates ϕ_{e_j} uniformly distributed between 0% and 1%; bad links have loss rates uniformly distributed between 5% and 10%. Taking the threshold between good and bad path transmission rates as 0.95, this model is separable if each path does not contain more than 5 links.

In a separable model, a path is bad if and only if at least one of its constituent links is bad. If we use the variable y_i to represent whether the path P_i is good ($y_i = 0$) or bad ($y_i = 1$) and the variable x_j is used to represent whether the network link e_j is good ($x_j = 0$) or bad ($x_j = 1$), we then have:

$$y_i = \bigvee_{j=1}^{n_l} x_j \cdot D_{ij} \text{ for all } i, \tag{2}$$

where "$\vee$" denotes the binary max operation, and "$\cdot$" denotes the usual multiplication operation.

Let us now introduce some concepts of Boolean vector spaces that are useful for the analysis of the Boolean model. Let $\mathcal{D} = \{d_i\}_{1 \le i \le h}$ be a set of binary vectors of equal length, and let $I = \{1, ..., h\}$ be the index set of $\mathcal{D}$. A vector span $\mathcal{S}$ can be defined on $\mathcal{D}$ as follows.

Definition 1. *[Vector span] The vector span of $\mathcal{D}$ is*

$$\mathcal{S} = <\mathcal{D}> = \{\bigvee_{i \in I} \alpha_i \cdot d_i \mid \alpha_i \in \{0, 1\}, d_i \in \mathcal{D}\}$$

Vectors in $\mathcal{D}$ are called the *generator vectors* of $\mathcal{S}$.

It was shown in [2] that each vector span $<\mathcal{D}>$ has a unique basis $\mathcal{B}$, which is the smallest set of vectors in $\mathcal{D}$ such that all other vectors in $<\mathcal{D}>$ can be written as a linear combination of vectors in $\mathcal{B}$. Let $b = |\mathcal{B}|$, b is called the dimension of $<\mathcal{D}>$. Without loss of generality, we assume that the first b rows of D form a basis of the row span $<\mathcal{D}>$ of D. Because every row vector d_i , $i > b$, of D can be represented by a linear combination of the first b independent row vectors, we can write that $d_i = \bigvee_{j=1}^{b} \alpha_j \cdot d_j$ for some α_j. The value y_i of the path P_i can be obtained from $y_1, ..., y_b$ as: $y_i = \bigvee_{j=1}^{b} \alpha_j \cdot y_j$. Therefore, only b independent equations of the n_p equations in (2) are needed to compute all elements of y, and hence we only need to measure b paths, which form a basis of $<\mathcal{D}>$, to determine whether any path in the overlay system is good or bad.

2.3   A Brief Comparison of the Two Models

If we are only interested in the classification of end-to-end paths as "good" or "bad", the continuous and Boolean models bring the same information. However, as we will show in this section, the two models use different algebraic structures, and as a result, they usually need different sets of end-to-end measurements to obtain this same information.

General properties relating network topologies with the dependency between measurement paths in the continuous and Boolean models are kept for future work. In this section, we are interested in the conditions under which a set of linearly dependent/independent vectors in the continuous model is dependent/independent in the Boolean model and vice versa. The observations in this section are useful to explain the results of our simulation and experimental studies in Sections 3 and 4. We first show that a linearly dependent set of vectors in $(+, \times)$ algebra is not necessarily linearly dependent in Boolean algebra and vice versa. This assertion can be verified in the following examples. The set of four vectors: $d_1 = \{1, 1, 1, 1\}, d_2 = \{0, 0, 1, 1\}, d_3 = \{1, 1, 1, 0\}$ and $d_4 = \{0, 1, 0, 1\}$ is linearly dependent in Boolean algebra as $d_1 = d_2 \vee d_3 \vee d_4$, but is linearly independent in $(+, \times)$ algebra; whereas the set of four vectors: $d_1 = \{1, 0, 0, 1\}, d_2 = \{0, 0, 1, 1\}, d_3 = \{1, 1, 0, 0\}$ and $d_4 = \{0, 1, 1, 0\}$ is linearly dependent in $(+, \times)$ algebra as $d_1 = d_2 + d_3 - d_4$ but not in Boolean algebra. It is not difficult to verify that for the Boolean model, the necessary and also sufficient condition for a set of vectors to be dependent is that one vector has entries of 1s at all the positions where other vectors have entries of 1s. Clearly, this statement does not apply for the continuous model as shown in the first example.

3   Numerical Evaluations

We conducted a series of numerical studies in order to obtain a preliminary comparison of the efficiency of the continuous and Boolean models in monitoring end-to-end network properties. We perform our investigations on three backbone ISP topologies with sizes ranging from small (Exodus: 80 nodes and 147 links) to medium (Telstra: 115 nodes and 153 links), and large (Tiscali: 164 nodes and 328 links). For the sake of simplicity, we assume that all the ISPs use shortest path routing to route traffic. In this section, we summarize our findings and provide explanations for the results in the context of the Rocketfuel topologies.

Recall that n is the number of nodes in the network. In our experiments, the number of beacons $|\mathcal{V}_B|$ is varied from $n/50$ to $n/2$. We select the beacon candidates randomly by picking a random permutation of the set of nodes in the network. After building the routing matrix as in Section 2, we first calculate the rank of the routing matrix D to obtain the number of end-to-end measurements for the continuous model, and then use the PS algorithm in [2] to find the number of end-to-end measurements for the Boolean model. For each topology, we plot the percentage of independent paths returned by the PS algorithm (for the Boolean model) and the rank of the matrix D (for the continuous model) for different numbers of beacons.

Fig. 1 shows the results for the Exodus topology. We observe that for a small number of beacons $|\mathcal{V}_B|$ (less than 5%) the Boolean model requires fewer direct measurements, whereas for a larger number of beacons $|\mathcal{V}_B|$ (between 10%-40%) the continuous model requires fewer direct measurements. However, when the

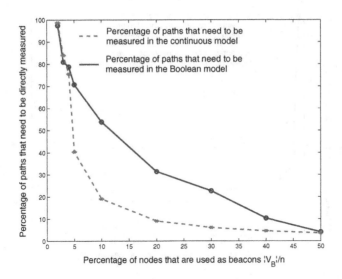

Fig. 1. Percentage of paths that need to be measured directly for complete determination of the quality of all paths in the Exodus topology with 80 nodes and 147 links

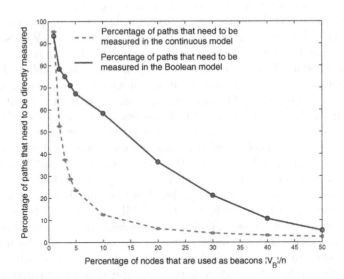

Fig. 2. Percentage of paths that need to be measured directly for complete determination of the quality of all paths in the Telstra topology with 115 nodes and 153 links

percentage of beacons is above 50%, both models require the same number of direct measurements.

Similar observations can be found for the Telstra and Tiscali topologies, as shown in Fig. 2 and 3, even though the exact percentages at which the two curves representing the binary and continuous model cross are different in each topol-

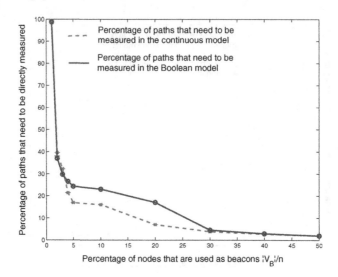

Fig. 3. Percentage of paths that need to be measured directly for complete determination of the quality of all paths in the Tiscali topology with 164 nodes and 328 links

ogy. The results suggest that although all topologies exhibit the same tendencies for the number of direct measurements in the continuous and Boolean model, exact results in each network are sensitive to the network topology. We also observe that for all three topologies, the percentages of required direct measurements for both continuous and Boolean models drop rapidly with the number of beacons when the latter is small. However, these percentages decrease slowly when the number of beacons is large. This observation suggests that the benefit of introducing additional beacons lessens with the increase of the number of beacons.

The above observations can be explained by the structure of a basis of a vector space in $(+, \times)$ algebra and the basis of a vector span in Boolean algebra. In the Boolean model, and contrary to the continuous model, if all links in a path are also present in some other paths, then the first path is redundant because it is a linear combination of the other paths. When the number of paths is small, this situation frequently occurs in the studied ISP networks. In $(+, \times)$ algebra, the rank of a matrix is upper bounded by the minimum of the number of rows and columns in the matrix. Hence, when we initially increase the number of paths (i.e., the number of rows) the dimension of the basis rapidly increases. But when the rank of the matrix approaches the number of columns, which stays almost constant, it increases only slowly. When there are a lot of paths, both the Boolean vector span and continuous vector space have the same basis that contains mostly unit vectors (vectors that have only one 1 entry).

4   Internet Evaluation

In this section, we compare the performance of the continuous and the binary models on the data collected from the NLANR's AMP (Active Measurement

Project) [14] measurement infrastructure, which performs site-to-site active measurements between campuses connected by high performance networks (most of them at NSF-funded HPC-sites). The NLANR measurement infrastructure consists of approximately 140 monitors and about 15,000 end-to-end paths. Each monitor performs RTT, topology and throughput tests to all other monitors. More details can be found at [14]. We have analyzed the data collected by AMP on the 23rd of January, 2004. After grouping the common edges and nodes in all end-to-end paths, we have a sample of 133 monitors and 17556 end-to-end paths that cover 9405 nodes and 36674 directed edges. Note here that the number of available monitors (133 monitors) is significantly smaller than the number of nodes (9405 nodes) in the systems. The AMP system is therefore operating in the regime where only a small percentage (below 1.4%) of nodes are used as beacons. To evaluate the effect of the number of beacons on the two models, we vary the number of beacons from 1 to 133, which corresponds to 0.01%-1.4% of the total number of nodes. We then construct the routing matrix D and calculate the rank of D to obtain the number of end-to-end measurements for the continuous model. We also calculate the number of independent paths in the Boolean model using the *PS* algorithm in [2]. The results are plotted in Fig. 4.

We observe that the results reflect the behaviors that we have already seen for the Rocketfuel topologies. That is, in the regime where the percentage of nodes that are used as beacons is very small (below 1% in this case) the Boolean model requires fewer direct measurements if we are only interested in separable "good" or "bad" performance of paths. However, the difference between the two models is very small for this range, which can be explained by the fact that in this case most of the end-to-end paths in the network are independent in both

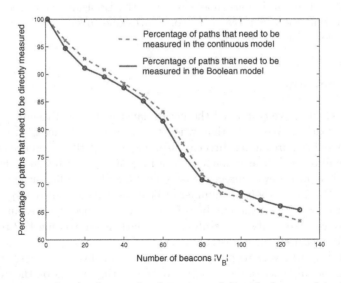

Fig. 4. Percentage of paths that need to be measured directly for complete determination of the quality of all paths in the NLANR's AMP monitoring system

Table 1. Accuracy of the continuous and Boolean models. CV is the coverage rate and FP is the false positive rate

Number of beacons	The continuous model	The Boolean model
2	CV = 99.7%, FP = 0.3%	CV = 98.7%, FP = 0.9%
100	CV = 97.2%, FP = 3.4%	CV = 95.6%, FP = 4.7 %

models. We suspect that the difference becomes more important, and in favor of the continuous model, for larger numbers of beacons. Furthermore, for both the continuous and Boolean models the percentage of direct measurements drops rapidly as the number of beacons increases.

We also evaluate the accuracy of the two approaches on predicting the quality of unmeasured paths. For this purpose, we proceed as follows. First, we fix the number of beacons. We then determine the set of independent paths in the continuous and Boolean models. We called these independent paths the *directly measured paths*. Using performance data from these directly measured paths, we calculate the performance on the unmeasured paths. That is, in the continuous model, we calculate the loss rates on all unmeasured paths and compare them against a loss threshold of 5%, which is the threshold between "tolerable loss" and "serious loss" as defined in [15], to determine whether the path is good or bad. In the Boolean model, we first determine the quality of the directly measured paths and then calculate the quality on the unmeasured paths. We compare the inferred results of the two models against the real measurement data. The results are given in Table 1. We observe that both models achieve a high coverage rate and a low false positive rate. The Boolean model is less accurate than the continuous model because the Boolean model relies on the assumption that the loss rates of network paths and links are separable, which sometimes does not hold in practice.

5   Conclusion

In this paper, we have compared the performance of two end-to-end path monitoring approaches. We show that when the number of beacons is small, the Boolean model requires fewer direct measurements; but the difference appear to be very small on real measurements. For a large number of beacons the continuous model requires fewer direct measurements, and the different can be quite significant. However, when the number of beacons is significantly large, there is no difference in terms of the number of paths that we need to measure directly in both models. We verify the results by simulations on existing ISP topologies and on real measurement infrastructure.

We are currently working on various extensions of this work. First, we are investigating the influence of the structure of the routing matrix on the differences between the number of probes required for the continuous and Boolean models. Second, so far in this work we have taken the restriction that nodes in the monitoring systems can send probes only to other nodes in the monitoring systems. However, since probes can be sent to many other nodes in the network, we are investigating how sending probes to other nodes in the network would affect the efficiency of the monitoring system in both the continuous and Boolean models.

Acknowledgements

Hung X. Nguyen's work is financially supported by grant DICS 1830 of the Hasler Foundation, Bern, Switzerland.

We would like to thank the researchers at the NLANR's Active Measurement Project for access to their raw data and the anonymous reviewer who gave us a pointer to these data.

References

1. Chen, Y., Bindel, D., Song, H., Katz, R.H.: An algebraic approach to practical and scalable overlay network monitoring. In: Proceedings of the ACM SIGCOMM, Portland (2004)
2. H.X.Nguyen, Thiran, P.: Active measurement for failure diagnosis in IP networks. In: Proceedings of the Passive and Active Measurment Workshop, Juan-les-Pins, France (2004) 185–194
3. Braynard, R., Kostic, D., Rodriguez, A., Chase, J., Vahdat, A.: Opus: an overlay peer utility service. In: Proceedings of the 5th International Conference on Open Architectures and Network Programming (OPENARCH). (2002)
4. Chen, Y.: Toward a Scalable, Adaptive and Network-aware Content Distribution Network. PhD thesis, University of carlifornia at Berkeley (2003)
5. Andersen, D.G., Balakrishnan, H., Kaashoek, M.F., Morris, R.: Resilient overlay networks. In: Proceeding of the 18th ACM Symp. on Operating System Priciples. (2001) 131–145
6. Coates, M., Hero, A., Nowak, R., Yu, B.: Internet tomography. IEEE Signal Processing Magazine **19** (2002)
7. Tang, C., McKinley, P.: On the cost-quality tradeoff in topology-aware overlay path probing. In: Proceedings of the IEEE ICNP. (2003)
8. Chua, D.B., Kolaczyk, E.D., Crovella, M.: Efficient monitoring of end-to-end network properties. private communication (2004)
9. Padmanabhan, V.N., Qiu, L., Wang, H.J.: Server-based inference of internet performance. In: Proceedings of the IEEE INFOCOM'03, San Francisco, CA (2003)
10. N.Duffield: Simple network perormance tomography. In: Proceedings of the IMC'03, Miami Beach, Florida (2003)
11. Bejerano, Y., Rastogi, R.: Robust monitoring of link delays and faults in IP networks. In: Proceedings of the IEEE INFOCOM'03, San Francisco (2003)
12. Horton, J., Lopez-Ortiz, A.: On the number of distributed measurement points for network tomography. In: Proceedings of IMC'03, Florida (2003)
13. Spring, N., Mahajanand, R., Wetherall, D.: Measuring ISP topologies with Rocketfuel. In: Proceedings of the ACM SIGCOMM. (2002)
14. AMP web site: http://watt.nlanr.net/. (Accessed January 2005)
15. Zhang, Y., Duffield, N., Paxson, V., Shenker, S.: On the constancy of internet path properties. In: Proceedings of ACM SIGCOMM Internet Measurement Workshop, San Francisco (2001)

Exploiting the IPID Field to Infer Network Path and End-System Characteristics

Weifeng Chen[1], Yong Huang[2], Bruno F. Ribeiro[1], Kyoungwon Suh[1],
Honggang Zhang[1], Edmundo de Souza e Silva[3], Jim Kurose[1],
and Don Towsley[1]

[1] Department of Computer Science,
University of Massachusetts at Amherst, MA 01002, USA
{chenwf, ribeiro, kwsuh, honggang, kurose, towsley}@cs.umass.edu
[2] Department of Electrical and Computer Engineering,
University of Massachusetts at Amherst, MA 01002, USA
yhuang@ecs.umass.edu
[3] COPPE and Computer Science Department,
Federal University of Rio de Janeiro (UFRJ),
Cx.P. 68511 Rio de Janeiro, RJ 21945-970 Brazil
edmundo@land.ufrj.br

Abstract. In both active and passive network Internet measurements, the IP packet has a number of important header fields that have played key roles in past measurement efforts, e.g., IP source/destination address, protocol, TTL, port, and sequence number/acknowledgment. The 16-bit identification field (IPID) has only recently been studied to determine what information it might yield for network measurement and performance characterization purposes. We explore several new uses of the IPID field, including how it can be used to infer: (a) the amount of internal (local) traffic generated by a server; (b) the number of servers in a large-scale, load-balanced server complex and; (c) the difference between one-way delays of two machines to a target computer. We illustrate and validate the use of these techniques through empirical measurement studies.

Keywords: IPID field, one-way delay difference, traffic activity, load-balanced server counting, estimation.

1 Introduction

In both active and passive network Internet measurements, the fundamental unit of measurement - the IP packet - includes a number of important header fields that have played key roles in past measurement efforts: IP source/destination address, protocol, TTL, port, and sequence number/acknowledgment. The 16-bit identification field (referred to here as the IPID field) has only recently been used to determine what information it might yield for network measurement and performance characterization purposes [3, 6, 9, 11, 7]. In this paper, we explore several new uses of the IPID field, including how it can be used to infer: (a)

C. Dovrolis (Ed.): PAM 2005, LNCS 3431, pp. 108–120, 2005.

the amount of internal (local) traffic generated by a server; (b) the number of servers in a large-scale, load-balanced server complex and; (c) the difference between one-way delays of two machines to a target computer. We illustrate and validate the use of these techniques through empirical measurement studies.

The remainder of this paper is structured as follows. In the following section we classify and discuss past work that has examined the use of the IPID field, and place our current work in this context. In Section 3, we describe a technique to infer the amount of a host's traffic that remains internal to its local network, and the complement amount of traffic that passes through a measured gateway link. In Section 4, we describe a technique to identify the number of load-balancing servers behind a single IP address. In Section 5, we introduce a technique to infer the difference between one-way delays. Section 6 concludes this paper with a discussion of future work.

2   Uses of the IPID Field

We begin with a brief description of the IPID field and the generation of IPID values, and then classify previous measurement work, as well as our current efforts, into three categories based on their use of the IPID field.

The 16-bit IPID field carries a copy of the current value of a counter in a host's IP stack. Many commercial operating systems (including various versions of Windows and Linux versions 2.2 and earlier) implement this counter as a global counter. That is, the host maintains a *single* IPID counter that is incremented (modulo 2^{16}) whenever a new IP packet is generated and sent. Other operating systems implement the IPID counter as a per-flow counter (as is done in the current version of Linux), as a random number, or as a constant, e.g., with a value of 0 ([3]).

2.1   Global IPID

In this paper, we only consider hosts that use a single *global* counter to determine the IPID value in a packet. To infer whether a host implements a global IPID counter, we probe the host from two different machines by sending http requests. IPID values in the packets returned from the host can be obtained by running tcpdump on the two probing machines separately. If the host uses a global IPID counter, these replying IPID values will belong to a unique sequence. By synchronizing the two probing machines, we are able to compare the replying IPID values, as presented in Figure 1. This figure clearly shows that the IPID values of the packets returned to the two probing machines belong to a unique sequence, and consequently, we can infer that this host uses a global IPID counter.

Instead, if the host does not implement a global IPID counter we obtain a different result. Figure 2 shows the result of a host implementing the IPID counter as a per-flow counter. The IPID values of the replying packets to the two different probing machines consist of two independent sequences, each corresponding to one probing machine. Note that the slopes of these two sequences

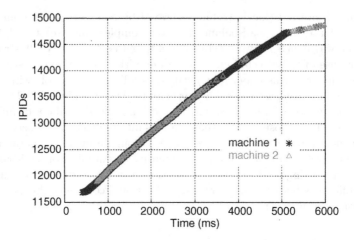

Fig. 1. IPID values returned from a global-IPID host

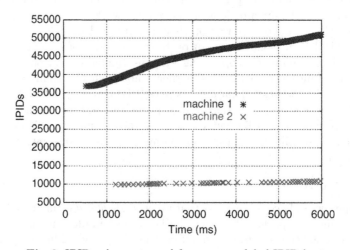

Fig. 2. IPID values returned from a non-global-IPID host

are different because of the different speeds of the probing packets sent from the probing machines.

Using this process, we probed the web-servers of the top 50 companies ranked by *Fortune* magazine [1] and found that 18 (36%) of them have a global IPID counter. Among the top 101 web sites ranked by *PC Magazine* [2], 40 of these web sites were found to have a global IPID counter.

2.2   Classifications of Using IPID Fields

We can broadly classify previous efforts, as well as our current efforts, using IPID sequences into three categories:

Application 1: Measuring traffic activity. Suppose that we observe a subset of the packets generated by a server, and consider the $(i-1)$-st and i-th observed packets. Let $T(i)$ denote the timestamp of the i-th packet and ΔIPID(i) the difference between the IPID values of the $(i-1)$-st and i-th packets[1]. In this case, $\sum_{i=1}^{n} \Delta$IPID(i) represents the number of packets sent by this server in the interval $(T(1), T(n))$. The use of IPID values to infer the total amount of outgoing server traffic is noted in [7]. We additionally note that, for stub networks with a single outbound connection, this also allows us to infer the relative amount of traffic sent to destinations within the network, and to destinations outside of the network. From this single measurement point, we can thus infer one aspect (local/remote) of the spatial distribution of traffic destinations. We consider this approach in Section 3.

Application 2: Clustering of sources. These applications make use of the fact that different hosts have independent (and thus generally different) IPID values, and that IPID values are incremented for each outgoing IP packet sent by a host. We denote the difference in the values of the IPID field of two successively observed packets as ΔIPID. Thus, if we observe two packets generated by the same host within a "short" interval of time, we will generally observe a small ΔIPID value. By identifying small ΔIPID values among a set of IP packets that were generated within a short interval of time from multiple sources, it is then often possible to identify packets coming from the same source. It is important to note that IPID-based source-identification is thus possible without actually examining the source IP address, which itself may have been aliased. Router alias detection [11], host alias detection and load-balanced multiplexed-server counting [7], and NATed host counting [3] all exploit this observation. Our work in Section 4 builds on initial suggestions in [7] by considering a specific algorithm for identifying the number of servers behind a load-balancer using only observed IPID values.

Application 3: Identifying packet loss, duplication and arrival order. Since a packet generated later in time by a host will carry a larger IPID (modulo 2^{16}) than a packet generated earlier in time by that host, it is possible (after solving the wrap-around problem) to determine the order in which packets are generated by a host. Previous work on detecting packet reordering and loss between a probing host and a router [9] and duplicate packet-detection and re-ordering at a passive monitor [8] exploit this observation. In Section 5, we use the fact that the IPID value of a packet generated in response to a received packet indicates the order in which received packets arrived to develop a new approach for inferring the absolute differences in one-way delays between a set of machines and a target host.

Several technical challenges must be met when using IPIDs in measurement studies. The most important regards wrap-around between two consecutively

---

[1] We may obtain a negative value for ΔIPID(i) due to wrap-around. We address this problem later in this paper.

observed packets from the same source. Correction is easy if we know that only a single wrap-around has occurred. With active probing techniques (where the measurement point sends active probes to a host and observes the IPID of the returned packet), multiple wrap-arounds can be avoided by choosing an appropriate probing interval. In a passive monitoring framework, a more sophisticated method is needed to deal with multiple wrap-arounds, as discussed in the following section.

3   Outbound Traffic from a Server

In this section, we present a simple technique for measuring the outbound traffic from a server (i.e., the number of packets sent by a server) by passively observing the IPIDs of packets generated by that server at a gateway. The use of active probes to infer the total amount of outgoing server traffic was suggested in [7]. Passive measurement avoids the overhead of active probing, and the attention that active probing may bring (indeed, several of our active probing experiments resulted in our measurement machines being black-listed at a number of sites!). We will see shortly, however that it is valuable to augment passive probing by occasionally sending active probes in order to handle IPID wrap-around.

Suppose that, at a gateway, we observe a subset of the packets generated by a server, and consider the $(i-1)$-st and i-th packets observed. Let $T(i)$ denote the timestamp of i-th packet and $\Delta \mathrm{IPID}(i)$ denote the difference between the IPID values of the $(i-1)$-st and i-th packets. In this case, $\sum_{i=1}^{n} \Delta \mathrm{IPID}(i)$ represents the total number of packets sent by this server during the interval $(T(1), T(n))$. Furthermore, if the server accesses the larger Internet only through this gateway, we know that all other packets generated between the $(i-1)$-st and i-th observed packets must have been sent to destinations within the network - providing an easy means to determine the amount of network-internal traffic being generated by a server.

We performed experiments on several popular web servers in our campus. One result is plotted in Figure 3. Since we could not instrument the server, we validated our measurements using periodic active probes. As shown in Figure 3, this result is consistent with that obtained using active probes. Figure 4 shows the amount of network-internal traffic from the server as determined by our proposed passive approach.

With a purely passive approach to measuring server activity, it can be difficult to detect IPID wrap-around if the amount of traffic observed at the monitor point is very small compared to the amount of network-internal traffic generated by the server. Indeed, in our experiments, we observed popular web servers in our campus that did not serve clients outside of our campus for long periods of time. To solve this problem, we adopt a *hybrid* approach in which adaptively-activated active measurement is used to supplement passive measurement. Specifically, we use an Exponential Weighted Moving Average (EWMA) to estimate the rate of IPID increase. Using this estimate, we can then estimate the next IPID wrap-around time, $T^{*}(\mathrm{msec})$, and start a timer with that value. Whenever we observe

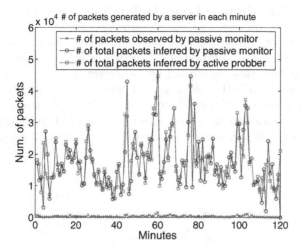

Fig. 3. Comparison between passively measured and actively measured outbound traffic from a server

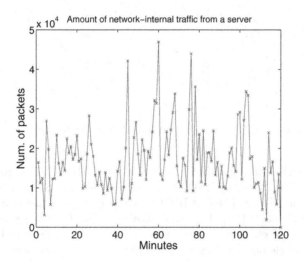

Fig. 4. Amount of network-internal traffic from a server

a new packet before this timer expires, we reset the timer based on the current estimated IPID rate. If the timer expires, we launch an active probe and reset the timer. We are currently performing additional work to evaluate this hybrid approach.

4   Inferring Number of Load-Balancing Servers

If each load-balancing server behind a single IP address has an independent global IPID counter, packets generated by one server have a sequence of IPID values that differs from those generated by a different server. As discussed below, using these observed IPID values, we can classify the packets into distinct sequences, with the number of distinct sequences being an estimate for the number of servers. Figure 5 shows the observed IPID values of the packets generated from a large commercial web server in response to the 5000 probing packets we sent to the server.

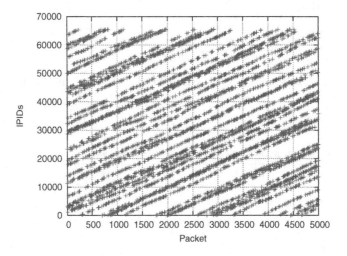

Fig. 5. IPIDs of the packets returned from the web server

We next describe an algorithm to classify the packet IPID sequences. Let $\{I_1, I_2, \ldots, I_{5000}\}$ be the set of IPIDs shown in Figure 5 and $\mathcal{S}$ the set of distinct sequences. Initially, $\mathcal{S} = \emptyset$. The first IPID I_1 is appended to sequence S_1 (namely, $S_1 = \{I_1\}$) and $\mathcal{S} = \mathcal{S} \cup \{S_1\}$. For each following IPID I_j ($2 \leq j \leq 5000$), I_j is compared to the tail element of all sequences in $\mathcal{S}$. If the difference between I_j and all of the tail elements is larger than a threshold T, a new sequence $S_{|\mathcal{S}|+1}$ is created and $S_{|\mathcal{S}|+1} = \{I_j\}$. Additionally, $\mathcal{S} = \mathcal{S} \cup \{S_{|\mathcal{S}|+1}\}$. Otherwise, I_j is appended to the sequence whose tail element has the smallest difference with I_j. Given T, the algorithm returns the number of sequences, i.e., $|\mathcal{S}|$, and the corresponding sequence sizes, i.e., the number of packets in each sequence.

Our algorithm will return a different number of sequences of different sizes for different values of T. Ideally, the sequence sizes should be equal, with probing packets being forwarded at equal rates to the servers. In practice, however, these rates are close but not equal, due to the mixing of probing packets with other traffic. For this experiment the interval between two successive probing packets was set to 3ms to minimize these effects.

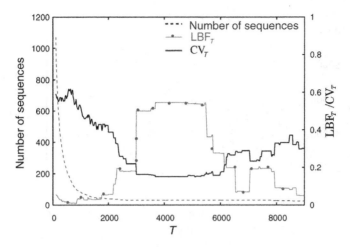

Fig. 6. Number of sequences, LBF_T and CV_T vs. T

To determine an appropriate T, we introduce two parameters: load balancing factor (LBF) and coefficient of variation (CV) of sequence size. We define LBF_T as $LBF_T = P_{min}/P_{max}$, where P_{min} (resp. P_{max}) is the number of packets in the smallest (resp. largest) sequence returned by the algorithm with a given T. Ideally, for a well balanced server, an appropriate T should produce a LBF_T very close to 1. The second parameter, CV_T, is defined as $CV_T = \sigma_T/\mu_T$, where σ_T and μ_T are the standard deviation and the mean of the sequence sizes respectively for a given T. Intuitively, an appropriate T results in a small CV_T.

Figure 6 shows LBF_T, CV_T and the number of sequences as a function of T. A T is *appropriate* when LBF_T achieves the maximum and CV_T achieves the minimum. The figure indicates that a $T \approx 4000$ is appropriate, resulting in 30 sequences. That is, we estimate that the web server has 30 load-balancing servers. Table 1 shows the numbers of packets in these 30 sequences.

Table 1. Number of packets in classified sequences

162, 165, 180, 155, 156, 131, 188, 136, 178, 186
170, 162, 167, 228, 208, 193, 158, 177, 144, 169
145, 145, 168, 192, 177, 124, 173, 129, 202, 132

Based on this value of T, the algorithm described above divides the IPID values shown in Figure 5 into 30 sequences. We plot these 30 sequences in different colors in Figure 7 where wraparounds of each sequence were removed by adding 64K to the values so that every sequence is always monotonically increasing. We observe that the slopes of all of these 30 sequences are almost the same, which suggests that each load-balancing server receives a comparative number of probing packets.

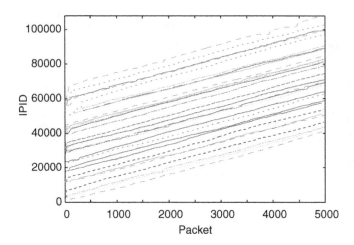

Fig. 7. IPID sequences of the load-balancing servers of a commercial web-server

5   Inferring One-Way Delay Differences

In this section, we present a simple technique that uses the IPID field to infer the
differences in one-way delays from a set of GPS-synchronized probing sources
to an "IPID order capable" destination target. By "IPID order capable" we
mean that the destination has a global IPID counter and that wrap-arounds
can be detected. Importantly, we do *not* require the destination to be GPS-
synchronized. Such delay differences can be used to infer shared path segments
using recently developed network tomograpic techniques [10, 4, 5]. In addition,
if one of the sources is able to determine (or accurately estimate) the absolute
magnitude of its one-way delay to the destination, then all other nodes can
determine the absolute values of their one-way delays as well. Knowledge of
one-way delay can be valuable in many circumstances.

[10] presents a methodology for estimating one-way delay differences from
non GPS-synchronized (or coarsely synchronized) sources to common destinations using a semi-randomized probing strategy and the packet arrival ordering
collected at destinations. Given GPS-synchronized source clocks, the deterministic probing strategy we study is considerably simpler. As in [10], the key idea
is for sources to send probes (e.g., ICMP echo packets) to a remote host, and use
the observed arrival ordering to infer path characteristics. Our approach differs
from [10] in the way we obtain arrival order information. In [10] all destination
machines must be instrumented. Using IPID, we are able to obtain the packet
arrival orders without instrumenting any destination machine. In the following,
we consider only two source nodes; the approach easily generalizes to the case
of additional source nodes.

Our goal is to infer the one-way delay difference from two GPS-synchronized
sources A and B to a destination D, i.e., the difference between path delays
d_{AD} and d_{BD}. Consider two packets p_1 and p_2 sent from A and B to D at the

same time. If p_1 arrives before p_2, the IPID, I_1, of the packet returned by D in response to p_1 will be smaller (modulo 2^{16}) than the IPID, I_2, of the packet responding to p_2.

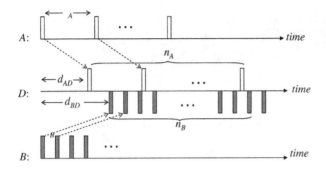

Fig. 8. Arriving orders of packets

We exploit this ordering of returned IPID values as follows. As illustrated in Figure 8, A and B begin simultaneously probing D using different probing intervals δ_A and δ_B, respectively. The n_A-th packet sent from A arrives at D between the $(n_B - 1)$-st packet and the n_B-th packet sent from B. If the delay does not change significantly during the measurement interval, we have:

$$d_{BD} + (n_B - 1)\delta_B \le d_{AD} + n_A\delta_A \le d_{BD} + n_B\delta_B$$
$$\Rightarrow (n_B - 1)\delta_B - n_A\delta_A \le d_{AD} - d_{BD} \le n_B\delta_B - n_A\delta_A$$

Note that the difference between the upper- and lower-bounds depends on δ_B. Thus by reducing δ_B, we can improve the accuracy of the inferred delay difference $d_{AD} - d_{BD}$. We conjecture that we can extend these techniques to handle the case of varying delays during the measurement interval as well.

We have validated the approach in a simple test scenario. In our experiments we send *ICMP echo* packets from source machine A (at Unifacs, a university in Brazil) and B (a machine at the University of Minnesota) to a destination machine D at the University of Massachusetts. Machine A sends one packet per second and machine B sends one packet every 3ms. Our measurements indicate that the IPID-inferred delay difference, namely, $d_{AD} - d_{BD}$, is around 230ms. We also send probes from A and B to a GPS-equipped machine, D', at the University of Massachusetts that was close to D. Based on the recorded data on D', we can measure $d_{AD} - d_{BD}$. Figure 9 shows the difference of the measured values and the IPID-inferred values as a function of time. From the figure, one can see that the inferred values are very close to the measured values. Furthermore, it should be noticed that most of the differences are within 3ms for $\delta_B = 3$ms. Figure 10 shows the relative error of the IPID-inferred values (I) to the measured values (M), where the relative error is defined as $(I - M)/M$.

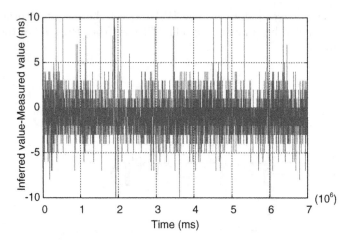

Fig. 9. Difference of IPID inferred $d_{AD} - d_{BD}$ and measured $d_{AD} - d_{BD}$

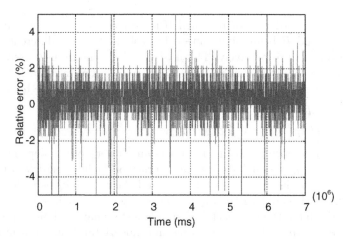

Fig. 10. Difference of IPID inferred $d_{AD} - d_{BD}$ and measured $d_{AD} - d_{BD}$

Table 2. Statistical results for several one-way delay inference experiments

Experiment ID	Time of day (EST)[1]	Mean of error[2] (ms)	Standard deviation of error, σ (ms)	Mean of measured values, $\bar{M}$ (ms)	$\sigma/\bar{M}$
1	2004-10-14 11pm	2.02	9.73	232.17	0.042
2	2005-01-14 9pm	3.37	18.57	354.49	0.052
3	2005-01-15 6am	3.53	22.27	361.14	0.062
4	2005-01-17 2pm	11.44	23.16	356.24	0.065
5	2005-01-17 5pm	9.53	20.38	361.11	0.056

[1] The beginning time when an experiment was conducted.
[2] error=$|I - M|$.

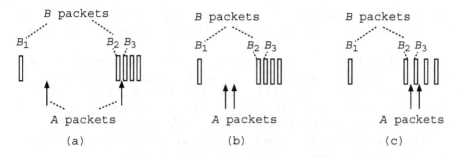

Fig. 11. Jitter examples

Statistical results from the experiment shown in Figure 9 are presented in the first line of Table 2, which includes the results of four other experiments. This table contains a broad set of experimental scenarios. We run experiment 2 and 3 with $\delta_B = 4$ms (Figure 8). Experiments 4 and 5 also used $\delta_B = 4$ms but during the busiest traffic hours. From the results we can see that during the busiest hours the one-way delay inference error becomes larger. This effect is due to the higher inter-packet jitter in our measures. Figure 11 depicts how higher jitters affect inter-arrival times of A and B packets at D (Figure 8). A packet from A is more likely to arrive between packets B_1 and B_2 than between packets B_2 and B_3. Thus the probability that packets from A will arrive between two given consecutive B packets increases with the jitter delay value. This problem could be ameliorated by sending two A packets with smaller sending intervals, i.e., a pair of A packets. These pairs can eliminate samples where both packets in a pair arrived between two identical B packets (Figure 11(b)). Samples where two packets of a pair interleaved with two B packets (Figure 11(c)) will produce more accurate inferences.

6   Conclusions

In this paper, we explored several uses of the IPID field for inferring network path and end-system characteristics. We classified previous IPID-related measurement efforts into three general application areas, and showed that, by using the IPID field, it is possible to infer: (a) the amount of internal (local) traffic generated by a server; (b) the number of servers in a large-scale, load-balanced server complex and; (c) the difference between one-way delays of two machines to a target computer. We illustrated and validated the use of these techniques through empirical measurement studies.

As with previous measurement techniques exploiting other packet header fields, header fields (such as the TTL and IPID fields) can be exploited for measurement purposes not initially envisioned in the design of IP. We hope that our work will add to the toolkit of network measurement techniques. We also hope that future measurement studies can build on this work, and that additional clever ways will be found to exploit the IPID field for measurement purposes.

Acknowledgments. The authors would like to thank the anonymous reviewers for their helpful comments. This research has been supported in part by the NSF under grant awards EIA-0131886, EIA-0080119, ANI-0085848, ANI-0070067, ANI-0319871, and ANI-0240487 and by the ARO under DAAD19-01-1-0610 and by CAPES (Brazil). Any opinions, findings, and conclusions or recommendations expressed in this material are those of the author(s) and do not necessarily reflect the views of the National Science Foundation.

References

1. "List of the fortune 500," http://www.usatoday.com/money/companies/2004-03-22-fortune-500-list_x.htm.
2. "Top 101 wet sites: Fall 2004," http://www.pcmag.com/category2/0,1738,7488,00.asp.
3. S. Bellovin. A technique for counting NATed hosts. In *Proc. ACM Internet Measurement Workshop(IMW)*, November 2002.
4. M. Coates and R. Nowak. Network tomography for internal delay estimation. In *Proc. IEEE International Conference on Acoustics, Speech and Signal Processing*, May 2001.
5. F. Lo Presti, N. Duffield, J. Horowitz, and D. Towsley. Multicast-based inference of network-internal delay distributions. *IEEE/ACM Trans. Networking*, 10:761–775, 2002.
6. A. Hussain, J. Heidemann, and C. Papadopoulos. A framework for classifying denial of service attacks. In *Proc. ACM SIGCOMM*, August 2003.
7. Insecure.org. Idle scanning and related IPID games. http://www.insecure.org/nmap/idlescan.html.
8. S. Jaiswal, G. Iannaccone, C. Diot, J. Kurose, and D. Towsley. Measurement and classification of out-of-sequence packets in a tier-1 IP backbone. In *Proc. IEEE INFOCOM*, April 2003.
9. R. Mahajan, N. Spring, D. Wetherall, and T. Anderson. User-level internet path diagnosis. In *Proc. ACM Symp. on Operating Systems Principles (SOSP)*, October 2003.
10. M. Rabbat, M. Coates, and R. Nowak. Multiple source, multiple destination network tomography. In *Proc. IEEE INFOCOM*, March 2004.
11. N. Spring, R. Mahajan, and D. Wetherall. Measuring ISP topologies with Rocketfuel. In *Proc. ACM SIGCOMM*, August 2002.

New Methods for Passive Estimation of TCP Round-Trip Times

Bryan Veal, Kang Li, and David Lowenthal

Department of Computer Science,
The University of Georgia,
Athens, GA 30602, USA
{veal, kangli, dkl}@cs.uga.edu

Abstract. We propose two methods to passively measure and monitor changes in round-trip times (RTTs) throughout the lifetime of a TCP connection. Our first method associates data segments with the acknowledgments (ACKs) that trigger them by leveraging the TCP timestamp option. Our second method infers TCP RTT by observing the repeating patterns of segment clusters where the pattern is caused by TCP self-clocking. We evaluate the two methods using both emulated and real Internet tests.

1 Introduction

Round-trip time (RTT) is an important metric in determining the behavior of a TCP connection. Passively estimating RTT is useful in measuring the the congestion window size and retransmission timeout of a connection, as well as the available bandwidth on a path [1]. This information can help determine factors that limit data flow rates and cause congestion [2]. When known at a network link along the path, RTT can also aid efficient queue management and buffer provisioning. Additionally, RTT can be used to improve node distribution in peer-to-peer and overlay networks [3].

Our work contributes two new methods to passively measure RTT at an interior measurement point. The first method works for bidirectional traffic through a measurement point. It associates segments from the sending host with the ACK segments that triggered their release from the sender. Our method uses TCP timestamps to associate data segments with the acknowledgments that trigger them. Since the other direction is easy—associating acknowledgments with the data segments they acknowledge—we can obtain a three-way segment association. Thus, we have a direct and simple solution that can collect many RTT samples throughout the lifetime of the connection.

There is no guarantee that the network route is symmetric, so only one direction of flow may be available to the measurement point. We introduce a second method to monitor a data stream and detect cyclical patterns caused by TCP's self-clocking mechanism. Because of self-clocking, a TCP connection's segment arrival pattern within one RTT is very likely to repeat in the next RTT. We use

C. Dovrolis (Ed.): PAM 2005, LNCS 3431, pp. 121–134, 2005.

algorithms that employ autocorrelation to find the period of the segment arrival pattern, which is the RTT. As with our previous method, we can take samples throughout the lifetime of a TCP session.

We show both methods to be accurate by evaluating them using both emulated and real network traces. For the emulated traces, we tested RTT estimates with network delays ranging from 15ms to 240ms, as well as with competing traffic over a bottleneck link using 0–1200 emulated Web users. The average RTT estimate for each delay tested was always within 1ms of the average RTT reported by the server. The maximum coefficients of variation (standard deviation/mean) were 3.79% for the timestamp based method and 6.69% for the self-clocking based method. Average RTT estimates for the tests with competing traffic were all within 1ms for the timestamp based method and 5ms for the self-clocking based method.

We also tested our RTT estimation methods with downloads from Internet FTP servers. Out of seven servers, the maximum coefficient of variation was 0.11% for the timestamp based method. For five of those servers, all RTT estimates for each server were within 1ms of each other using the self-clocking based method, and their average estimates were within 2.2ms of the average estimates from the timestamp based method.

2   Related Work

The method [4] uses segment association during the three-way handshake that initiates a TCP connection, as well as during the slow start phase. This takes advantage of the fact that the number of data segments sent can be easily predicted in advance. However, during the congestion avoidance phase, it is hard to predict the RTT based on the number of segments. Our method can associate a data segment with the ACK that triggered it, and thus it can follow changes in the RTT throughout the lifetime of a TCP session.

There is a method [5] to associate, throughout the lifetime of a session (including during congestion avoidance), a data segment with the ACK segment that triggered it. This method first generates a set of all possible candidate sequences of ACKs followed by data segments. Sequences that can be determined to violate basic TCP properties are discarded. The method then uses maximum-likelihood estimation to choose from the remaining possible sequences. This method is complex and would be cumbersome to implement as a passive estimation method at a device such as a router. Our method of using TCP timestamps to associate segments is simpler and more direct.

A previous work [6] introduces a method to passively measure RTT by mimicking changes in the sender's congestion window size. The measurement point must accurately predict the type of congestion control used: Tahoe, Reno, or NewReno. The accuracy of the estimate is affected by packet loss, the TCP window scaling option, and buggy TCP implementations. Our method avoids these difficulties by directly detecting the associations between segments.

3   TCP Timestamps

Both our RTT estimation methods use the TCP timestamp option. The original purpose of the option was to estimate the RTT at the sender for the purpose of deriving the retransmission timeout. The option adds two fields to the TCP header: timestamp value (*TSval*) and timestamp echo reply (*TSecr*). *TSval* is filled with the time at which the segment was sent, and *TSecr* is filled with the *TSval* of most recently received segment, with some exceptions. If a segment is received before a segment previous to it in the sequence arrives, leaving a hole, then the timestamp of the segment previous to the hole in the sequence is echoed. When this hole is filled by an out-of-order segment or a retransmission, the timestamp of the segment that fills the hole is echoed rather than the timestamp of a segment later in the sequence.

3.1   Timestamp Deployment

For timestamps to be useful for passive RTT measurement, the option should have a wide deployment and its implementation should be consistent across different hosts. We have developed a tool that can test the timestamp option on remote Web servers. This tool was run on 500 servers taken from the Alexa Global 500 list [7]. Of these, 475 servers responded to HTTP requests from our tool.

The tool tests for timestamp deployment by sending SYN segments with the timestamp option enabled and checking the SYN/ACK response for timestamps. Of the 475 responding servers, 76.4% support the TCP timestamp option. We expect timestamp deployment to increase over time. Furthermore, the self-clocking based RTT estimation method does not have to rely on TCP timestamps as the time unit used to associate segments into clusters. Other time units are possible, such as arrival time at the measurement point. We will address this possibility in future work.

3.2   Implementation Consistency

The tool also tests for implementation consistency. It tests the exceptions to echoing the most recent timestamp, described above. The tool sends three data segments with the last two out of order in sequence. The server should indicate the hole by sending a duplicate ACK with the timestamp of the first segment. When the client sends the last segment that fills the hole, the server should echo its timestamp. Of the servers tested that support TCP timestamps, 100% echoed the correct timestamp in both cases.

Another possible implementation error is to echo the timestamps of only data segments, disregarding ACKs that carry no data. Our tool tests for this possibility by sending an HTTP request to the server, receiving a data segment, sending an acknowledgment, and receiving more data. The congestion window is throttled to one byte to ensure that one segment is sent at a time. The second data segment from the server should echo the timestamp of the ACK and not the timestamp of the HTTP request. Of the servers tested, 99.4% correctly echo the timestamp of the the ACK.

3.3   Timestamp Granularity

The granularity chosen for TCP timestamps is implementation dependent. A fine granularity increases the accuracy and usefulness of both our RTT estimation methods, as shall be explained in later sections. Our tool tests granularity by sending data segments to the server at a known interval and then measuring the difference between the timestamps of the ACKs the server sends in response. Table 1 shows the distribution of timestamp granularity across the servers tested that support the timestamp option.

Table 1. Distribution of timestamp granularity

Granularity	Percent of Servers
500ms	0.6%
476ms	0.6%
100ms	36.9%
10ms	54.8%
1ms	7.2%

4   RTT Estimation Using Timestamps

Our first RTT estimation method method requires finding *associations* between TCP segments at an interior point along the route between the sender and receiver. The first segment in an association is a data segment from the sending end of a TCP connection. The second is the ACK segment from the receiving end that acknowledges receipt of the data segment. The third segment in the association is the next data segment from the sender, which is triggered when it receives the ACK. This assumes that the sender always has enough data ready to fill the congestion window as soon as more room becomes available.

Since multiple data and ACK segments may be in transmission concurrently, it is not obvious at an interior point which segments from one host have been triggered by the other. For the interior point to recognize an association, a segment must carry identification of the segment that triggered it. For the case of a data segment triggering an ACK, the acknowledgment number carried by the ACK is derived from the sequence number of the data segment. Thus the interior point can associate the two segments. However, the sequence numbers of ACK segments remain constant as long as the receiver sends no data. Because of this, it is impossible to use the acknowledgment number of a data segment to identify the ACK that triggered it.

The measurement point may use TCP timestamps instead of sequence numbers to associate segments. Timestamps are used only for association and not for calculating the RTT. Both the sender and receiver of a TCP session echo the most recently received timestamp, with minor exceptions in the cases of loss and segment reordering. The measurement point records the timestamps, their echoes, and arrival times of segments in each direction to estimate the RTT.

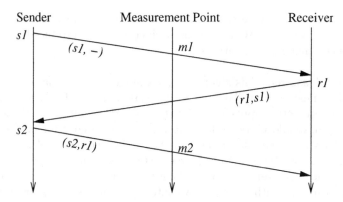

Fig. 1. Association of segments using TCP timestamps

Figure 1 provides an example. The sender transmits a segment at time $s1$. It arrives at the interior measurement point at time $m1$. The receiver responds with an ACK at time $r1$ and echoes the sender's timestamp, $s1$. The measurement point recognizes $s1$ in both segments and makes an association. Upon receiving the ACK, the sender transmits more data at time $s2$ and echoes the receiver's timestamp, $r1$. The measurement point receives this segment at time $m2$. It recognizes $r1$ in both segments and forms an association. Having associated all three segments, the measurement point estimates the RTT to be $m2 - m1$.

4.1   Constraints

Timestamp Granularity. The granularity of timestamps depends upon the TCP implementation of the sender. Even with a granularity as fine as 1ms, a burst of segments sent in a short interval may carry the same timestamp. The receiver may acknowledge parts of the burst at different times, but all the ACKs would carry the same timestamp echo. It would be difficult for the interior point to determine which data segments caused which ACKs. Since the first segment carrying a timestamp may be associated safely with the first segment carrying its echo, the algorithm only considers the first arriving segment with a particular timestamp and others with identical timestamps are discarded. However, a coarser timestamp granularity increases the the number of segments with identical timestamps, and thus allows for fewer measurements to be taken.

A side effect of preventing associations with ACKs containing old timestamps is that later ACKs containing the same timestamp echo as the discarded segment may be used to make an association, leading to an overestimate. To prevent this situation, only the first ACK with any particular timestamp echo is used to make associations.

Packet Loss. When the receiver is missing data due to packet loss, it sends duplicate ACKs. Since timestamp echoes are not updated when the receiver is missing data, this problem is automatically eliminated by discarding associations with ACKs that contain old timestamp echoes. However, when selective acknowledgments are enabled, overestimates can still occur. This problem is avoided by not considering selective ACK segments (which are only produced when loss is present), when making associations.

Interactive Sessions. This algorithm does not consider situations where the sender has no new data available when it receives and ACK. Such sessions are typically for interactive applications, such as ssh or telnet. Though not implemented here, it should be possible to obtain RTT estimates for interactive sessions based on some simple application heuristics. For example, in a typical session, when a user types a key, the character is sent to the server. Then the server echos the character back to the client to be displayed on the terminal. The client then responds with an ACK. An interior measurement point could take advantage of this to make an association for the three segments and estimate the RTT.

It is still possible that the sender has some delay in sending more data during a bulk transfer which could lead to an inflated RTT estimate at the measurement point. To filter such measurements, we have devised a method that tracks current maximum RTT for the session between the measurement point itself and each of the two hosts. These RTTs would be taken for only data-ACK pairs to avoid any possibility of sender delay. Any RTT estimates greater than the current sum of the two maximum delays would be discarded as an inflated estimate. We plan to evaluate this method as future work.

Asymmetric Routing. Though the RTT estimation algorithm requires both data and ACKs, there is no guarantee that both directions of traffic will follow the same route. However, it is still possible to obtain estimates using the second algorithm described in the next section.

5   RTT Estimation Using Self-Clocking Patterns

Our second algorithm detects patterns in a bulk data stream caused by a mechanism in TCP known as *self-clocking*. Capturing ACKs from the receiver is not required, so this algorithm maybe used for either asymmetric or symmetric routes. With self-clocking, the bulk data sender produces more data each time it receives an ACK, and the receiver sends an ACK each time it receives more data. Because of this, the the spacing between bursts of segments is likely preserved from one round trip to the next. Although packet losses and competing traffic could change the spacing and cause bursts to split or merge, the changes do not always happen frequently, and the bursts tend to persevere for at least a few round trips after each change. There may be multiple bursts of segments per round trip, and their size and spacing generally repeat every RTT. This algorithm detects the repetition of these burst-gap patterns to find the RTT. An example of such a pattern is shown in Fig. 2.

Discrete autocorrelation measures how well a data set is correlated with itself at an offset determined by the lag (l). If the correlation is strong, then the data

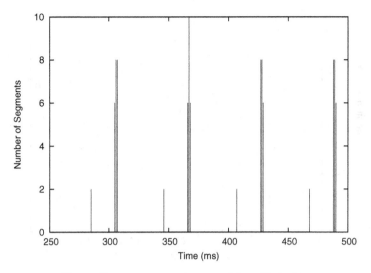

Fig. 2. Burst-gap pattern caused by self-clocking

matches its offset closely. Figure 3 shows the autocorrelation strengths for the data in Fig. 2. The strong correlation at 61ms corresponds closely to the RTT (which is 60ms).

Our algorithm uses autocorrelation to make RTT estimates. The algorithm repeats the RTT estimation once per *measurement interval*, T, which is supplied as a parameter. During this interval, the number of packets that arrive with timestamp t is stored in array $P[t]$ ranging from 0 to $T-1$. Once the count is complete, the discrete autocorrelation $A[l]$ is computed for each lag l from 1 to $l/2$. The RTT estimate is computed as $\max(A)$.

This process is repeated to produce multiple estimates throughout the session. The number of estimates depends upon the duration of the measurement interval and the duration of the session. However, more estimates may be taken by allowing measurement intervals to overlap.

5.1 Constraints

Timestamp Granularity. According to a theoretical limit known as the *Nyquist period*, it is only possible to measure RTTs at least twice the TCP timestamp granularity. For instance, if the granularity is 10ms, we can only detect RTTs of at least 20ms. This is a problem with timestamp granularities of 100ms or more. Although we do not explore it in this paper, a possible solution is to use arrival times at the measurement point rather than TCP timestamps from the sender.

Harmonic Frequencies. A consequence of a burst-gap pattern that repeats every RTT is a strong autocorrelation at multiples of the RTT that is sometimes stronger than that of the actual RTT. Rather than assuming that the strongest

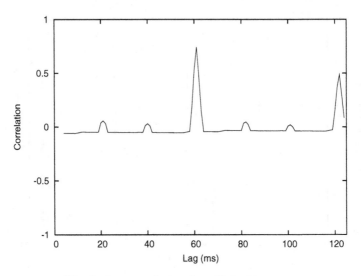

Fig. 3. Autocorrelation for self-clocking pattern

correlation corresponds to the RTT, the algorithm starts with the lag at the strongest correlation, s, and compares it to $A[\frac{s}{2}], A[\frac{s}{3}], A[\frac{s}{4}], \ldots$, until a certain limit is reached. If the correlation at the fractional lag is at least a certain percent of the actual lag, then that lag is considered the RTT instead. The limit of fractional lags and the percent of the maximum correlation are both provided as parameters to the algorithm.

Measurement Interval. The measurement interval chosen places an upper bound on the maximum RTT that can be measured. Autocorrelation becomes unreliable at a lag of half the measurement interval, since two complete round trips are needed to fully compare one round trip with its offset.

Delay Variation. While a smaller measurement interval may miss large RTTs, a larger measurement interval decreases the amount RTT variation that can be detected. If multiple strong correlations exist at different lags, this may indicate different RTTs at different times within a measurement interval. Our algorithm can report multiple candidate RTT estimates within an interval along with their correlation strengths.

Noise Effects on Self-Clocking Patterns. Congestion caused by competing traffic or other network conditions may disrupt the burst-gap pattern caused by self-clocking. One consequence of this is a high correlation at very small lags. This problem can be corrected by allowing a lower bound on the RTT to be specified as a parameter. We evaluate the effects of competing traffic on our algorithm in the next section.

6   Evaluation of RTT Estimation Methods

To evaluate these two passive RTT estimation methods, we have implemented them in a toolkit called TCPpet (TCP Passive Estimation Toolkit). Our implementations take traces captured by tcpdump and generate RTT estimates. The implementation of the timestamp method takes a trace with bidirectional TCP traffic and generates as many RTT estimates as the timestamp granularity will allow.

The implementation of the self-clocking method can use a trace with bidirectional or unidirectional traffic. Harmonic frequency detection is enabled for all experiments. If $\frac{1}{2}$, $\frac{1}{3}$, or $\frac{1}{4}$ of the lag having the strongest correlation has at least 75% of that correlation, it is taken as the RTT measurement. Note that 75% is measured from the minimum correlation in the measurement interval, which may be negative, instead of from zero. These values were chosen for good overall performance with the FTP downloads described later. Additionally, RTTs are assumed to be at least 10ms. Lags less than 10ms are not considered in order to correct for strong autocorrelations caused by noise effects. Note that if the Nyquist period is higher than 10ms, it becomes the minimum instead.

6.1   Emulation Test with a Single Flow

We have evaluated both RTT estimation methods with different network delays over an emulated network. The network consists of four machines: a sender, a receiver, and two routers, creating a three-hop route between the sender and receiver. Both the sender and receiver have timestamp granularities of 1ms. NIST Net [8] was used to add delay along the route by adding delay to both directions of traffic on each of the two routers. Thus, there are a total of four sources of delay along the route. If, for instance, a total round trip delay of 100ms is desired, then a delay of 25ms is added to each of the four points.

Traces were taken with tcpdump on the router closest to the sender. These traces were taken from the network interface that connects to the receiver's router, so that segments from the sender are delayed both before and after being recorded by tcpdump.

TCP data transfers were generated by the ttcp utility, which has been instrumented to report RTT estimates from the server's TCP implementation. All transfers were 16MB of data generated by ttcp.

Figure 4 shows all the RTT estimates for a network trace with a 60ms delay. Although the trace is longer, we only show the first 2 seconds for clarity. It includes estimates made by the server as well those made by the two passive estimation methods. A 250ms measurement interval was heuristically chosen for the self-clocking method. We plan to find a general default measurement interval as future work. As shown in the figure, nearly all the RTT measurements are close to those reported by the server. Note that the first few estimates for the server were influenced by preexisting state in the TCP implementation. After the 1s mark, the server estimates level off throughout the duration of the trace.

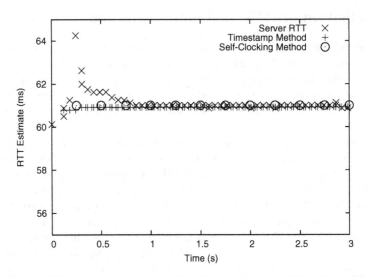

Fig. 4. RTT estimates for an emulated network trace with a 60ms delay

RTT estimates were taken for traces with 15, 30, 60, 120, and 240ms delays. Figure 5 shows the average of all the RTT estimates reported by each method for each trace. Here, a 500ms measurement interval was chosen for the self-clocking estimation method to accommodate possible RTTs up to 250ms. As shown in the figure, the averages are nearly identical. The largest coefficients of variation (standard deviation/mean) occurred for the 15ms delay trace, which were 3.79% for the timestamp based method and 6.69% for the self-clocking based method.

6.2   Emulation Test with Competing Flows

While the previous experiments show that our methods work well with different delays, real networks have conditions such as bottlenecks and competing traffic. The emulated network used previously was modified using NIST Net to limit the segments sent from the sender to the receiver to 10mb/s with a queue length of 13 packets. Delay was only added to the receiver side router, and it was set to 30ms for either direction of traffic. Thus, there was a total 60ms delay for the round trip.

To generate competing traffic, an Apache web server was run on the sender. Surge [9] was run on the receiver to generate HTTP requests to be served by the sender. Surge was configured with the default settings based on analyses in [10]. Traces were generated with `ttcp` while Surge was concurrently generating traffic. The traces were captured by `tcpdump` as described previously. A 250ms measurement interval was chosen for the self-clocking based method.

Figure 6 shows average RTT estimates when Surge is generating requests from 0, 200, 400, 600, 800, 1000, and 1200 emulated Web users. The error bars show the standard deviation of each set of estimates. Note that the initial 1s

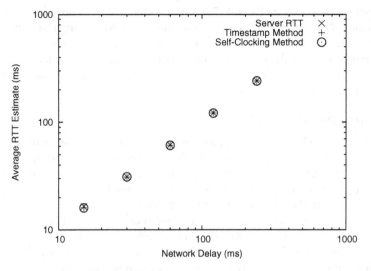

Fig. 5. Emulated network traces with varying delay

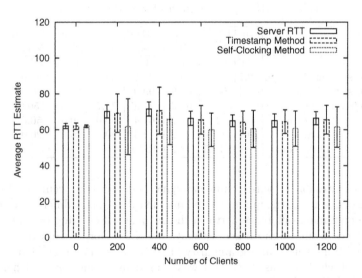

Fig. 6. RTT estimates with competing traffic from emulated Web users

period of server RTT estimates for each trace is discarded because the initial estimates were influenced by preexisting state in the TCP implementation.

The average RTT estimates from the timestamp based method are all within 1ms of that of the server. They are all within 5ms for the self-clocking method. High loss and retransmission rates were the principle causes of variation in the estimates, especially in the traces with 200 and 400 emulated users. In fact, for the self-clocking based method, 9.7% of the 250ms measurement intervals could not produce RTT estimates because no segments arrived during that time. Other intervals had one or two segments–too few to produce accurate estimates. We are currently investigating why traces with more than 600 emulated users had lower loss rates. Considering the severity of the effects of congestion, our average RTT estimates were very accurate.

6.3   Real Network Tests

To evaluate our RTT estimation methods on real networks, we performed FTP downloads from seven sites (Table 2). Traces were captured for the FTP data streams on the client with `tcpdump`. A measurement interval of 500ms was chosen for the self-clocking based method to accommodate possible RTTs up to 250ms. Note that RTT estimates for the servers' TCP implementations are not available since we do not have access to these machines. ICMP ping times were captured for reference at a later date than the traces. The RTT for the Sun server appears to have changed after the traces were taken. For the Intel server, the 100ms timestamp granularity was too coarse, so no valid RTT estimate could be produced. The Intel server also did not respond to ping requests.

Table 2. RTT estimates in ms for FTP downloads

Site	Ping	Timestamps Method			Self-Clocking Method		
		Avg.	Min.	Max.	Avg.	Min.	Max.
ftp10.us.freebsd.org	97.6	98.7	97.9	134.5	98.1	98.0	99.0
ftp.cs.washington.edu	59.7	62.0	60.7	87.0	60.0	60.0	60.0
ftp.cs.stanford.edu	62.4	65.3	64.0	99.9	65.1	65.0	66.0
ftp.jriver.com	60.5	75.3	60.6	101.6	75.7	40.0	150.0
docs-pdf.sun.com	90.7	52.2	51.7	61.4	50.0	50.0	50.0
ftp.cs.uiuc.edu	20.7	21.5	21.2	24.4	20.0	20.0	20.0
download.intel.com	—	110.6	98.2	127.1	—	—	—

The estimates for the self-clocking based method were very accurate. With the exception of one trace, the minimum and maximum estimates differed by at most 1ms. The maximum difference between the average self-clocking based estimate and the average timestamp based estimate was 2.2ms for the Sun trace, which had a high 10ms timestamp granularity. The 150ms estimate for JRiver trace from the self-clocking based method was caused by a single missed harmonic frequency. Similarly, the low 40ms was caused by two consecutive measurements during a series of half-RTT bursts.

Despite some of the high maximum estimates for the timestamp based method, the *number* of large estimates were few. The highest coefficient of variation was only 0.11% for the JRiver trace. While many of the other RTT estimates from the timestamp based method are affected by a few outliers, estimates for this trace have a more scattered distribution in both directions from the mean. This suggests that our method is detecting actual small-scale variations in RTT caused by conditions in the network route.

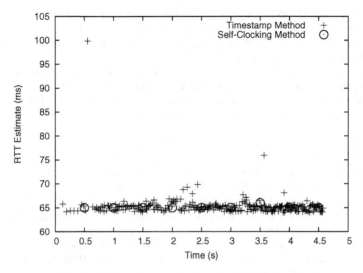

Fig. 7. RTT estimates for Stanford FTP download

Figure 7 shows the passive RTT estimates for the FTP download from Stanford as an example. Note that the timestamp based method is able to take many samples, while samples from the self-clocking based method are limited by the size of the measurement interval. All the estimates from the self-clocking based method are either 65 or 66ms. Nearly all the estimates from the timestamps based method are within 5ms of the average self-clocking estimate. Only two estimates larger than 70ms exist, which were likely caused by sender delay.

7   Conclusions

We have presented two new methods for passive estimation of round-trip times for bulk TCP transfers. These RTT estimations can be made at an interior point along the network route. One method uses TCP timestamps to locate segments from a bulk data sender that arrive one RTT apart, while the other detects patterns caused by self-clocking that repeat every RTT. Both methods can be used throughout the lifetime of a TCP session. The timestamp based method can be used for symmetric routes, while the self-clocking based method works for both symmetric and asymmetric routes.

References

1. Jain, M., Dovrolis, C.: End-to-end available bandwidth: measurement methodology, dynamics, and relation with tcp throughput. In: SIGCOMM, ACM (2002)
2. Zhang, Y., Breslau, L., Paxson, V., Shenker, S.: On the characteristics and origins of Internet flow rates. In: SIGCOMM, ACM (2002)

3. Ratnasamy, S., Handley, M., Karp, R., Shenker, S.: Topologically-aware overlay construction and server selection. In: INFOCOM, IEEE (2002)
4. Jiang, J., Dovrolis, C.: Passive estimation of TCP round-trip times. ACM Computer Communication Review **32** (2002)
5. Lu, G., Li, X.: On the correspondency between tcp acknowledgment packet and data packet. In: Internet Measurement Conference, ACM (2003)
6. Jaiswal, S., Iannaccone, G., Diot, C., Kurose, J., Towsley, D.: Inferring TCP connection characteristics through passive measurements. In: INFOCOM, IEEE (2004)
7. Alexa Internet: Alexa Web search—top 500. http://www.alexa.com/site/ds/top_sites?ts_mode=global&lang=none (2004)
8. Carson, M., Santay, D.: NIST Net: a Linux-based network emulation tool. ACM Computer Communication Review **33** (2003) 111–126
9. Barford, P., Crovella, M.: Generating representative web workloads for network and server performance evaluation. In: Measurement and Modeling of Computer Systems. (1998) 151–160
10. Barford, P., Bestavros, A., Bradley, A., Crovella, M.: Changes in web client access patterns: Characteristics and caching implications. World Wide Web **2** (1999)

Detecting Duplex Mismatch on Ethernet

Stanislav Shalunov and Richard Carlson*

{shalunov, rcarlson}@internet2.edu

Abstract. IEEE 802.3 Ethernet networks, a standard LAN environment, provide a way to auto-negotiate the settings of capacity (10, 100, or 1000 Mb/s) and duplex (full- or half-). Under certain conditions described below, the auto-negotiation protocol fails to work properly. The resultant configuration problem, *duplex mismatch*, appears to be common; when this problem occurs, the connectivity is impaired, but not completely removed. This can result in performance problems that are hard to locate.

This paper describes a work in progress aimed at (i) studying the condition of duplex mismatch in IEEE 802.3 Ethernet networks, (ii) producing an analytical model of duplex mismatch, (iii) validating the model, (iv) studying the effects of duplex mismatch on TCP throughput, (v) designing an algorithm for duplex mismatch detection using data from active testing, and (vi) incorporating the detection algorithm into an existing open-source network troubleshooting tool (NDT).

1 Introduction

Ethernet duplex mismatch is a condition that occurs when two devices communicating over an IEEE 802.3 Ethernet link (typically, a host and a switch) do not agree on the duplex mode of the direct Ethernet connection between them; the switch might operate as if the connection were full-duplex with the host operating in half-duplex mode, or *vice versa* (section 2 describes the condition in more detail). Duplex mismatch causes some packets to be lost (see section 3) and, in many cases, leads to serious performance degradation (see section 4.2). Section 5 discusses soft failures similar to duplex mismatch. Section 6 describes an algorithm to detect a duplex mismatch condition by observing the behavior of the network when test traffic is injected, and describes integrating the detection algorithm into our existing network testing system.

2 Problem Description

The IEEE 802.3 Ethernet specifications [1] define a set of Media Access Control (MAC) protocols that are the standard for Local Area Networks (LANs) used

---

* This work was supported by contract 467-MZ-401761 from the National Library of Medicine.

C. Dovrolis (Ed.): PAM 2005, LNCS 3431, pp. 135–148, 2005.

around the world. The original protocol was developed for communications over shared media and uses the Carrier Sense Multiple Access with Collision Detection (CSMA/CD) protocol to arbitrate access to the shared media. That the media is shared means that a host could either transmit or receive data at any given time (in other words, devices operate in a *half-duplex mode*). Later enhancements define how devices can communicate over a wide variety of co-axial, twisted-pair, and fiber-optic cables. The twisted-pair and fiber-optic cable protocols allow for development of switched network environments—enabling a device to simultaneously transmit and receive data (thus, operating in a *full-duplex mode*). All IEEE 802.3-compliant implementations (10-, 100-, and 1000-Mb/s) must support half-duplex operations and may support full-duplex operations.

2.1 Half-Duplex Ethernet Operation

To reliably operate in the half-duplex mode ([1], section 4.1), it is necessary to specify how a device should operate when several devices attempt to transmit simultaneously (a *collision event* is said to occur when this happens). The CSMA/CD protocol is used to resolve collisions: in half-duplex mode, an interface will monitor the Carrier Sense (CS) flag at all times to determine if any device is transmitting data on the network; when the local device has data to send, these data pass through one or more upper-layer protocols and finally arrive at the local interface layer where they are encapsulated into an Ethernet frame. The transmitting interface checks the CS flag and, if it is *true*, holds the frame in the transmit queue. When the CS flag becomes *false*, an inter-frame gap (IFG) timer is started. When this timer expires, the frame is transmitted onto the network. While the frame is being transmitted, the interface monitors the outgoing data to detect if another device has also started to transmit. This Collision Detection (CD) algorithm improves network efficiency by stopping transmissions when they are obviously corrupted. The CSMA/CD protocol also prescribes a way to retransmit frames after the collision is detected.

Two types of collisions may be detected by the transmitting interface. A slot time is defined as the time it takes to transmit the first 64 octets (512 octets for 1000-Mb/s interfaces) of the frame. A normal collision will occur within the slot time period, a *late collision* will occur from the end of the slot time to the end of the frame. If no collision is detected within the slot time, the transmitting interface is said to have captured the network. If a normal collision is detected, an interface will halt its transmission, broadcast a jamming signal to ensure the collision is detected by other devices, and then wait a pseudo-random time before attempting to transmit again. A host that detects a late collision will still stop transmitting its frame and discard the incoming frame. At speeds up to 100 Mb/s, the standard does not make it mandatory to retransmit the outgoing frame (this is implementation-specific); at 1000 Mb/s, the frame shall be discarded. Late collisions should not occur on a properly operating network.

2.2 Full-Duplex Ethernet Operation

With the introduction of the 10BASE-T specification for 10-Mb/s twisted-pair media ([1], section 14.1.2), full-duplex operation over switched network connections became possible. In switched mode, a network link capable of transmitting and receiving data simultaneously is dedicated to each device in the network. This usually means that two twisted-pair copper or fiber-optic cables are used create a bidirectional network link. This operating mode offers several advantages over half-duplex operation: a device can transmit data whenever they are available, thus improving performance; in addition, the need for CSMA/CD vanishes, thus simplifying the operation and making it ostensibly more robust.

When a device has data to send in full-duplex mode ([1], section 4.1), an Ethernet frame is created and placed in the transmit queue. The frame at the head of the queue is transmitted after the expiration of the IFG timer. Since collisions are impossible, a higher throughput can be achieved; the interface can also stop monitoring the CS and CD flags.

2.3   Auto-Configuration on Ethernet

In addition to operating in full- or half-duplex mode, the IEEE 802.3 specifications ([1], section 28.2) describe how a device can configure itself to operate in an unknown environment by use of a negotiation protocol that detects the capacity and duplex settings. The protocol uses an out-of-band pulse-code sequence based on the 10BASE-T link integrity test pulses. This string of closely spaced pulses, the Fast Link Pulse (FLP) burst, encodes the operating modes that each interface is capable of supporting into a Link Code Word (LCW).

The protocol operates thusly: at power-on, after a physical cable is connected, or after a management command is received, the interface enters the auto-negotiation phase. At this time, the interface starts sending FLP bursts to the remote interface and listening for incoming FLP bursts. An old 10BASE-T-only interface will begin sending only single Normal Link Pulses (NLP), indicating it does not support the auto-negotiation function. The receiving interface measures the interval between NLP pulses to differentiate between the NLP and FLP bursts. If the FLP bursts are detected, then the Auto-Negotiate flag in the Media Independent Interface (MII) control register is set to *true* and the auto-negotiation process continues. If the FLP bursts are not detected, auto-negotiation is disabled and the interface is brought up in its default state.

The default parameter settings ([1], section 22.2.4) determine how the interface will operate if auto-negotiation fails (the interface will set the interface to the highest possible speed and half-duplex).

Upon receipt of three consecutive and consistent LCWs, the receiving interface will respond by setting the ACK bit in the transmitted LCWs and the MII control register to *true*. The receipt of three consecutive and consistent LCWs with the ACK bit set indicates that the peer interface has received the negotiation parameters and is ready to complete the negotiation process. The mode priority ([1], annex 28b.3) settings determine how the auto-negotiation process determines which speed and duplex setting to use. This process prefers the highest performance capability that both interfaces support, so 100 Mb/s is preferred over 10 Mb/s and full-duplex is preferred over half.

In addition to the auto-negotiation protocol, an auto-sense protocol ([1], section 28.2.3.1) can detect 100BASE-TX, 100BASE-T4, and 10BASE-T interfaces that do not support the auto-negotiation protocol. In this situation, an interface that receives an NLP signal will use the link integrity test functions to detect the speed of the peer interface. While speed and media type (TX or T4) is detected, only half-duplex operations are supported by this protocol.

Some older (or cheaper) network interface cards support only half-duplex mode; most modern cards support both auto-negotiation and manual configuration modes. It should be noted that the speed and duplex settings chosen by an interface are never explicitly communicated to the peer interface. We believe this lack of robustness makes the auto-negotiation protocol susceptible to errors that can cause operational problems on real networks.

2.4   Duplex Mismatch

While auto-configuration makes connecting to the network easier, it can lead to major performance problems when it fails, as it sometimes does [2]. If the two sides of a Fast Ethernet connection disagree on the duplex mode (*i.e.*, one is using full- and the other is using half-), a *duplex mismatch* is said to occur.

A duplex mismatch can happen in one of these ways (among others), but the symptoms seen by the hosts will be the same regardless of the cause:

1. One card is hard-coded to use full-duplex and the other is set to auto-negotiate: the hard-coded side will not participate in negotiation and the auto-negotiating side will use its half-duplex default setting;
2. The two cards are hard-coded to use different duplex modes;
3. Both cards are set to auto-negotiate, but one or both of them handles auto-negotiation poorly [3, 4]; note that, in this case, the problem can occur sporadically and rebooting or resetting the interface on either side could clear the condition.

3   Predicted Behavior

When duplex mismatch happens, a peculiar breakdown in communication occurs. Denote the interface that thinks that the connection is in full-duplex mode F and the interface that thinks that the connection is in half-duplex mode H.

3.1   Model of the Pattern of Layer-3 Packet Loss

Denote the propagation delay between H and F by δ (in seconds),[1] the period of time after starting to send during which H will retransmit in case of collision by ξ (in seconds),[2] and the capacity of the link by c (in bits/second). Let us consider the two directions separately.

---

[1] The standard requires that cables not exceed 100 m in length; this means that $\delta \leq 0.5\,\mu$s for 100-Mb/s Ethernet.

[2] Note that ξ is determined by a specific interface card model. The standard guarantees that for 10- and 100-Mb/s Ethernet, $\xi \geq 512/c$; for 1000-Mb/s Ethernet, $\xi = 4096/c = 4096/10^9 \approx 4.1\,\mu$s.

$F \to H$ Duplex mismatch can cause loss in this direction (but no extra delays can be introduced, as F never delays frames in its transmit queue and never retransmits frames). Suppose that a frame is being sent in this direction at time t_F. It will be lost if and only if it starts arriving at H while H is sending: *i.e.*, there exists a frame of size m (in bits) sent by H at time t_H such that

$$t_H < t_F + \delta < t_H + m/c. \tag{1}$$

$H \to F$ Duplex mismatch can cause loss in this direction for three reasons:

(i) Non-recoverable collision loss. Suppose a frame of size m (in bits) is being sent by H at time t_H. This frame will be lost if and only if there exists a frame sent by F at time t_F such that

$$t_H + \xi < t_F + \delta < t_H + m/c. \tag{2}$$

(ii) Buffer overrun loss. If the link $F \to H$ is not idle long enough for H to drain its transmit queue, the queue will overflow and packets will be lost. If the average rates at which F and H are sending are c_F and c_H, loss will always happen if

$$c_F + c_H > c. \tag{3}$$

Note: Since H can spend considerable time idle due to exponential back-off,[3] buffer overrun loss can occur even if condition 3 does not hold.

(iii) Excessive collision loss. A frame will be lost if a collision occurs more than 16 times when H attempts to retransmit the frame.

In addition, delays can be introduced by H waiting for the network to go idle before sending and by retransmitting. The more traffic goes in the $F \to H$ direction, the more traffic going in the $H \to F$ direction will be delayed.

3.2   Manifestation of Duplex Mismatch in the Case of UDP

Using UDP *per se* imposes no particular sending schedule. For the purposes of producing verifiable predictions made by our model, the case of Poisson streams is considered in this section; this case is easy to analyze and, therefore, the verification of the predictions will help validate the model.

Assume that two Poisson streams of frames are injected into the system for transmission by the interfaces. The average rate of the streams are c_F and c_H (in bits/second, as above). Since F never delays packets, the stream that leaves F is Poisson. Let us consider the situation when $c_F \ll c$ and $c_H \ll c$. In this case, the stream leaving H is not disturbed to a large extent and can be approximated by a Poisson stream.

Our model then predicts loss in $F \to H$ direction, p_F, to be

$$p_F = \frac{c_H}{c}. \tag{4}$$

---

[3] For attempt n, the delay is a uniformly distributed pseudo-random integer between 0 and $2^{\min(n,10)} - 1$, multiplied by slot time.

In the $H \to F$ direction, since $c_F + c_H \ll c$, buffer overrun loss will never occur (cf. condition 3). Excessive collision loss rate, $(c_F/c)^{16}$, will be negligible. Further, denote the size of packets that leave H by m. We have:

$$p_H = \frac{c_H}{c} \max\left(0, 1 - \frac{c\xi}{m}\right). \tag{5}$$

Note: Formula 5 allows one to measure ξ externally by observing $\max(0, 1 - c\xi/m)$ (the proportion of bits in frames sent by H that are transmitted later than ξ seconds after start of frame transmission).

3.3   Manifestation of Duplex Mismatch in the Case of TCP

TCP is a reliable transport protocol that uses ACK packets to control the flow of data between two Internet nodes. Only unidirectional TCP data flows with ACKs flowing in the opposite direction are considered. Denote the interface on whose side the TCP sender is located S, and the interface on whose side the TCP receiver is located R.[4] In bulk transfer mode, a stream of MTU-sized packets will go from S to R and a stream (with typically half as many packets per second when delayed ACKs are used) of small packets containing TCP ACKs will go from R to S. ACKs are cumulative: if an ACK is lost and a subsequent ACK is delivered, the same information is conveyed (and the effect of the missed increment of the TCP congestion window is minimized). Denote the period of time it takes the receiver to generate an ACK, and for this ACK to reach interface R, by Δ.

Since, with TCP, packets flow in both directions, there is a potential for collisions and loss during normal TCP operation on a link with duplex mismatch. Consider the case when TCP is not in a timeout state and congestion window is not too small (at least a few packets). If the network path between the sender and the receiver does not contain any tight links, then the arrival of several back-to-back TCP data packets should cause a collision and a loss event will occur.

For simplicity, consider two cases where a single duplex mismatch condition exists on the last hop of the network path (*e.g.*, next to the user's computer):[5]

1. $S = F$, $R = H$: The interface R will obey the CSMA/CD protocol and refrain from transmitting while a frame is being received. It will also detect and re-transmit frames when collisions occur using the proper collision slot time (defined in 2.1). The interface S will follow the full-duplex protocol and transmit frames whenever they become available without checking for frames on the receive circuit. Collisions will be ignored by S and the entire packet transmission time m/c will always be used.

---

[4] Note that S and R are *not* the sender and receiver, but rather the two interfaces on the sides of a link with duplex mismatch. Often, S could be on the sender or R could be on the receiver, but for both to be true, the network path would need to consist of exactly one layer-2 hop.

[5] Only data and ACK packets from a single TCP flow are considered.

When gaps between data packets are wider than Δ, ACK packets will be transmitted and will arrive at the sender for processing by the TCP stack. When gaps between data packets are less than Δ, a collision could occur. Consider the case during slow start when the congestion window (CWND) on the sender reaches four packets; the sender could have four data packets to transmit back-to-back. After receiving two data packets, the receiver would generate an ACK. The interface R will receive this ACK in time Δ, attempt to transmit it to S, and find that S is currently transmitting the third data packet, so R will delay the ACK's transmission. When this frame's transmission completes, both R and S will wait an IFG time and simultaneously begin to transmit (the fourth data packet and the ACK packet) causing a collision to occur. R will detect this collision and re-schedule another retransmission, but S will continue to blindly send the entire frame. This frame will be discarded by R as corrupted due to the collision. The sender will detect and handle this loss event.

In general, whenever CWND on the sender increases enough to allow for a burst of at least four packets to arrive at the receiver, the ACK generated in response to the first or second data packet in the burst will allow the next packet to be delivered, but will cause all subsequent packets in the burst to be lost; the ACK itself will be delivered to S after the burst. TCP would thus suffer from both inability to raise CWND and timeouts caused by multiple packet losses in a single RTT (whenever CWND becomes large enough). Empirically, TCP infrequently enters slow start in this case, since CWND remains small enough; the goodput obtained is thus better than that in case 2.

The large number of lost data packets will cause the receiver to generate a large number of duplicate ACKs. Thus, the TCP source node will receive a larger number of ACK packets than would normally be expected in a delayed ACK environment.

2. $S = H$, $R = F$: Consider the case where a burst of packets arrives at S so that the next frame is ready to begin transmission before the previous frame ends. The first two data packets will arrive at the receiver, which will generate an ACK packet. This packet will be transmitted during the receipt of the third data packet if Δ is small enough. The switch will detect the collision and abort the data transer, but the ACK packet will be lost. If $\Delta < \xi$, then the data packet will be resent by S; otherwise, it will be lost. This loss of ACKs not only has a detrimental effect on TCP throughput by reducing the number of increments of CWND, but also creates a situation when the last ACK for a burst is not received, thus causing the sender to timeout and reenter slow start.

The large number of lost ACK packets will mean that the sender will see fewer ACKs then would otherwise be expected (about one ACK per burst).

4   Validation of Predictions

To validate the model of section 3, we create the duplex mismatch condition artificially, send test traffic, and compare the results with predictions.

4.1   Validation with UDP Streams

Since the loss pattern created by duplex mismatch is complex, it is easier to analyze first the results with simple synthetic traffic consisting of UDP packets. Our model makes the following predictions (see section 3.2):

1. Unidirectional traffic (in either direction) will not suffer any loss;
2. When a small amount of traffic is sent in each direction as a Poisson stream, loss is given by formulae 4 and 5.

Our UDP tests were run with a program THRULAY (using option -u), which had UDP mode added for the purposes of conducting these experiments.[6] A duplex mismatch condition was artificially created between a Linux host and an Ethernet switch with MII-TOOL. A series of bidirectional 10-second UDP tests were conducted between the host with the duplex mismatch condition and another host connected to the same Ethernet switch. Both hosts were connected at 100 Mb/s. The host without duplex mismatch was connected using full duplex. The sending rate was varied from 1 Mb/s to 9 Mb/s on each side in 1-Mb/s increments; 81 tests were run.

The results of these experiments are presented in figures 1 and 2. In figure 1, the line predicted by equation 4, on which data points are expected to fall, is shown; as can be seen, the match is quite good. In figure 2, data points for any given value of c_H are expected (in accordance with equation 5) to fall on a horizontal line, which generally agrees with the observations. Prediction 2 thus holds. In might be of interest that indirectly, the value of ξ (inherent in the Ethernet implementation of the HP switch we used in this experiment) is measured here and it appear that the value is quite small—perhaps as small as allowed by the standard.[7]

In addition, we conducted tests in each direction without opposing traffic; no loss was observed during these tests, thus verifying prediction 1.

Formula 4 was only proven for $c_H \ll c$ and $c_F \ll c$. However, empirically (data not presented in this paper), this formula extends to values of c_F/c as large as 0.3 or even 0.4 and values of c_H/c as large as 0.5.

---

[6] Many network testing programs allow the use of a UDP mode. However, we found no program that would generate a Poisson streams of packets that are sent with no correlation with operating system's time slice boundaries. Our tool can use a busy-wait loop on the TSC timing register to effect a true Poisson stream. The program is made freely available at http://www.internet2.edu/~shalunov/thrulay/.

[7] The best match for the value of the term $\max\left(0, 1 - \frac{c\xi}{m}\right)$ is about 0.93. For these tests, we had $m = 1518\,B$; therefore, ξ was large enough to cover about 106 B of each packet.

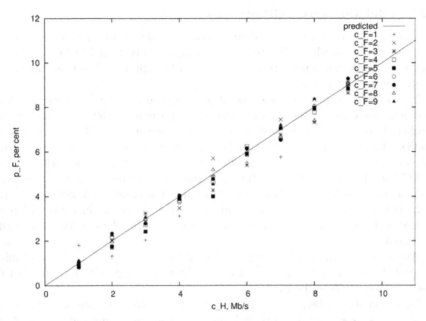

Fig. 1. Loss probability p_F in the stream in $F \to H$ direction as a function of c_H, the rate of the stream going in the opposite direction. The prediction is based on equation 4

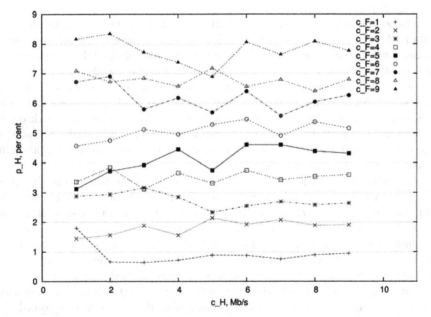

Fig. 2. Loss probability p_H in the stream in $H \to F$ direction as a function of c_H, the rate of the stream

4.2   Validation with TCP Streams

To validate the model developed in section 3.3, two types of tests were run. One test used a pseudo-TCP stream consisting of a stream of UDP packets sent back to back, with a small stream returning in the opposite direction; the other involved actual TCP streams.

The TTCP test program was modified to perform these experiments. The modifications involved having one host send a stream of 50 UDP packets to another. Each packet contained a unique sequence number. The receiver was modified to return a small UDP packet for every second packet received. This simulates the typical mode of operation of the delayed ACK algorithm found in TCP implementations. This returned packet contained a unique sequence number and a copy of the data packet sequence number that generated this ACK. As before, the MII-TOOL program was used to force one link into different normal and mismatch states. The raw data was captured using the TCPDUMP command.

An analysis of the resulting traces showed that there were two different mismatch behaviors and two flavors of each behavior. These behaviors match the predicted models described in section 3.3.

One behavior is when $S = H$ and $R = F$. In this situation, most of the returning ACK packets are lost. Even though the receiver is generating and transmitting them, only the final ACK packet is recieved by the sender. The rest are lost due to collisions with the arriving data packets. Depending on how late collisions are handled, the data packets may be retransmitted or discarded. Thus, two flavors of this behavior are observed.

The second behavior is when $S = F$ and $R = H$. In this situation, the arriving data packets are discarded due to collisions. The ACK packets are delayed due to numerous retransmission attempts. When the receiver's local link is mismatched, Δ is small, so the first ACK is given back to the interface while data packet 3 is being received. As predicted in section 3.3, data packets 4, 5, 6, 7 and 8 are lost due to collisions as this frame is retransmitted over and over. Eventually, the retransmit delay increases to a point where several data packets do get through, causing more ACKs to be generated. These ACKs are queued behind the first ACK and can only be transmitted after a successful transmission (or if the maximum retransmission count is exceeded).

A slightly different flavor of this behavior occurs when the mismatch link is not the last link in the path. In this case the first collision will occur when the ACK has propagated back along the path to the point of the mismatch. In the tests, the sender and receiver roles were exchanged but no changes were made to the link with duplex mismatch. This resulted in data packets 6, 7, 8, 9, and 10 being lost. This shift exactly matches the extra time needed for the ACK to propagate back along the network path to the mismatched switch port.

For comparison purposes, the normal full-duplex and half-duplex operating conditions were also tested. No exceptional loss or delay values were recorded.

While this simple test can demonstrate how the returning ACK packets can collide with data packets, it does not explain the complex dynamics of a real TCP

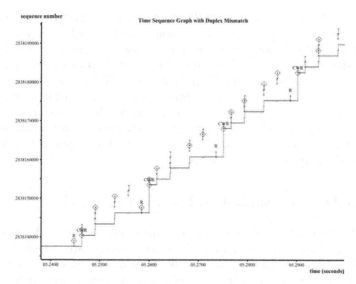

Fig. 3. TCP trace with duplex mismatch, $S = F$ and $R = H$

flow. We need to capture the retransmission and congestion control dynamics to more completely describe the TCP behavior.

To examine the various normal and mismatch cases, a series of tests were run using the existing NDT client/server program. The client's link was configured using the MII-TOOL program to simulate the expected configuration: where the NDT server is properly configured but the client may have a mismatch condition. The results of these test are described below.

Figure 3 shows the mismatch case with $S = F$. We begin the trace after the NDT server has successfully sent several packets and a timeout occurred. Following the retransmission, an ACK is received causing the RWIN value to increase and allowing more data to be sent. Two data packets are transmitted, and, following an RTT, the ACKs are returned. Note that after two RTTs a hole appears due to a lost packet. The sender continues to inject new packets as ACK arrive. After three duplicate ACKs have arrived, the sender retransmitts the missing data packet. The receipt of this packet fills in the hole and causes a large jump in RWIN. This process repeats for the entire 10 second test time.

The result is that throughput is limited by CWND and the RTT of the end-to-end path. Slight variations in packet arrival times can cause multiple packets to be lost at the client. This increases the probability of a TCP RTO event, further decreasing in throughput.

When the mismatch condition is reversed such that $S = H$, an even greater impact is seen on TCP throughput. As noted above, this condition will result in ACK packets being discarded. A typical trace shows that the NDT server sends several back-to-back packets to the receiver. These packets are received, but the returning ACK is lost. The lack of an ACK prevents the sender from sending more packets into the network, thus preventing any more ACKs from being generated. The connection stalls until the TCP RTO event occurs. This causes the sender to resend what it thought were lost packets. When the first retransmitted data packet is received, the receiver recognizes it as a duplicate and immediately sends a duplicate ACK. This ACK causes two more packets to be sent, another duplicate and a single new packet. When the receiver receives the duplicate data packet, it sends another duplicate ACK, which collides with the third data packet. Thus, the ACK is lost, but the data packet is successfully received after a small delay. The loss of this ACK causes another RTO event to occur, starting the entire process over again. TCP throughput in this situation is limited by the numerous RTO events.

5   Soft Failures Similar to Duplex Mismatch

Our goal is to develop a duplex mismatch detection algorithm that treats the network as a black box. Both false positives and false negatives must be considered. A particularly harmful and insidious mode of false positive would be the characterization of another soft failure condition as duplex mismatch: not only would users waste their time not solving their real problem, but they might, while responding to perceived duplex mismatch diagnostics, change settings to introduce duplex mismatch in a network that previously did not have one. For example, when a copper twisted-pair cable is subtly damaged, *cross-talk* can occur; *i.e.*, a signal transmitted on one wire induces a spurious signal on another. Bit corruption during bidirectional transmission that occurs thusly could be confused with duplex mismatch. Chief differentiators of cross-talk from duplex mismatch are:

1. Cross-talk can affect unidirectional traffic; duplex mismatch cannot;
2. Duplex mismatch occurs deterministically; cross-talk corrupts bits randomly.

6   Detecting Duplex Mismatch in NDT

While other packet losses can cause the number of duplicate ACK packets to increase, the asymmetric throughput results are not observed. Thus we believe that the combination of these two conditions is a reliable indicator of duplex mismatch.

The Network Diagnostic Tool (NDT) [5] is an open-source software program that uses a client/server approach to detect common network problems. The server process communicates with a Java-based client program to generate a series of TCP unidirectional data flows. The server also captures Web100 data allowing it to delve into the depths of the TCP connection. By analyzing this data, it is possible to determine if a duplex mismatch condition existed over some portion of the network path.

The original NDT duplex mismatch detection heuristic was created after running a series of experiments in a testbed network environment. This environment was expanded to encompass a campus network. This heuristic used the amount of time the connection spent in the congestion window limited case, the number of packets retransmitted per second, and the throughput predicted by the MTU/RTT$\sqrt{p}$ formula. A later modification was made to ignore cases when the client was located behind a tight link (cable modem or DSL circuit).

A new algorithm that takes advantage of the analytical model described in this paper has now been incorporated into the NDT tool. Web100 [6] variables can be used to perform duplex mismatch detection thusly: The NDT server generates two test streams sequentially, one on each direction, to measure TCP throughput. Each stream is monitored to determine the path capacity. Thus, we have two TCP throughput measurements and an estimate of the tight link in the end-to-end path.

Duplex mismatch causes major disruptions to a TCP flow. The generation and transmission of ACK packets in direct response to received data packets increases the probability of a collision. Once this happens, either the data or the ACK packet will be lost and the other delayed by the CSMA/CD algorithm. However, TCP is a reliable transport protocol, so it will retransmit lost data packets. These retransmission will cause the receiver to generate more duplicate ACKs per data packet compared to one delayed ACK for every other data packet.

The Web100 variables are captured on the NDT server only in the case where the NDT server is sending data to the client. The two duplex mismatch cases can now be examined.[8]

If the client is the receiver and the mismatch condition is such that $R = H$, then numerous data packets will be lost. The original transmission and an subsequent retransmissions will cause the Web100 DataPktsOut counter to increment. In addition, the loss of individual packets will cause the client to generate a large number of duplicate ACKs. The Web100 AckPktsIn counter will be incremented every time a new or duplicate ACK is received. Thus, the ratio of data packets transmitted *vs* ACK packets received will skew away from 2:1 towards more ACKs. As noted above, throughput will be a function of CWND and RTT.

If the client is the receiver, $R = F$, and $S = H$, then numerous ACK packets will be lost. In addition, a large number of packets will be retransmitted and a large number of timeouts will occur. This will skew the data packet *vs* ACK packet ratio in the opposite direction from that described above. Thus, a ratio of more than 2:1 is expected. TCP throughput will be dramatically affected due to the large number of RTO events.[9]

This means that we can create a reliable duplex mismatch detection algorithm by combining the asymmetric throughput with the skewed ACK:data packet ratio.

At present, we are collecting data to validate these predictions and results. NDT servers at several locations are gathering data from production environments. We will analyze the log files produced and compare the results with the observations from the NDT administrator. Our presentation will describe the results of this effort.

---

[8] In each case, the TCP flow in the opposite direction will exhibit the other mismatch behavior.

[9] The NDT tool will display both conditions as it runs two unidirectional tests.

7   Conclusions

Duplex mismatch is an insidious condition, undetectable with such simple network operation tools as *ping*. It can affect a user's performance for weeks or months before it is detected. A model of duplex mismatch is described and a detection algorithm is proposed. The algorithm is implemented in the NDT.

References

1. IEEE: Part 3: Carrier sense multiple access with collision detection (CSMA/CD) access method and physical layer specifications. IEEE Standards, 802.3 (2002)
2. Apparent Networks: Duplex conflicts: How a duplex mismatch can cripple a network. White Paper (2002)
3. Eggers, J., Hodnett, S.: Ethernet autonegotiation best practices. Sun BluePrints OnLine (2004)
4. Hernandez, R.: Gigabit Ethernet auto-negotiation. Dell Power Solutions 1 (2001)
5. Carlson, R.A.: Developing the Web100 based network diagnostic tool (NDT). In: Proc. Passive and Active Measurement Workshop (PAM), San Diego (2003)
6. Mathis, M., Heffner, J., Reedy, R.: Web100: Extended TCP instrumentation for research, education, and diagnosis. ACM Computer Communications Review **33** (2003)

Improved Algorithms for
Network Topology Discovery

Benoit Donnet[1], Timur Friedman[1], and Mark Crovella[2,*]

[1] Université Pierre & Marie Curie, Laboratoire LiP6-CNRS
[2] Boston University Department of Computer Science

Abstract. Topology discovery systems are starting to be introduced in
the form of easily and widely deployed software. However, little consideration has been given as to how to perform large-scale topology discovery efficiently and in a network-friendly manner. In prior work, we
have described how large numbers of traceroute monitors can coordinate
their efforts to map the network while reducing their impact on routers
and end-systems. The key is for them to share information regarding the
paths they have explored. However, such sharing introduces considerable
communication overhead. Here, we show how to improve the communication scaling properties through the use of Bloom filters to encode a
probing stop set. Also, any system in which every monitor traces routes
towards every destination has inherent scaling problems. We propose capping the number of monitors per destination, and dividing the monitors
into clusters, each cluster focusing on a different destination list.

1   Introduction

We are starting to see the wide scale deployment of tools based on *traceroute* [1]
that discover the Internet topology at the IP interface level. Today's most extensive tracing system, *skitter* [2], uses 24 monitors, each targeting on the order of
one million destinations. Other well known systems, such as *RIPE NCC TTM* [3]
and *NLANR AMP* [4], conduct a full mesh of traceroutes between on the order
of one- to two-hundred monitors. An attempt to scale either of these approaches
to thousands of monitors would encounter problems from the significantly higher
traffic levels it would generate and from the explosion in the data it would collect.
However, larger scale systems are now coming on line.

If a traceroute monitor were incorporated into screen saver software, following
an idea first suggested by Jörg Nonnenmacher (see Cheswick et al. [5]), it could

---

* The authors are participants in the traceroute@home project. Mr. Donnet and Mr.
Friedman are members of the Networks and Performance Analysis research group
at LiP6. This work was supported by: the RNRT's Metropolis project, NSF grants
ANI-9986397 and CCR-0325701, the e-NEXT European Network of Excellence, and
LiP6 2004 project funds. Mr. Donnet's work is supported by a SATIN European
Doctoral Research Foundation grant. Mr. Crovella's work at LiP6 is supported by
the CNRS and Sprint Labs.

C. Dovrolis (Ed.): PAM 2005, LNCS 3431, pp. 149–162, 2005.

lead instantaneously to a topology discovery infrastructure of considerable size, as demonstrated by the success of other software distributed in this manner, most notably *SETI@home* [6]. Some network measurement tools have already been released to the general public as screen savers or daemons. *Grenouille* [7] was perhaps the first, and appears to be the most widely used. More recently we have seen the introduction of *NETI@home* [8], and, in September 2004, the first freely available tracerouting monitor, *DIMES* [9].

In our prior work [10], described in Sec. 2 of this paper, we found that standard traceroute-based topology discovery methods are quite inefficient, repeatedly probing the same interfaces. This is a concern because, when scaled up, such methods will generate so much traffic that they will begin to resemble distributed denial of service (DDoS) attacks. To avoid this eventuality, responsibly designed large scale systems need to maintain probing rates far below that which they could potentially obtain. Thus, skitter maintains a low impact by maintaining a relatively small number of monitors, and DIMES does so by maintaining a low probing rate. The internet measurement community has an interest in seeing systems like these scale more efficiently. It would also be wise, before the more widespread introduction of similar systems, to better define what constitutes responsible probing.

Our prior work described a way to make such systems more efficient and less liable to appear like DDoS attacks. We introduced an algorithm called Doubletree that can guide a skitter-like system, allowing it to reduce its impact on routers and final destinations while still achieving a coverage of nodes and links that is comparable to classic skitter. The key to Doubletree is that monitors share information regarding the paths that they have explored. If one monitor has already probed a given path to a destination then another monitor should avoid that path. We have found that probing in this manner can significantly reduce load on routers and destinations while maintaining high node and link coverage.

This paper makes two contributions that build on Doubletree, to improve the efficiency and reduce the impact of probing. First, a potential obstacle to Doubletree's implementation is the considerable communication overhead entailed in sharing path information. Sec. 3 shows how the overhead can be reduced through the use of Bloom filters [11]. Second, any system in which every monitor traces routes towards every destination has inherent scaling problems. Sec. 4 examines those problems, and shows how capping the number of monitors per destination and dividing the monitors into clusters, each cluster focusing on a different destination list, enables a skitter-like system to avoid appearing to destinations like a DDoS attack. We discuss related and future work in Sec. 5.

2   Prior Work

Our prior work [10] described the inefficiency of the classic topology probing technique of tracing routes hop by hop outwards from a set of monitors towards a set of destinations. It also introduced Doubletree, an improved probing algorithm.

Data for our prior work, and also for this paper, were produced by 24 skitter [2] monitors on August 1st through 3rd, 2004. Of the 971,080 destinations towards which all of these monitors traced routes on those days, we randomly selected a manageable 50,000 for each of our experiments.

Only 10.4% of the probes from a typical monitor serve to discover an interface that the monitor has not previously seen. An additional 2.0% of the probes return invalid addresses or do not result in a response. The remaining 87.6% of probes are redundant, visiting interfaces that the monitor has already discovered. Such redundancy for a single monitor, termed *intra-monitor redundancy*, is much higher close to the monitor, as can be expected given the tree-like structure of routes emanating from a single source. In addition, most interfaces, especially those close to destinations, are visited by all monitors. This redundancy from multiple monitors is termed *inter-monitor redundancy*.

While this inefficiency is of little consequence to skitter itself, it poses an obstacle to scaling far beyond skitter's current 24 monitors. In particular, inter-monitor redundancy, which grows in proportion to the number of monitors, is the greater threat. Reducing it requires coordination among monitors.

Doubletree is the key component of a coordinated probing system that significantly reduces both kinds of redundancy while discovering nearly the same set of nodes and links. It takes advantage of the tree-like structure of routes in the internet. Routes leading out from a monitor towards multiple destinations form a tree-like structure rooted at the monitor. Similarly, routes converging towards a destination from multiple monitors form a tree-like structure, but rooted at the destination. A monitor probes hop by hop so long as it encounters previously unknown interfaces. However, once it encounters a known interface, it stops, assuming that it has touched a tree and the rest of the path to the root is also known.

Both backwards and forwards probing use stop sets. The one for backwards probing, called the *local stop set*, consists of all interfaces already seen by that monitor. Forwards probing uses the *global stop set* of (interface, destination) pairs accumulated from all monitors. A pair enters the stop set if a monitor visited the interface while sending probes with the corresponding destination address.

A monitor that implements Doubletree starts probing for a destination at some number of hops h from itself. It will probe forwards at $h + 1$, $h + 2$, etc., adding to the global stop set at each hop, until it encounters either the destination or a member of the global stop set. It will then probe backwards at $h - 1$, $h - 2$, etc., adding to both the local and global stop sets at each hop, until it either has reached a distance of one hop or it encounters a member of the local stop set. It then proceeds to probe for the next destination. When it has completed probing for all destinations, the global stop set is communicated to the next monitor.

The choice of initial probing distance h is crucial. Too close, and intra-monitor redundancy will approach the high levels seen by classic forward probing techniques. Too far, and there will be high inter-monitor redundancy on destinations. The choice must be guided primarily by this latter consideration to avoid having probing look like a DDoS attack.

While Doubletree largely limits redundancy on destinations once hop-by-hop probing is underway, its global stop set cannot prevent the initial probe from reaching a destination if h is set too high. Therefore, we recommend that each monitor set its own value for h in terms of the probability p that a probe sent h hops towards a randomly selected destination will actually hit that destination. Fig. 1 shows the cumulative mass function for this probability for skitter monitor `apan-jp`. For example, in order to restrict hits on destinations to just 10% of

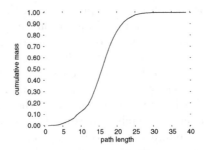

Fig. 1. Cumulative mass plot of path lengths from skitter monitor `apan-jp`

initial probes, this monitor should start probing at $h = 10$ hops. This distance can easily be estimated by sending a small number of probes to randomly chosen destinations.

For a range of p values, Doubletree is able to reduce measurement load by approximately 70% while maintaining interface and link coverage above 90%.

3 Bloom Filters

One possible obstacle to successful deployment of Doubletree concerns the communication overhead from sharing the global stop set among monitors. Tracing from 24 monitors to just 50,000 destinations with $p = 0.05$ produces a set of 2.7 million (interface, destination) pairs. As 64 bits are used to express a pair of IPv4 addresses, an uncompressed stop set based on these parameters requires 20.6 MB. This section shows that encoding the stop set into a Bloom filter [11] can reduce the size by a factor of 17.3 with very little loss in node and link coverage. Some additional savings are possible by applying the compression techniques that Mitzenmacher describes [12]. Since skitter traces to many more than 50,000 destinations, a skitter that applied Doubletree would employ a larger stop set. Exactly how large is difficult to project, but we could still expect to reduce the communication overhead by a factor of roughly 17.3 by using Bloom filters.

A Bloom filter encodes information concerning a set into a bit vector that can then be tested for set membership. An empty Bloom filter is a vector of all zeroes. A key is registered in the filter by hashing it to a position in the vector and setting the bit at that position to one. Multiple hash functions may be used, setting several bits set to one. Membership of a key in the filter is tested by checking if all hash positions are set to one. A Bloom filter will never falsely return a negative result for set membership. It might, however, return a false positive. For a given number of keys, the larger the Bloom filter, the less likely is a false positive. The number of hash functions also plays a role.

To evaluate the use of Bloom filters for encoding the global stop set, we simulate a system that applies Doubletree as described in Sec. 2. The first monitor initializes a Bloom filter of a fixed size. As each subsequent monitor applies Doubletree, it sets some of the bits in the filter to one. A fixed size is necessary because, with a Bloom filter, the monitors do not know the membership of the stop set, and so are unable to reencode the set as it grows.

Our aim is to determine the performance of Doubletree when using Bloom filters, testing filters of different sizes and numbers of hash functions. We use the skitter data described in Sec. 2. A single experiment uses traceroutes from all 24 monitors to a common set of 50,000 destinations chosen at random. Hashing is emulated with random numbers. We simulate randomness with the Mersenne Twister MT19937 pseudorandom number generator [13]. Each data point represents the average value over fifteen runs of the experiment, each run using a different set of 50,000 destinations. No destination is used more than once over the fifteen runs. We determine 95% confidence intervals for the mean based, since the sample size is relatively small, on the Student t distribution. These intervals are typically, though not in all cases, too tight to appear on the plots.

We first test p values from $p = 0$ to $p = 0.19$, a range which our prior work identified as providing a variety of compromises between coverage quality and redundancy reduction. For these parameters, the stop set size varies from a low of 1.7 million pairs ($p = 0.19$) to a high of 9.2 million ($p = 0$). We investigate ten different Bloom filter sizes: 1 bit (the smallest possible size), 10, 100, 1,000, 10,000, 100,000, 131,780 (the average number of nodes in the graphs), 279,799 (the average number of links), 1,000,000, 10,000,000 and, finally, 27,017,990. This last size corresponds to ten times the average final global stop set size when $p = 0.05$. We test Bloom filters with one, two, three, four, and five hash functions. The aim is to study Bloom filters up to a sufficient size and with a sufficient number of hash functions to produce a low false positive rate. A stop set of 2.7 million pairs encoded in a Bloom filter of 27 million bits using five hash functions, should theoretically, following the analysis of Fan et al. [14–Sec. V.D], produce a false positive rate of 0.004.

3.1   Bloom Filter Results

The plots shown here are for $p = 0.05$, a typical value. Each plot in this section shows variation as a function of Bloom filter size, with separate curves for varying numbers of hash functions. The abscissa is in log scale, running from 100,000 to 30 million. Smaller Bloom filter sizes are not shown because the results are identical to those for size 100,000. Curves are plotted for one, two, three, four, and five hash functions. Error bars show the 95% confidence intervals for the mean, but are often too tight to be visible.

Fig. 2 shows how the false positive rate varies as a function of Bloom filter size. Ordinate values are shown on a log scale, and range from a low of 0.01% to a high of 100%. The figure displays two sets of curves. The upper bound is the

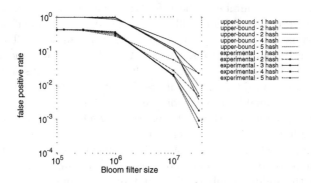

Fig. 2. Bloom filter false positive rate

false positive rate that one would obtain from a Bloom filter of the given size, with the given number of hash functions, encoding the global stop set at its final size, and presuming that any possible key is equally likely to be tested. This is an upper bound because the number of elements actually in the stop set varies. The first monitor's Bloom filter is empty, and it conducts the most extensive exploration, never obtaining a false positive. Successive monitors encounter higher false positive rates, and the value that results from an experiment is the rate over all monitors. The false positive rate should also differ from the theoretic bound because all keys are not equally likely. We would expect a disproportionate number of set membership tests for interfaces that have high betweenness (Dall'Asta et al. work [15] point out the importance of this parameter for topology exploration).

Looking at the upper bounds, we see that false positives are virtually guaranteed for smaller Bloom filters, up to a threshold, at which point false positive rates start to drop. Because of the abscissa's log scale, the falloff is less dramatic than it might at first appear. In fact, rates drop from near one hundred percent to the single digit percentiles over a two order of magnitude change in the size of the Bloom filter, between 2.8×10^5 and 2.7×10^7. The drop starts to occur sooner for a smaller number of hash functions, but then is steeper for a larger number of hash functions.

Based on an average of 2.7×10^6 (interface, destination) pairs in a stop set, we find that the decline in the false positive rate starts to be perceptible at a Bloom filter size of approximately $1/10$ bit per pair encoded, and that it drops into the single digit percentiles at approximately ten bits per pair encoded. This translates to a range of compression ratios from 640:1 to 6.4:1.

Looking at the experimental results, we find, as we would expect, that the false positive rates are systematically lower. The experimental curves parallel the corresponding upper bounds, with the false positive rate starting to decline noticeably beyond a somewhat smaller Bloom filter size, 1.3×10^5, rather than 2.8×10^5. False positive rates are below one percent for a Bloom filter of 2.7×10^7 bits. We would expect to find variation in performance over the same range, as the subsequent figures bear out.

The main measure of performance for a probing system is the extent to which it discovers what it should. Fig. 3 shows how the node and link coverage varies as a function of Bloom filter size. The ordinate values are shown on linear scales, and represent coverage proportional to that discovered by skitter. A value 1.0, not

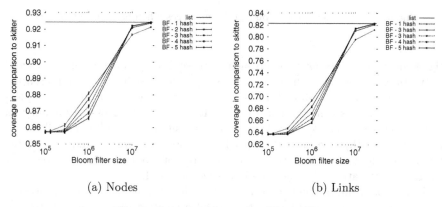

(a) Nodes   (b) Links

Fig. 3. Coverage when using Bloom filters

shown on these scales, would mean that application of Doubletree with the given Bloom filter had discovered exactly the same set of nodes or links as had skitter. The introduction of Doubletree, however, implies, as we found in our prior work, a reduction in coverage with respect to skitter, and this is irrespective of whether Bloom filters are introduced or not. The straight horizontal line labeled *list* in each plot shows the coverage that is obtained with a list of (interface, destination) pairs instead of a Bloom filter, and thus no false positives. For the parameters used here, $p = 0.05$ and 50,000 destinations, the coverage using a list is 0.924 for nodes and 0.823 for links.

The lowest level of performance is obtained below a Bloom filter size of 10^5, the point at which the false positive rate is at its maximum. Note that the lowest level of performance is not zero coverage. The first monitor conducts considerable exploration that is not blocked by false positives. It is only with subsequent monitors that a false positive rate close to one stops all exploration beyond the first probe. Baseline coverage is 0.857 for nodes and 0.636 for links.

The goal of applying Doubletree is to reduce the load on network interfaces in routers and, in particular, at destinations. If the introduction of Bloom filters were to increase this load, it would be a matter for concern. However, as

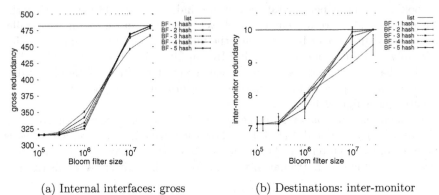

(a) Internal interfaces: gross   (b) Destinations: inter-monitor

Fig. 4. Redundancy on 95th percentile interfaces when using Bloom filters

Fig. 4 shows, there seems to be no such increase. For both router interfaces and destinations, these plots show the 95th percentile of redundancy, representing the extreme values that should prompt the greatest concern. Ordinates are plotted on linear scales. The ordinates in Fig. 4(a) specify the gross redundancy on router interfaces: that is, the total number of visits to the 95th percentile interfaces. In Fig. 4(b), the ordinates specify the inter-monitor redundancy on the 95th percentile destinations: that is, the number of monitors whose probes visit the given destination, the maximum possible being 24.

That Bloom filters seem to add no additional redundancy to the process is a good sign. It is also to be expected, as false positive results for the stop set would tend to reduce exploration rather than increase it, as Fig. 3 has already shown. However, it was not necessarily a foregone conclusion. False positives introduce an element of randomness into the exploration. The fact of stopping to explore one path artificially early could have the effect of opening up other paths to more extensive exploration. If this phenomenon is present, it does not have a great impact.

4   Capping and Clustering

The previous section focused on one potential obstacle to the successful deployment of the Doubletree algorithm for network topology discovery: the communication overhead. This section focuses on another: the risk that probe traffic will appear to destinations as a DDoS attack as the number of monitors scales up. Doubletree already goes some way towards reducing the impact on destinations. However, it cannot by itself cap the probing redundancy on all destinations. That is why we suggest imposing an explicit limit on the number of monitors that target a destination. This section proposes a manner of doing so that should also reduce communication overhead: grouping the monitors into clusters, each cluster targeting a subset of the overall destination set.

As we know from our prior work, Doubletree has the effect of reducing the redundancy of probing on destinations. However, we have reason to believe that the redundancy will still tend to grow linearly as a function of the number of monitors. This is because probing with Doubletree starts at some distance from each monitor. As long as that distance is not zero, there is a non-zero probability, by definition of p (see Sec. 2), that the monitor, probing towards a destination, will hit it on its first probe. There is no opportunity for the global stop set to prevent probing before this first probe. If there are m monitors probing towards all destinations, then the average per-destination redundancy due to these first probes will tend to grow as mp. To this redundancy will be added any redundancy that results from subsequent probes, though this would be expected to grow sublinearly or even be constant because of the application of Doubletree's global stop set.

There is a number of approaches to preventing the first probe redundancy on destinations from growing linearly with the number of monitors. One would be to only conduct traceroutes forward from the monitors. However, as discussed in our prior work, this approach suffers from considerable inefficiency. Another approach would use prior topological knowledge concerning the location of the monitor and the destination in order to set the initial probing distance so as to avoid hitting the destination. Such an approach indeed seems viable, and is a subject for our future work. However, there are numerous design issues that would need to be worked out first: Where would the topology information be stored, and how frequently would it need to be updated? Would distances be calculated on the basis of shortest paths, or using a more realistic model of routing in the internet? Would there still be a small but constant per-monitor probability of error? A simpler approach, and one that in any case could complement an approach based on topology, is to simply cap the number of monitors that probe towards each destination.

If we are to cap the number of monitors per destination, we run the risk of reduced coverage. Indeed, the results presented here show that if skitter were to apply a cap of six monitors per destination, even while employing all 24 monitors and its full destination set, its node coverage would be 0.939 and its link coverage just 0.791 of its normal, uncapped, coverage. However, within a somewhat higher range of monitors per destination, the penalty associated with capping could be smaller. Our own experience has shown that in the range up to 24 monitors, there is a significant marginal utility in terms of coverage for each monitor added. We also find that the marginal utility decreases for each additional monitor, a phenomenon described in prior work by Barford et al. [16], meaning that a cap at some as-yet undefined point would be reasonable.

Suppose, for the sake of argument, that skitter's August 2004 level of 24 monitors per destination is sufficient for almost complete probing of the network between those 24 monitors and their half million destinations. If that level were imposed as a cap, then it would suffice to have 806 monitors, each probing at the same rate as a skitter monitor, in order to probe towards one address in each of the 16.8 million potential globally routable /24 CIDR [17] address prefixes. Most

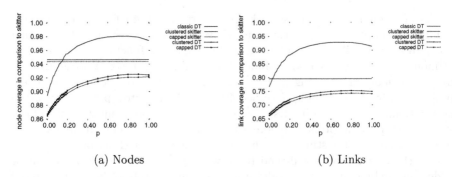

(a) Nodes   (b) Links

Fig. 5. Coverage when capping and clustering

of the additional discovery would presumably take place near the new monitors and new destinations, rounding out an overall map of the network.

If capping is a reasonable approach, then the question arises of how to assign monitors to destinations. It could be done purely at random. Future work might reveal that a topologically informed approach provides better yield. However, one straightforward method that promises reductions in communication overhead is to create clusters of monitors within which all monitors target a common destination set. This would allow the Doubletree global stop sets to be encoded into Bloom filters, as described in Sec. 3, and shared within each cluster. There would be no need to share between clusters, as no destinations would overlap.

We evaluate capping and clustering through experiments similar to those described in Sec. 3. Using the same data sets as described in Sec. 2, we cap the number of monitors per destination at 6. This means that each monitor traces towards 1/4 of the destinations, or 12,500 destinations per monitor. We investigate the effects on redundancy and coverage of the capping. We also investigate the difference between capping with and without the clustering of monitors around common destination sets.

Experiments for clustering and capping employ the methodology that is described in Sec. 3. However, for capping, six monitors are chose at random for each destination. For clustering, six monitors and 12,500 destinations are chosen at random for each cluster. Each monitor appears in only one cluster, and each destination appears in only one cluster.

4.1   Capping and Clustering Results

In these plots, we vary Doubletree's single parameter, p, over its entire range, from $p = 0$ to $p = 1$, with more measurements being taken in the range $p < 0.2$, where most change occurs. The abscissa is in linear scale. Error bars, where visible, show the 95% confidence intervals for the mean.

Fig. 5 shows how the average node and link coverage varies as a function Doubletree's parameter p. The ordinate values are shown on linear scales, and represent coverage proportional to that which is discovered by skitter. A value of 1.0 would mean that application of the given approach had discovered exactly the same set of nodes or links as had skitter.

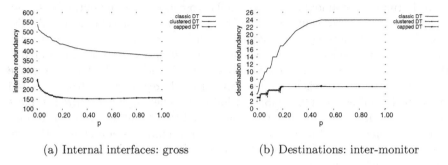

(a) Internal interfaces: gross   (b) Destinations: inter-monitor

Fig. 6. Redundancy on 95th percentile interfaces when capping and clustering

The straight horizontal line labeled *capped skitter* in each plot shows the coverage that is obtained by a hypothetical version of skitter in which each destination is assigned to just six monitors. The lines show the cost of capping skitter at this level, as already discussed. The straight horizontal line labeled *clustered skitter* in each plot shows what would be obtained by skitter if its monitors were to be divided into four clusters. In both plots, the results are very close. Clustered skitter has slightly better coverage than capped skitter, so there is a small effect of promoting exploration due to restricting the global stop sets to within clusters.

The curve labeled *classic DT* in each plot shows how uncapped, unclustered Doubletree performs, and can be compared to the curves labeled *capped DT* and *clustered DT*. As for skitter, the coverage for capping and clustering is slightly better than for simply capping. While it appears that capping imposes significant coverage costs when compared to an uncapped version, these plots alone do not tell the entire story. To better understand the tradeoff we need to look at the redundancy plots as well.

Fig. 6 show the 95th percentile of redundancy for internal interfaces and destinations, in the same manner as in Fig. 4. Fig. 6(b) is of particular interest, because the purpose of capping is to constrain redundancy on destinations. We see that the maximum redundancy for the 95th percentile destination is indeed maintained at six. But this was a foregone conclusion by the design of the experiment. Much more interesting is to compare the parameter settings at which both uncapped and capped Doubletree produce the same redundancy level. To obtain a redundancy of six or less on the 95th percentile destination, uncapped Doubletree must operate at $p = 0.015$. Capped Doubletree can operate at any value in the range $0.180 \leqslant p \leqslant 1$.

If the goal is to maintain a constant level of redundancy at the destinations, the performance, in terms of coverage, of capped and uncapped Doubletree is much closer than it initially appeared. Capped Doubletree can use a value of $p = 0.800$ to maximise both its node and link coverage, at 0.920 and 0.753, respectively. Uncapped Doubletree must use a value of $p = 0.015$, obtaining values of 0.905 and 0.785. Capping, in these circumstances, produces a slightly better result on nodes and a slightly worse on on links.

If destination redundancy results similar to capping can be obtained simply by operating at a lower value of p, then what is the advantage of capping? As discussed earlier, there is a penalty associated with conducting forward traceroutes starting close to the monitor. The same router interfaces are probed repeatedly. We see the effects in Fig. 6(a). The gross redundancy on the 95th percentile router interface is 510 visits for uncapped Doubletree at $p = 0.015$. It is 156 for capped Doubletree at $p = 0.800$. Additional benefits, not displayed in plots here, come from reduced communication costs. If Bloom filters are used to communicate stop sets, and monitors are clustered, then filters of a quarter the size are shared within sets of monitors that are a quarter the size, compared to the uncapped, unclustered case.

5   Conclusion

This paper addresses an area, efficient measurement of the overall internet topology, in which very little related work has been done. This is in contrast to the number of papers on efficient monitoring of networks that are in a single administrative domain (see for instance, Bejerano and Rastogi's work [18]). The two problems are extremely different. An administrator knows their entire network topology in advance, and can freely choose where to place their monitors. Neither of these assumptions hold for monitoring the internet with screen saver based software. Since the existing literature is based upon these assumptions, we need to look elsewhere for solutions.

Some prior work has addressed strategies for tracing routes in the internet. Govindan and Tangmunarunkit [19] proposed the idea of starting traceroutes far from the source, and incorporated a heuristic based on it into the *Mercator* system. No results on heuristic's performance have been published.

A number of papers have examined the tradeoffs involved in varying the number of monitors used for topological exploration of the internet. As previously mentioned, Barford et al. [16] found a low marginal utility for added monitors for the purpose of discovering certain network characteristics, implying that a small number of monitors should be sufficient. However, Lakhina et al. [20] found that this depends upon the parameters under study, and that small numbers of monitors could lead to biased estimates. These biases have been further studied by Clauset and Moore [21], Petermann and De Los Rios [22], and Dall'Asta et al. [15]. Guillaume and Latapy [23] have extended these studies to include the tradeoff between the number of monitors and the number of destinations.

We believe that, employing the heuristics described here, a system such as skitter can be safely extended to a more widely deployed set of monitors, or a system such as DIMES could safely increase its rate of probing. The next prudent step for future work would be to test the algorithms that we describe here on an infrastructure of intermediate size, on the order of hundreds of monitors. We have developed a tool called *traceroute@home* that we plan to deploy in this manner. While we have seen the potential benefits of capping and clustering, we are not yet prepared to recommend a particular cluster size. Data from traceroute@home should allow us better to determine the marginal benefits and costs of adding monitors to clusters.

We also plan further steps to reduce communication overhead and increase probing effectiveness. One promising means of doing this would be to make use of BGP [24] information to guide probing. We are collaborating with Bruno Quoitin to incorporate his C-BGP simulator [25] into our studies.

Acknowledgments

Without the skitter data provided by kc claffy and her team at CAIDA, this research would not have been possible. They also furnished much useful feedback. Marc Giusti and his team at the Centre de Calcul MEDICIS, Laboratoire STIX, Ecole Polytechnique, offered us access to their computing cluster, allowing faster and easier simulations. Finally, we are indebted to our colleagues in the Networks and Performance Analysis group at LiP6, headed by Serge Fdida, and to our partners in the traceroute@home project, José Ignacio Alvarez-Hamelin, Alain Barrat, Matthieu Latapy, Philippe Raoult, and Alessandro Vespignani, for their support and advice.

References

1. Jacobsen, V., et al.: traceroute. man page, UNIX (1989) See source code: ftp://ftp.ee.lbl.gov/traceroute.tar.gz, and NANOG traceroute source code: ftp://ftp.login.com/pub/software/traceroute/.
2. Huffaker, B., Plummer, D., Moore, D., claffy, k: Topology discovery by active probing. In: Proc. Symposium on Applications and the Internet. (2002) See also the skitter project: http://www.caida.org/tools/measurement/skitter/.
3. Georgatos, F., Gruber, F., Karrenberg, D., Santcroos, M., Susanj, A., Uijter-waal, H., Wilhelm, R.: Providing active measurements as a regular service for ISPs. In: Proc. PAM. (2001) See also the RIPE NCC TTM service: http://www.ripe.net/test-traffic/.
4. McGregor, A., Braun, H.W., Brown, J.: The NLANR network analysis infrastructure. IEEE Communications Magazine **38** (2000) 122–128 See also the NLANR AMP project: http://watt.nlanr.net/.
5. Cheswick, B., Burch, H., Branigan, S.: Mapping and visualizing the internet. In: Proc. USENIX Annual Technical Conference. (2000)
6. Anderson, D.P., Cobb, J., Korpela, E., Lebofsky, M., Werthimer, D.: SETI@home: An experiment in public-resource computing. Communications of the ACM **45** (2002) 56–61 See also the SETI@home project: http://setiathome.ssl.berkeley.edu/.
7. Schmitt, A., et al.: La météo du net (ongoing service) See: http://www.grenouille.com/.
8. Simpson, Jr., C.R., Riley, G.F.: NETI@home: A distributed approach to collecting end-to-end network performance measurements. In: Proc. PAM. (2004) See also the NETI@home project: http://www.neti.gatech.edu/.
9. Shavitt, Y., et al.: DIMES (ongoing project) See: http://www.netdimes.org/.
10. Donnet, B., Raoult, P., Friedman, T., Crovella, M.: Efficient algorithms for large-scale topology discovery. Preprint (under review). arXiv:cs.NI/0411013 v1 (2004) See also the traceroute@home project: http://www.tracerouteathome.net/.

11. Bloom, B.H.: Space/time trade-offs in hash coding with allowable errors. Communications of the ACM **13** (1970) 422–426
12. Mitzenmacher, M.: Compressed Bloom filters. In: Proc. Twentieth Annual ACM Symposium on Principles of Distributed Computing. (2001) 144–150
13. Matsumoto, M., Nishimura, T.: Mersenne Twister: A 623-dimensionally equidistributed uniform pseudorandom number generator. ACM Trans. on Modeling and Computer Simulation **8** (1998) 3–30 See also the Mersenne Twister home page: http://www.math.sci.hiroshima-u.ac.jp/~m-mat/MT/emt.html.
14. Fan, L., Cao, P., Almeida, J., Broder, A.Z.: Summary cache: A scalable wide-area web cache sharing protocol. In: Proc. ACM SIGCOMM. (1998)
15. Dall'Asta, L., Alvarez-Hamelin, I., Barrat, A., Vázquez, A., Vespignani, A.: A statistical approach to the traceroute-like exploration of networks: theory and simulations. In: Proc. Workshop on Combinatorial and Algorithmic Aspects of Networking (CAAN). (2004) Preprint: arXiv:cond-mat/0406404.
16. Barford, P., Bestavros, A., Byers, J., Crovella, M.: On the marginal utility of network topology measurements. In: Proc. ACM SIGCOMM Internet Measurement Workshop (IMW). (2001)
17. Fuller, V., Li, T., Yu, J., Varadhan, K.: Classless inter-domain routing (CIDR): an address assignment and aggregation strategy. RFC 1519, IETF (1993)
18. Bejerano, Y., Rastogi, R.: Robust monitoring of link delays and faults in IP networks. In: Proc. IEEE Infocom. (2003)
19. Govindan, R., Tangmunarunkit, H.: Heuristics for internet map discovery. In: Proc. IEEE Infocom. (2000)
20. Lakhina, A., Byers, J., Crovella, M., Xie, P.: Sampling biases in IP topology measurements. In: Proc. IEEE Infocom. (2003)
21. Clauset, A., Moore, C.: Why mapping the internet is hard. Technical report. arXiv:cond-mat/0407339 v1 (2004)
22. Petermann, T., De Los Rios, P.: Exploration of scale-free networks. Eur. Phys. J. B **38** (2004) Preprint: arXiv:cond-mat/0401065.
23. Guillaume, J.L., Latapy, M.: Relevance of massively distributed explorations of the internet topology: Simulation results. In: Proc. IEEE Infocom. (2005)
24. Rekhter, Y., Li, T., et al.: A border gateway protocol 4 (BGP-4). RFC 1771, IETF (1995)
25. Quoitin, B., Pelsser, C., Bonaventure, O., Uhlig, S.: A performance evaluation of BGP-based traffic engineering. International Journal of Network Management (to appear) See also the C-BGP simulator page: http://cbgp.info.ucl.ac.be/.

Using Simple Per-Hop Capacity Metrics to Discover Link Layer Network Topology

Shane Alcock[1], Anthony McGregor[1,2], and Richard Nelson[1]

[1] WAND Group, University of Waikato*
[2] NLANR Measurement and Network Analysis Group,
San Diego Supercomputer Center**
{spa1, tonym, richardn}@cs.waikato.ac.nz

Abstract. At present, link layer topology discovery methodologies rely on protocols that are not universally available, such as SNMP. Such methodologies can only be applied to a subset of all possible networks. Our goal is to work towards a generic link layer topology discovery method that does not rely on SNMP. In this paper, we will present a new link layer topology discovery methodology based on variable packet size capacity estimation. We will also discuss the problems that arose from preliminary testing where different brands of network cards affected the capacity estimates used to detect serializations. As a result, topologically equivalent links fail to be classified as such by the capacity estimation tool. To circumvent this issue, non-VPS methods of capacity estimation that utilise back to back packet pairs have been investigated as a calibration technique.

1   Introduction

Most topology discovery research focuses on the network (or IP) layer, dealing with host machines and routers. At present, there are a number of tools that can successfully perform topology discovery in this capacity. By contrast, there are few effective topology discovery tools that operate at the link layer. Most existing tools utilise the Simple Network Management Protocol (SNMP) to perform link layer topology discovery [1][2][3]. This is an effective and straightforward method but it requires that SNMP agents are running on every node in the target network and that the appropriate access strings are known. This is not always possible. Other tools that operate at the link layer are manufacturer specific and, as a result, are even more restricted than SNMP-based tools. The goal is to create

* The University of Waikato Network Research Group (WAND) Measurement and Simulation project is supported by the New Zealand Foundation for Research Science and Technology under the New Economy Research Fund, Contract UOWX0206.
** NLANR Measurement and Network Analysis Group (NLANR/MNA) is supported by the National Science Foundation (NSF) under cooperative agreement no. ANI-0129677.

C. Dovrolis (Ed.): PAM 2005, LNCS 3431, pp. 163–176, 2005.

a methodology for generic link layer topology discovery that does not rely on SNMP.

Despite the lack of tools to automate the discovery process, knowledge of a network's topology at the link layer is important. Large scale Ethernet networks linked by switches are becoming increasingly commonplace and it is difficult for network operators to accurately keep track of all the devices in their network. A link layer topology discovery tool could form part of a larger network management suite that would create and maintain an accurate picture of the network's topology. The maintenance portion of the suite would be particularly useful for spotting unauthorized devices and machines being added to the network, and for troubleshooting by identifying failing link layer devices. The discovery portion could be used to check for bottlenecks and redundancy (or a lack thereof), or to generate a map of the network, which is very difficult to do manually for a large network.

The lack of link layer topology discovery tools is due to the inherent transparency of the link layer. Link layer devices cannot be communicated with, queried, or interacted with by a remote computer (except via SNMP). This means there is no *direct* way of discovering the existence of these devices. However, this does not rule out the use of a less direct method. An indirect method would rely on the link layer devices affecting the performance of the network in a manner that is detectable and consistent.

Fortunately, one particular class of link layer device imparts a detectable effect upon any link in which such devices are present. Prasad, Dovrolis, and Mah's paper [4] on the effect of store and forward switches on variable packet size (VPS) capacity estimation tools showed that switches caused VPS capacity estimation tools to consistently underestimate link capacity. This effect can be used to detect not only the presence of store and forward switches, but also the quantity of switches and their capacities in the measured link. This effect is limited to store and forward devices so the focus of the remainder of this paper will be on this particular class of devices. In modern Ethernet networks, store and forward switches are the most common type of link layer device so limiting the scope to switches only is not unreasonable. Other link layer devices, such as hubs, require a different method to discover and are the subject of future research.

Initial testing showed that link layer topology can be inferred from VPS capacity estimates, but it also introduced a more practical difficulty with the methodology. Different brands of network interface cards cause VPS capacity estimation tools to provide different estimates for otherwise equivalent links. This is due to each brand of card having a slightly different processing delay for differently sized packets, which appears as a variation in the initial serialization delay at both ends of the link. To alleviate this problem, we have attempted to use non-VPS capacity estimation techniques as a means of calibrating the topology discovery system so that the variation in delay at the network cards is factored out.

The paper is organised as follows. Section 2 introduces variable packet size capacity estimation and describes the underlying theories and techniques associated with it. The effect of switches on VPS tools and how that effect can be used to generate link layer topology information will also be discussed. Section 3 presents the results of putting the theory into practice. The practical problems that arose from testing are described in Sect. 4 and the solutions we have investigated are detailed in Sect. 5. Section 6 concludes the paper with a discussion of the current state of the methodology and future work in the area of link layer topology discovery.

2   Variable Packet Size Capacity Estimation

Variable packet size (VPS) capacity estimation techniques utilize the relationship between packet size, serialization delay, and capacity. The basic premise is that serialization delay is proportional to the size of the packet being sent. The larger the packet, the longer the serialization delay. The capacity of a link is the rate at which bits can be inserted onto the physical medium and, hence, is directly related to serialization delay. By measuring round trip times for different sized packets, it is possible to calculate the ratio of change in packet size to change in serialization delay. This ratio will describe the capacity of the measured link.

A potential problem with the VPS method is the possibility of other delays, such as queuing, increasing the round trip time by a significant amount. The potential effects of queuing are presented in Fig. 1. Any queuing is going to result in a round trip time measurement that is not solely affected by serialization delay. As such, an accurate capacity estimate cannot be made based on such skewed round trip time measurements. To alleviate this, for each packet size numerous packets are sent and the minimum round trip time is assumed to have been unaffected by queuing or other delays. This technique has been standard in VPS capacity estimation since *pathchar* [5]. VPS capacity estimation tools typically allow the user to specify the number of packets to be sent for each packet size. In situations where it is difficult to observe a minimum RTT, the number of probes may be increased to compensate.

Propagation and processing delays are assumed to be constant. For the propogation delay, a change would normally indicate a change in path requiring restarting of the topology discovery process. In the test network used, the distances are only tens of meters at most so the propagation delay is negligable. The processing delay is assumed to be deterministic and constant. This has held true for the switches we have tested, but not the end stations as we will discuss in section 4.

Tools that use the VPS methodology to generate capacity estimates include Van Jacobson's *pathchar* [5], Mah's *pchar* [6], and Downey's *clink* [7]. Each tool uses the same basic algorithm. A packet size is selected at random from a series of possible packet sizes. A packet of that size that will generate a response from the destination machine is created and sent. The round trip time for the packet is then recorded. Once a packet from each possible size has been sent a certain

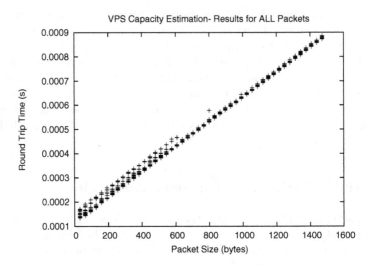

Fig. 1. This graph shows the complete results of a typical *pchar* probe of a link. 32 packets were sent at each packet size and each packet size was 32 bytes apart. Note the differences in round trip times between equally sized packets, often around half a millisecond for many of the smaller packet sizes. These differences are often caused by queuing. Taking the minimum round trip time at each packet size produces a smooth straight line which describes the capacity of the link

number of times (this is usually specified by the user), the minimum round trip time is calculated for each packet size. Using linear regression, the gradient of a line that shows packet size versus round trip time can be calculated. Inverting that gradient will give the estimated capacity of the link.

One major flaw with variable packet size capacity estimation is the fact that such a method will significantly underestimate the capacity of any link that contains store and forward link layer devices. As described in a paper by Prasad, Dovrolis, and Mah [4], this effect is due to each store and forward device adding an extra serialization into the link that is not accounted for by the capacity estimation tool (see Fig. 2). VPS tools use the TTL field of a packet to determine the number of hops in a link but link layer devices do not decrement the TTL counter due to their transparency. From the perspective of the VPS tool, each hop only contains one serialization, regardless of how many switches might be present. However, the round trip time is multiplied by the number of extra unnoticed serializations, making the link appear a lot slower than it really is.

For example, a link between two machines contains two store and forward switches. Both the switches and the Ethernet adaptors on the machines are operating at the same capacity. This link contains three serialization delays: one at the originating host, and one for each switch. Each serialization delay increases the round trip time of a packet sent across the link. Hence, the round trip time for the link is approximately three times what it would be if there were no switches in the link. This makes the gradient of the packet size versus

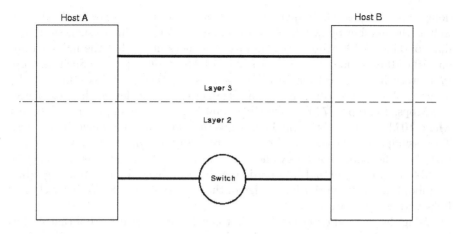

Fig. 2. A demonstration of the effect of a switch on a link between two hosts. At the network layer, it appears that the hosts are directly connected. At the link layer, a switch basically splits the hop into two, creating an extra serialization delay. Since VPS tools operate at the network layer, they do not take that serialization into account when making capacity estimates, resulting in underestimation

round trip time line three times what it would normally be. This gradient, when inverted, would produce a capacity estimate one-third the value of the nominal capacity.

Although, this underestimation effect is a problem for capacity estimation, it provides information regarding the presence of store and forward devices. If the nominal capacity is known (or accurately estimated by another method) prior to calculating a VPS capacity estimate, the degree of underestimation can be used to infer the number of extra serializations and, as a result, the number of link layer devices, within the link. The equation to convert a capacity estimate into a quantity of serializations is:

$$\text{Serializations} = \frac{\text{nominal capacity}}{\text{estimated capacity}} \qquad (1)$$

Using the above equation gives the number of serializations *including* the original serialization at the sending host. This equation works best when all the devices are operating at the same capacity. If some of the serialization delays are of different lengths, it is no longer a simple case of comparing the VPS estimate to the nominal capacity. Fortunately, most Ethernet switches have capacities that make it easier to detect the differing serialization delays.

The serialization delay of a 10 Mbps device is ten times that of a 100 Mbps device. This relationship also holds between 100 Mbps devices and 1 Gbps devices. As it is unlikely that a single hop will contain ten switches of the same capacity, it is reasonable to assume that every ten serializations suggested by a VPS estimate are actually a single serialization for a lower capacity device. By doing this, it is possible to quantify the capacities of the switches in the link, as well as the number of switches. For example, if a VPS estimate suggests that a link contains 11 100 Mbps serializations, it is more likely that the link contains one 100 Mbps serialization and a single 10 Mbps serialization. Such assumptions work in practice because the capacities of Ethernet devices all differ by a factor of 10. This would not be as straightforward if devices had capacities of 5 Mbps, 14 Mbps, 37 Mbps, 96 Mbps, and 112 Mbps, as opposed to Ethernet where 10 Mbps, 100 Mbps and 1 Gbps are the most common device capacities. The assumption that there will be no more than ten switches in any given network layer hop does not always hold true in practice. In such cases, it may be possible to use contextual information such as topology information regarding neighbouring links or prior knowledge of the network layout to correctly classify links with many switches.

Using the method described above, it is possible to create a tool that probes links using a VPS capacity estimation algorithm, infers the number of serializations in the link using the above equation and calculates the number of switches present in those links. Rather than adapt an existing VPS capacity estimation utility, such as *pchar*, a Python implementation of the VPS algorithm called *pychar* was written to perform the probing of links. This has the advantage of providing a tool that supports easy modification and expansion to suit the specific purposes of this project. *pychar* is also designed to be integrated into a future topology discovery suite in an efficient and straightforward manner. *pychar* is capable of using either ICMP or UDP to perform the probing, at least one of which should be available on any Ethernet network.

3   *pychar* Results

Using the WAND emulation network [8], the performance of the *pychar* tool and the validity of the underlying theory has been tested. The test network (see Fig. 3) consists of seven host machines, all running Linux 2.4.20. Three of the host machines are using Mikrotik 4 port Intel Pro100 Ethernet cards, while the remaining four are using single port DSE Realtek 8139 based Ethernet cards. All the cards are operating at 100 Mbps. The machines are connected via three Gigabyte brand 5-port mini switches which are also operating at 100 Mbps. Hence, all the links have a nominal capacity of 100 Mbps. Each machine is within a single transport layer hop of each other, but the number of switches in each link is between one and three. There are no hosts directly connected without at least one switch between them. It is significant that all the devices at both the link and network layers are operating at the same capacity, creating a straightforward situation for initial testing. There is very little traffic operating on this network at any given time, making it easy to gather minimum round trip time data that is free of queuing delays.

Table 1 contains the results of sending probes from Machine 1 to all the other machines in the test network. What these results show is that the basic theory does prove to be correct in practice and that it is possible to approximate the

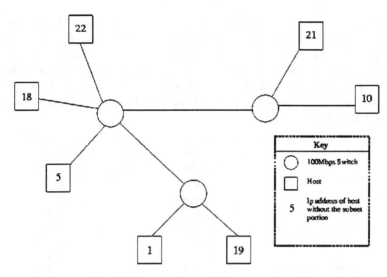

Fig. 3. The topology of the test network

number of switches in any of the links using the technique described earlier. The serialization estimates are not perfect, as there is a fractional component to each of the estimates seen in the table. This is not an issue in this case, as there is prior knowledge that all the devices are operating at the same capacity so any fractional remainders can be ignored. In an environment where device capacities are more varied, this could become a problem. This situation will be discussed in Sect. 4.

Table 1. The results of sending *pychar* probes from Machine 1 to all the other machines in the test network. When rounded to the nearest whole number, the estimated number of serializations is equal to the actual number of serializations. This is ideal in situations where all the serialization delays are the same length

Destination Machine	Estimated Capacity (Kbps)	Nominal Capacity (Kbps)	Estimated Serializations	Actual Serializations
5	31704	100000	3.15	3
10	24124	100000	4.15	4
18	30708	100000	3.26	3
19	44640	100000	2.24	2
21	23540	100000	4.25	4
22	30764	100000	3.25	3

It is also important that topologically equivalent links, i.e. links that contain the same number of serializations, produce equivalent capacity estimates when probed by *pychar*. Table 2 presents *pychar* estimates for a number of different

Table 2. *pychar* estimates for a selection of the two switch links in the test network. The "From" column displays the machine number of the sending host and the "To" column describes the receiving host. The letter in parentheses beside each machine number represents the type of network card used by that host. M represents a Mikrotik card whereas R represents a Realtek card. The rightmost column contains estimates calculated where DAG cards were placed at both ends of the link to perform the timing. This is discussed more in Sect. 5

From	To	Standard *pychar* Estimate (Kbps)	*pychar* Estimate Using DAG Cards (Kbps)
1 (M)	5 (M)	31726	49970
5 (M)	1 (M)	31860	50023
10 (M)	5 (M)	31689	50023
1 (M)	18 (R)	30721	50010
10 (M)	18 (R)	30707	50036
18 (R)	21 (R)	29786	49984
21 (R)	18 (R)	29738	49931

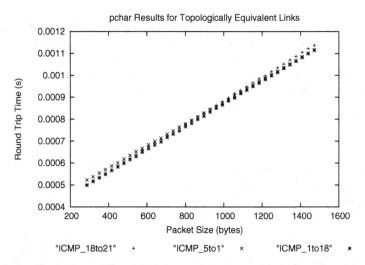

Fig. 4. A graphical illustration of different estimates provided by *pychar* for topologically equivalent links. Each link depicted in the graph contains two switches but the slopes of the lines are different, resulting in different capacity estimates

two switch links in the test network. The results show that topologically equivalent links are producing different capacity estimates when probed by *pychar*. A graphical view of this situation is presented in Fig. 4. Again, this is not an issue in this particular case as it is known that all the devices are operating at the same capacity. However, this problem will need to be dealt with to successfully create a generic link layer topology discovery tool.

4   *pychar* Problems

One problem with variable packet size capacity estimation is that the estimates produced are not exactly what one might expect for any given link. For example, given the theory of underestimation due to extra serializations presented earlier in this paper, it is expected that a link with a nominal capacity of 100000 Kbps that contains three serializations would produce a capacity estimate of 33333 Kbps (one third of the nominal capacity). Looking at Table 1, it is apparent that the three links that match this profile give slightly different estimates - 31704 Kbps, 30708 Kbps, and 30764 Kbps respectively.

This becomes significant when there is the possibility of higher capacity devices being present in the link. A 1 Gbps serialization is one-tenth that of a 100 Mbps device. This means that a serialization quantity estimate of 3.25 for a link with a nominal capacity of 100 Mbps, as seen in Table 1, not only suggests three 100 Mbps devices, but also two 1 Gbps devices (and another five 10 Gbps devices, if one has reason to believe devices of such a capacity might be present). As a result, the slight difference between the ideal estimate and the actual estimate can cause *pychar* to erroneously detect high capacity devices. In situations similar to the test network, where there is prior knowledge that all the devices are the same capacity, this is not a problem and the fractional component can be rounded off and ignored. However, most topology discovery takes place in an *unknown* environment so such an assumption cannot be made.

Table 2 highlights another problem that arose from the initial testing of *pychar*. Some links, despite having the same number of switches present, produce capacity estimates that vary. The links can be divided into three groups based on the capacity estimate given by *pychar*. One group contains links from a host with a Mikrotik Ethernet card to another host with a Mikrotik Ethernet card, one contains links from a host with a Realtek Ethernet card to a host with another Realtek card, and the final group consists of links that have a Realtek card at one end and a Mikrotik card at the other. In the latter case, it does not matter which is the sending host. The important factor is that there are two different cards involved.

This effect is explained by the notion that different brands of network interface card take different amounts of time to put the packet onto the wire, depending on the size of the packet. Equally sized packets have slightly different round trip times depending on the network interface cards involved in the link. For Mikrotik card to Mikrotik card links, smaller sized packets have comparatively longer round trip times. Similarly, larger sized packets have longer round trip times on Realtek to Realtek links. As a result, the variation in gradients as seen in Fig. 4 occurs. This variation in slope translates into a variation in capacity estimate. There can be a difference in excess of 1 Mbps between two estimates for topologically equivalent links, which can be enough to suggest the presence of an extra 1 Gbps serialization. The uncertainty is great enough that links that are identical from a topological standpoint can be classified as different by *pychar* simply due to different network interface cards in the end nodes.

5   Possible Solutions

The difficulty in trying to eliminate the variation in capacity estimates for topologically equivalent links is that the variation in minimum round trip times for
any given packet size is very small. For example, the difference between minimum
RTTs for a 288 byte packet on the two switch links described in Fig. 4 is only
20 microseconds. This is the difference in processing time for the packet on two
different Ethernet adaptors. The difference in RTTs is consistent across multiple
tests so the problem cannot be resolved by increasing the number of probes.
Instead, solutions that eliminate the processing time of the network cards from
the serialization estimates must be considered.

5.1   Hardware Measurement

The first possible solution investigated involves using DAG passive measurement
cards to capture timestamps immediately after the packets have been transmitted. Designed to capture and record details of every packet passing through
them, DAG cards are GPS synchronized network monitoring cards capable of
timestamping at a better than 100 nanosecond resolution [9]. By placing DAG
cards at each end of a link, the network card serializations can be bypassed,
eliminating the variability that causes topologically equivalent links to produce
different capacity estimates under *pychar*. DAG cards also offer more precise
timing than the operating system based timing used by *pychar*.

   If two hosts, X and Y, are both connected to DAG cards, as seen in Fig.
5, and a packet is sent between them, the outgoing timestamp is generated at
DAG 1 rather than at Host X. When the packet reaches DAG 2, an incoming
timestamp is generated. Host Y sends a response packet back to Host X which is
timestamped again as it passes through each of the DAG cards on the way back.
If the difference between the DAG 2 timestamps (the turnaround time at Host
Y) is subtracted from the difference between the DAG 1 timestamps, which is
effectively the round trip time minus the initial serialization delay, the result is
the round trip time for the packet travelling from DAG 1 to DAG 2. The hosts
are removed from the round trip time calculation without altering the link in
any significant way.

   The results of probing some two switch links using dual DAG cards are presented in the rightmost column in Table 2. Instead of the 1 Mbps difference
between estimates for links that had different network interface card configurations, the difference is reduced to less than 100 Kbps. Hence, using DAG cards
reduces the problems caused by different network cards to a negligible level. Applying this solution requires that there be a DAG card connected to both ends
of every probed link. It is neither practical nor economic to deploy DAG cards
on every host in a network of non-trivial size. However, deploying a single DAG
card on the host that would be initiating all the *pychar* probes remains practical. This will allow for much more accurate and consistent capacity estimation
due to the increased timing precision. It will also eliminate any variation at the
sending end of the link, meaning that only the receiving network card will affect
the *pychar* estimates. This means that there will only be two groups of links in

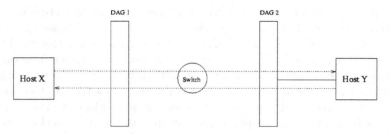

Fig. 5. This diagram demonstrates how DAG cards can be used to eliminate the effects of the network interface cards. Connecting both ends of a link to DAG cards means that timestamps can be captured after the initial serialization delays, removing any hardware variability from the round trip time measurement

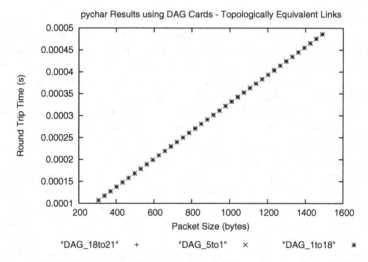

Fig. 6. A graphical illustration of how placing DAG cards at each end of a link can eliminate the variability in *pychar* estimates. The links depicted above are the same links seen in Fig. 4. On this occasion, each link produces exactly the same slope and, as a result, exactly the same capacity estimate

the test network instead of three: links that have a Mikrotik card at the receiving end and links that have a Realtek card at the receiving end.

5.2   Two Packet Probing

Another approach to factoring out the effects of different network interface cards involves the use of a different type of probe prior to running *pychar* to calibrate the system and provide a reference point that includes the adverse effect of the Ethernet card present at the receiving end of the link. To do this, an estimation technique that does not underestimate capacity due to the presence of switches but is affected by the network interface card is needed to provide a base estimate. This base estimate will act as a replacement for the nominal capacity required for the serialization calculation described above. As a result, this would also remove the need for the nominal capacity of the link to be known in advance.

Capacity estimation techniques that use back to back packets to estimate link capacity are typically unaffected by the presence of switches in the link [10]. Rather than using round trip times as the metric upon which the capacity estimates are based, these techniques involve sending two response-seeking packets back to back to the target machine and using the distance between the two response packets to calculate a estimate of link capacity. Similar to VPS capacity estimation techniques, back to back packet methods generate their estimate by inverting the slope of the packet size versus the distance between response packets line. Different back to back packet techniques apply the varying packet size differently.

The three back to back packet methods that will be investigated are packet pairing, packet tailgating and a leading packets method. Packet pairing involves sending two packets of the same size back to back across a link. The idea is that the trailing packet will queue behind the first packet at the link bottleneck but will not be able to catch up any further, so that upon arrival the gap between the two packets is equal to the serialization delay at the bottleneck. Under packet tailgating, the packet doublet consists of a packet that will vary in size and a trailing packet that is always very small in size. This technique has is similar conceptually to packet pairing. By contrast, a leading packet doublet is made up of a small fixed size leading packet and a variable size trailing packet. Unlike packet pairing and packet tailgating, the leading packets method is not a recognised capacity estimation technique and is a variation on packet tailgating created for the purposes of this project. The gap between the two packets is due to the longer serialization time of the second (larger) packet, but it includes the cumulative serializations of all devices on the link. The numbers produced by this method should be not be seen as capacity estimates. However, the leading packets method may produce results that describe characteristics of the link that the other two methods do not.

A prototype Linux kernel module has been written to enable experimentation with back to back packet theories. A kernel module was used because initial testing with a Python script has shown that a user space application requires too much overhead when sending packets, making it virtually impossible to send packets in an optimal back to back manner. Specifically, the aim is to investigate if there is any useful relationship between estimates produced by back to back packet methods and estimates produced by VPS methods that allows the variable effects of network cards to be factored out. Packet size selection is performed in the same manner as in *pychar*: random selection without replacement from a list of possible sizes. In the case of packet pairing, both packets are created to be the selected size. Under packet tailgating and the leading packets method, the first and trailing packets are set to be the selected size, respectively. To prevent inaccuracies due to queuing, each packet size is used multiple times, as in *pychar*, and the minimum distance between packets is used as the metric for each packet size.

None of these back to back methods have proven effective in eliminating the turnaround time at the receiving end of the link. Although there is not enough space for a detailed discussion of the results produced by those techniques here, they were either too inconsistent or failed to exhibit the effects of the receiving network card. As a result, the major problem with this technique remains unresolved. Until it is, it is not possible to produce a generic link layer topology discovery tool that uses variable packet size capacity estimation techniques to gather information about each link.

6   Future Work and Conclusion

The results produced when using dual DAG cards show that a tool such as *pychar* can be used for link layer topology discovery in more heterogenous environments, provided the variation in Ethernet adaptor serialization delay can be factored out.

While placing DAG cards at the end of every link in a network is impractical, it may still be possible to create a viable hardware-based solution. A device that can be inserted at each end of link could run *pychar* and discover the link layer topology between two such devices. Because the devices will all have the same hardware (and possibly DAG cards doing the timing) variation in network interface card is non-existent. The device can simply account for the known effects of the particular brand of card it uses, if necessary. The only drawback to such an approach is that only a single link can be dealt with at any given time, rather than an entire network. Probing multiple links will require manual movement of the devices to the appropriate endpoints.

However, even without hardware assistance, the progress that has been made up to this point still has some more specific uses. If a network is known to contain switches that are all the same capacity (our test network being a prime example, see Fig. 3), then simply rounding the serialization estimates will produce the correct results. Such a network is usually small enough that link layer topology discovery is not necessary but there may be some occasions where *pychar* could prove useful, especially for troubleshooting.

A number of further practical issues with the *pychar*-based technique not addressed in this paper will require future work. This includes finding a method for detecting cut-through devices such as hubs. Also, this paper has not detailed how the link layer information will be combined to create a topology map. Some rudimentary thought has been given to this problem without settling on a comprehensive solution. Finally, the emulated network that *pychar* has been tested on is very homogeneous with regard to operating systems, host hardware, and both the capacity and manufacturer of the switches. Further testing on more varied networks will be required to reveal problems similar to the network interface card problem. However, all these outstanding issues are irrelevant if the problem with differing serialization delays on the network interface cards cannot be resolved.

The next step for the VPS-based link layer topology discovery project is to investigate other measurement techniques that might be able to provide information that will allow the effects of the network interface cards to be factored out. At this stage, given the failure of back to back packet capacity estimation methods to provide such information, we do not know of any techniques that might be of use for this purpose. However, that seems the only way forward for a software-based solution that utilizes VPS capacity estimation.

Although variable packet size capacity estimation appears to be a viable tool for inferring link layer topology information, it is susceptible to the effects of different varieties of network interface cards. Back to back packet capacity estimation techniques have proven ineffective in factoring out these effects. As a result, although *pychar* can provide link layer topology information under specific conditions, it is currently not a viable link layer topology discovery technique in a generic environment.

References

1. D. T. Stott, "Layer-2 Path Discovery Using Spanning Tree MIBs," Avaya Labs Research, Avaya Inc, March 7 2002.
2. Y. Bejerano, Y. Breitbart, M. Garofalakis, and R. Rastogi, "Physical Topology Discovery in Large Multi-Subnet Networks," IEEE Infocom 2003.
3. B. Lowekamp, D. R. O'Hallaron, and T. R. Gross, "Topology Discovery for Large Ethernet Networks," in Proceedings of ACM SIGCOMM, San Diego, California, Aug. 2001.
4. R. S. Prasad, C. Dovrolis, and B. A. Mah, "The Effect of Layer-2 Store-and-Forward Devices on Per-Hop Capacity Estimation," http://citeseer.ist.psu.edu/prasad03effect.html/, 2003.
5. V. Jacobson, "Pathchar: A Tool to Infer Characteristics of Internet Paths," ftp://ftp.ee.lbl.gov/pathchar/, April 1997.
6. B. A. Mah, "pchar: a Tool for Measuring Internet Path Characteristics," http://www.employees.org/~bmah/Software/pchar/, February 1999.
7. A. Downey, "clink: a Tool for Estimating Internet Link Characteristics," http://allendowney.com/research/clink/, 1999.
8. http://www.wand.net.nz/~bcj3/emulation/
9. Endace Measurement Systems, http://www.endace.com/networkMCards.htm
10. K. Lai and M. Baker, "Measuring Link Bandwidths Using a Deterministic Model of Packet Delay," SIGCOMM 2000, pp. 283-294.

Revisiting Internet AS-Level Topology Discovery

Xenofontas A. Dimitropoulos[1], Dmitri V. Krioukov[2], and George F. Riley[1]

[1] School of Electrical and Computer Engineering,
Georgia Institute of Technology,
Atlanta, Georgia 30332–0250
{fontas, riley}@ece.gatech.edu
[2] Cooperative Association for Internet Data Analysis (CAIDA),
La Jolla, California 92093–0505
dima@caida.org

Abstract. The development of veracious models of the Internet topology has received a lot of attention in the last few years. Many proposed models are based on topologies derived from RouteViews [1] BGP table dumps (BTDs). However, BTDs do not capture all AS–links of the Internet topology and most importantly the number of the hidden AS–links is unknown, resulting in AS–graphs of questionable quality. As a first step to address this problem, we introduce a new AS–topology discovery methodology that results in more complete and accurate graphs. Moreover, we use data available from existing measurement facilities, circumventing the burden of additional measurement infrastructure. We deploy our methodology and construct an AS–topology that has at least 61.5% more AS–links than BTD–derived AS–topologies we examined. Finally, we analyze the temporal and topological properties of the augmented graph and pinpoint the differences from BTD–derived AS–topologies.

1   Introduction

Knowledge of the Internet topology is not merely of technological interest, but also of economical, governmental, and even social concern. As a result, discovery techniques have attracted substantial attention in the last few years. Discovery of the Internet topology involves passive or active measurements to convey information regarding the network infrastructure. We can use topology abstraction to classify topology discovery techniques into the following three categories: AS–, IP– and LAN–level topology measurements. In the last category, SNMP–based as well as active probing techniques construct moderate size networks of bridges and end-hosts. At the IP–level (or router–level), which has received most of the research interest, discovery techniques rely on path probing to assemble WAN router–level maps [2,3,4]. Here, the two main challenges are the resolution of IP aliases and the sparse coverage of the Internet topology due to the small number of vantage points. While the latter can be ameliorated by increasing the number of measurement points using overlay networks and distributed agents [5,6,7], the former remains a daunting endeavor addressed only partially thus far [8,9].

C. Dovrolis (Ed.): PAM 2005, LNCS 3431, pp. 177–188, 2005.
© Springer-Verlag Berlin Heidelberg 2005

AS–level topology discovery has been the most straightforward, since BGP routing tables, which are publicly available in RouteViews (RV) [1], RIPE [10] and several other Route Servers [11], expose parts of the Internet AS–map. However, the discovery of the AS–level topology is not as simple as it appears.

The use of BTDs to derive the Internet AS–level topology is a common method. Characteristically, the seminal work by Faloutsos *et al.* [12] discovered a set of simple power–law relationships that govern AS–level topologies derived from BTDs. Several followup works on topology modeling, evolution modeling and synthetic topology generators have been based on these simple power–law properties [13,14,15]. However, it is well–known among the research community that the accuracy of BTD–derived topologies is arguable. First, a BGP table contains a list of AS–paths to destination prefixes, which do not necessarily unveil all the links between the ASs. For example, assume that the Internet topology is a hypothetical full mesh of size n, then from a single vantage point, the shortest paths to every destination would only reveal $n - 1$ of the total $n(n-1)/2$ links. In addition, BGP policies limit the export and import of routes. In particular, prefixes learned over peering links[1] do not propagate upwards in the customer-provider hierarchy. Consequently, higher tier ASs do not see peering links between ASs of lower tiers. This is one reason BTD–based AS–relationships inference heuristics [16] find only a few thousands of peering links, while the Internet Routing Registries reveal tens of thousands [17]. Lastly, as analyzed comprehensively in [18], RV servers only receive partial views from its neighboring routers, since the eBGP sessions filter out backup routes.

The accuracy of AS–level topologies has been considered previously. In [19] Chang *et al.* explore several diverse data sources, i.e. multiple BTDs, Looking Glass servers and Internet Routing Registry (IRR) databases, to create a more thorough AS–level topology. They report 40% more connections than a BTD-derived AS–map and find that the lack of connectivity information increases for smaller degree ASs. Mao *et al.* [20] develop a methodology to map router–graphs to AS–graphs. However they are more concerned with the methodology rather then the properties of the resulting AS–graph. Finally, in [21] Andersen *et al.* explore temporal properties of BGP updates to create a correlation graph of IP prefixes and identify clusters. The clusters imply some topological proximity, however their study is not concerned with the AS–level topology, but rather with the correlation graph.

Our methodology is based on exploiting BGP dynamics to discover additional topological information. In particular we accumulate the AS–path information from BGP updates seen from RV to create a comprehensive AS–level topology. The strength of our approach relies on a beneficial side–effect of the problematic nature of BGP convergence process. In the event of a routing change, the so-called "path exploration" problem, [22], results in superfluous BGP updates, which advertise distinct backup AS–paths of increasing length. Labovitz

---

[1] "Peering links" refers to the AS–relationship, in which two ASs mutually exchange their customers' prefixes free of charge.

Table 1. Example of a simple BGP–update sequence that unveils a backup AS–link (2828 14815) not seen otherwise

Time	AS–path	Prefix
2003-09-20 12:13:25	(withdrawal)	205.162.1/24
2003-09-20 12:13:55	10876-1239-2828-14815-14815-14815-14815-14815	205.162.1/24
2003-09-20 12:21:50	10876-1239-14815	205.162.1/24

et al. [22] showed that there can be up to $O(n!)$ superfluous updates during BGP convergence. We analyze these updates and find that they uncover a substantial number of new AS–links not seen previously. To illustrate this process, consider the simple update sequence in Table 1, which was found in our dataset. The updates are received from a RV neighbor in AS10876 and pertain to the same prefix. The neighbor initially sends a withdrawal for the prefix 205.162.1/24, shortly after an update for the same prefix that exposes the unknown to that point AS–link 2828–14815, and finally an update for a shorter AS–path, in which it converges. The long AS–prepending in the first update shows that the advertised AS–path is a backup path not used at converged state. We explore the backup paths revealed during the path exploration phenomenon and discover 61.5% more AS–links not present in BTDs.

2   Methodology

Our dataset is comprised of BGP updates collected between September 2003 and August 2004 from the RV router `route-views2.oregon-ix.net`. The RV router has multihop BGP sessions with 44 BGP routers and saves all received updates in the MRT format [1]. After converting the updates to ASCII format, we parse the set of AS–paths and mark the time each AS–link was first observed, ignoring AS–sets and private AS numbers. There are more than 875 million announcements and withdrawals, which yield an AS–graph, denoted as G_{12}, of 61,134 AS–links and 19,836 nodes. Subscript 12 in the notation G_{12} refers to the number of months in the accumulation period. To quantify the extent of additional information gathered from updates, we collect BTDs from the same RV router on the 1st and 15th of each month between September 2003 and August 2004. For each BTD we count the number of unique AS–links, ignoring AS–sets and private AS–numbers for consistency. Figure 1 illustrates the comparison. The solid line plots the cumulative number of unique AS–links over time, seen in BGP updates. Interestingly, after an initial super–linear increase, the number of additional links grows linearly, much faster than the corresponding increase observed from the BTDs. At the end of the observation window, BGP updates have accumulated an AS–graph that has 61.5% more links and 10.2% more nodes than the largest BTD–derived graph G_{12}^{BTD}, which was collected on 08/15/2004. The notable disparity suggests that the real Internet AS topology may be different from what we currently observe from BTD–derived graphs, and merits

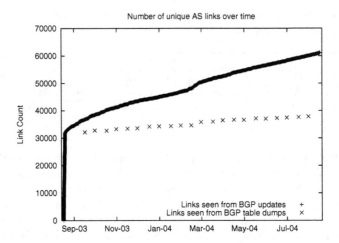

Fig. 1. Number of unique AS–links observed in BGP updates vs BTDs

further investigation. To gain more insight in the new information we analyze the temporal and topological properties of the AS–connectivity.

3   Temporal Analysis of Data

Identifying temporal properties of the AS–connectivity observed from BGP updates is necessary to understand the interplay between the observation of AS–links and BGP dynamics. In particular, we want to compare the temporal properties of AS–links present in BTDs with AS–links observed in BGP updates. To do so, we first introduce the concept of *visibility* of a link from RV. We say that at any given point in time a link is visible if RV has received at least one update announcing the link, and the link has not been withdrawn or replaced in a later update for the same prefix. A link stops been visible if all the prefix announcements carrying the link have been withdrawn or reannounced with new paths that do not use the link. We then define the following two metrics to measure the temporal properties of AS–links:

1. *Normalized Persistence* (NP) of a link is the cumulative time for which a link was visible in RV, over the time period from the first time the link was seen to the end of the measurements.
2. *Normalized Lifetime* (NL) of a link is the time period from the first time to the last time a link was seen, over the time period from the first time the link was seen to the end of the measurements.

The NP statistic represents the cumulative time for which a link was visible in RV, while the NL represents the span from the beginning to the end of the lifetime of the link. Both are normalized over the time period from the first time a link was seen to the end of the measurements to eliminate bias against links that were not seen from the beginning of the observation.

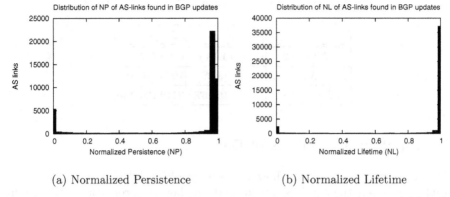

(a) Normalized Persistence   (b) Normalized Lifetime

Fig. 2. Distribution of Normalized Persistence and Normilized Lifetime of AS–links seen between September 2003 and January 2004 in BGP updates

To calculate the NP and NL statistics, we replicate the dynamics of the RV routing table using the BGP updates dataset. We implement a simple BGP routing daemon that parses BGP updates and reconstructs the BGP routing table keeping per–peer and per–prefix state as needed. Then for each link we create an array of time intervals for which the link was visible and calculate the NP and NL statistics. Unfortunately, the BGP updates cannot explicitly pinpoint the event of a session reset between RV and its immediate neighbors. Detection of session resets is necessary to flush invalid routing table entries learned from the neighbor and to adjust the NP and NL statistics. We implement a detection algorithm, described in the Appendix, to address the problem.

We measure the NP and NL statistics over a 5–month period, from September 2003 to January 2004, and plot their distributions in Figure 2. Figure 2(a) demonstrates that NP identifies two strong modes in the visibility of AS–links. At the lower end of the x axis, more than 5,000 thousand links have $NP \leq 0.2$, portraying that there is a significant number of links that only appear during BGP convergence turbulence. At the upper end of the x axis, almost 35,000 links have an NP close to 1. The distribution 2(b) of the NL statistic is even more modal, conveying that most of the links have a high lifetime span. At the end of the 5–month period, BGP updates have accumulated a graph G_5 that we decompose into two parts. One subgraph, G_5^{BTD}, is the topology seen in a BTD collected from RV at the end of the 5–month period and the second subgraph is the remaining $G_5 - G_5^{BTD}$. Table 2 shows the number of links with $NP \leq 0.2$, $0.2 < NP < 0.8$ and $NP \geq 0.8$ in G_5^{BTD} and in $G_5 - G_5^{BTD}$. Indeed, only 0.2% of the links in G_5^{BTD} have $NP \leq 0.2$, demonstrating that BTDs capture only the AS–connectivity seen at steady–state. In contrast, most links in $G_5 - G_5^{BTD}$ have $NP \leq 0.2$, exhibiting that most additional links found with our methodology appear during BGP turbulence.

Table 2. Normalized Persistence in G_5^{BTD} and $G_5 - G_5^{BTD}$

	G_5^{BTD}	$G_5 - G_5^{BTD}$
$NP \leq 0.2$	65 (0.2%)	6891 (57.5%)
$0.2 < NP < 0.8$	1096 (3.2%)	1975 (16.5%)
$NP \geq 0.8$	33141 (96.6%)	3119 (26.0%)

4   Topological Analysis of Data

Ultimately, we want to know how the new graph is different from the BTD graphs, e.g. where the new links are located, and how the properties of the graph change. A handful of graph theoretic metrics have been used to evaluate the topological properties of the Internet. We choose to evaluate three representative metrics of important properties of the Internet topology:

1. *Degree Distribution of AS–nodes.* The Internet graph has been shown to belong in the class of power–law networks [12]. This property conveys the organization principle that few nodes are highly connected.
2. *Degree–degree distribution of AS–links.* The degree–degree distribution of the AS–links is another structural metric that describes the placement of the links in the graph with respect to the degree of the nodes. More specifically, it is the joint distribution of the degrees of the adjacent ASs of the AS–links.
3. *Betweenness distribution of AS–links.* The betweenness of the AS–links describes the communication importance of the AS–links in the graph. More specifically, it is proportional to the number of shortest paths going through a link.

One of the controversial properties of the Internet topology is that the degree distribution of the AS–graph follows a simple power–law expression. This observation was first made in [12] using a BTD–derived AS–graph, later disputed in [23] using a more complete topology, and finally reasserted in [24] using an augmented topology as well. Since our work discovers substantial additional connectivity over the previous approaches, we re–examine the power–law form of the AS–degree distribution. For a power–low distribution the complementary cumulative distribution function (CCDF) of the AS–degree is linear. Thus, after plotting the CCDF, we can use linear regression to fit a line, and calculate the correlation coefficient to evaluate the quality of the fit. Figure 3 plots the CCDF of the AS–degree for the updates-derived graph, G_{12}, and for the corresponding BTD-derived graph, G_{12}^{BTD}. Due to the additional connectivity in G_{12}, the updates–derived curve is slightly shifted to the right of the G_{12}^{BTD} curve, without substantial change in the shape. Figures 4 and 5 show the CCDF of the AS–degree and the corresponding fitted line for G_{12} and G_{12}^{BTD}, accordingly. The correlation coefficient for G_{12}^{BTD} is 0.9836, and in the more complete AS–graph G_{12} it slightly decreases to 0.9722, which demonstrates that the AS–degree dis-

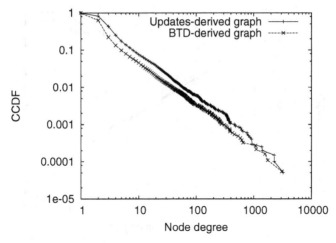

Fig. 3. CCDF of the AS–degree for the updates–derived AS–graph (G_{12}) and the largest BTD–derived AS–graph (G_{12}^{BTD})

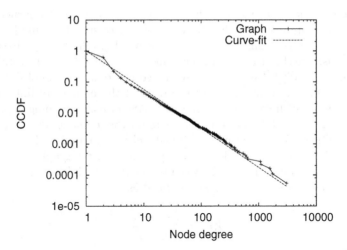

Fig. 4. CCDF of the AS–degree for the largest BTD–derived AS–graph (G_{12}^{BTD}) and linear regression fitted line

tribution in our updates–derived graph follows a power–law expression fairly accurately.

We then examine the degree–degree distribution of the links. The degree–degree distribution $M(k_1, k_2)$ is the number of links connecting ASs of degrees k_1 and k_2. Figure 6, compares the degree–degree distributions of the links in the full G_{12} graph and of the links present only in updates, $G_{12} - G_{12}^{BTD}$. The overall structure of the two contourplots is similar, except for the differences in the areas of links connecting low-degree nodes to low-degree nodes and links connecting

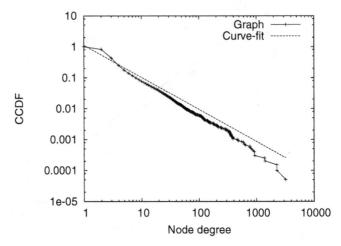

Fig. 5. CCDF of the AS–degree for the updates–derived AS–graph (G_{12}) and linear regression fitted line

medium-degree nodes to medium-degree nodes (the bottom-left corner and the center of the contourplots). The absolute number of such links in $G_{12} - G_{12}^{BTD}$ is smaller than in G_{12}, since $G_{12} - G_{12}^{BTD}$ is a subgraph of G_{12}. However, the contours illustrate that the ratio of such links in $G_{12} - G_{12}^{BTD}$ to the total number of links in $G_{12} - G_{12}^{BTD}$ is higher than the corresponding ratio of links in G_{12}. Figure 7 depicts the contourplot of the ratio of the number of links in G_{12}^{BTD} over the number of links in G_{12} connecting ASs of corresponding degrees. The dark region between 0.5 and 1.5 exponents on the x and y axes, signifies the fact that BGP updates contain additional links, compared to BTDs, between low and medium-degree ASs close to the periphery of the graph.

Finally, we examine the link betweenness of the AS–links. In graph $G(V,E)$, the betweenness $B(e)$ of link $e \in E$ is defined as

$$B(e) = \sum_{ij \in V} \frac{\sigma_{ij}(e)}{\sigma_{ij}},$$

where $\sigma_{ij}(e)$ is the number of shortest paths between nodes i and j going through link e and σ_{ij} is the total number of shortest paths between i and j. With this definition, link betweenness is proportional to the traffic load on a given link under the assumptions of uniform traffic distribution and shortest–path routing. Figure 8 illustrates the betweenness distribution of G_{12} and of G_{12}^{BTD} and reveals that our updates–constructed graph yields more links with small betweenness. Links with small betweenness have lower communication importance in a graph theoretic context, demonstrating that our methodology unveils backup links and links used for local communication in the periphery of the graph.

Overall, our topological analysis shows that our augmented graph remains a power-law network and has more links between low and medium–degree nodes

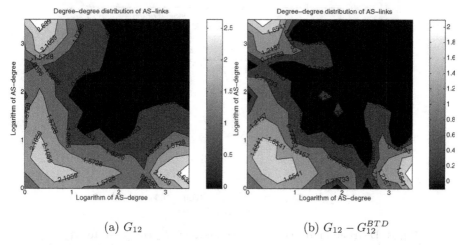

(a) G_{12}

(b) $G_{12} - G_{12}^{BTD}$

Fig. 6. Degree–degree distributions of AS–links. The x and y axes show the logarithms of the degrees of the nodes adjacent to a link. The color codes show the logarithm of the number of the links connecting ASs of corresponding degrees

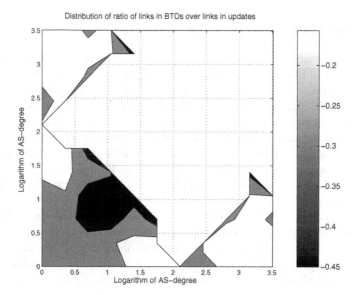

Fig. 7. Distribution of the ratio of the number of links in G_{12}^{BTD} over the number of links in G_{12} connecting ASs of corresponding degrees. The x and y axes show the logarithms of the degrees of the nodes adjacent to a link. The color codes show the logarithm of the above ratio

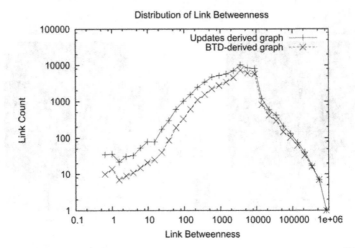

Fig. 8. Distribution of the link betweenness of G_{12} compared to G_{12}^{BTD}

and more links of lower communication importance compared to BTD–derived graphs.

5   Conclusions

In this work we exploit the previously unharnessed topological information that can be extracted from the most well–known and easily accessible source of Internet interdomain routing data. We evidence that the Internet topology is vastly larger than the common BTD–derived topologies and we show how an undesired aspect of the interdomain architecture can be used constructively. We find that our substantially larger AS–graph retains the power–law property of the degree distribution. Finally, we show that our method discovers links of small communication importance connecting low and medium–degree ASs, suggesting AS–links used for backup purposes and local communication in the periphery of the Internet.

Closing, we highlight that our work is a step forward showing a large gap in our knowledge of the Internet topology. For this reason, we pronounce the need to focus more on the perpetual problem of measuring Internet topology before accepting far–reaching conclusions based on currently available AS–level topology data, which are undeniable rich but substantially incomplete.

Acknowledgments

We thank Priya Mahadevan for sharing her betweenness scripts, Andre Broido, Bradley Huffaker and Young Hyun for valuable suggestions, and Spyros Denazis for providing computer resources.

Support for this work was provided by the DARPA N66002-00-1-8934, NSF award number CNS-0427700 and CNS-0434996.

References

1. Meyer, D.: University of Oregon Route Views Project (2004)
2. Govindan, R., Tangmunarunkit, H.: Heuristics for Internet map discovery. In: IEEE INFOCOM 2000, Tel Aviv, Israel, IEEE (2000) 1371–1380
3. k claffy, Monk, T.E., McRobb, D.: Internet tomography. Nature (1999) http://www.caida.org/tools/measurement/skitter/.
4. Spring, N., Mahajan, R., Wetherall, D.: Measuring ISP topologies with Rocketfuel. In: ACM SIGCOMM. (2002)
5. Spring, N., Wetherall, D., Anderson, T.: Scriptroute: A facility for distributed Internet measurement. In: USENIX Symposium on Internet Technologies and Systems (USITS). (2003)
6. Shavitt, Y.: Distributed Internet MEasurements and Simulations (DIMES). http://www.netdimes.org (2004)
7. Friedman, T.: Work in progress on traceroute@home. In: ISMA - Internet Statistics and Metrics Analysis Workshops. (2004)
8. Spring, N., Dontcheva, M., Rodrig, M., Wetherall, D.: How to Resolve IP Aliases (2004) UW CSE Technical Report 04-05-04.
9. Keys, K.: iffinder. http://www.caida.org/tools/measurement/iffinder/ (2002)
10. : RIPE. http://www.ripe.net (2004)
11. : A traceroute server list. http://www.traceroute.org (2004)
12. Faloutsos, M., Faloutsos, P., Faloutsos, C.: On power-law relationships of the Internet topology. In: Proceedings of the ACM SIGCOMM. (1999)
13. Aiello, W., Chung, F., Lu, L.: A random graph model for massive graphs. In: Proceedings of the 32nd Annual ACM Symposium on Theory of Computing (STOC), ACM Press (2000) 171–180
14. Chen, Q., Chang, H., Govindan, R., Jamin, S., Shenker, S.J., Willinger, W.: The origin of power laws in Internet topologies revisited. In: IEEE INFOCOM. (2002)
15. Tangmunarunkit, H., Govindan, R., Jamin, S., Shenker, S., Willinger, W.: Network topology generators: Degree-based vs structural (2002)
16. Gao, L.: On inferring autonomous system relationships in the Internet. In: Proc. IEEE Global Internet Symposium. (2000)
17. Siganos, G., Faloutsos, M.: Analyzing BGP policies: Methodology and tool. In: IEEE INFOCOM. (2004)
18. Teixeira, R., Rexford, J.: A measurement framework for pin-pointing routing changes. In: ACM SIGCOMM Network Troubleshooting Workshop. (2004)
19. Chang, H., Govindan, R., Jamin, S., Shenker, S.J., Willinger, W.: Towards capturing representative AS-level Internet topologies. Computer Networks Journal **44** (2004) 737–755
20. Mao, Z.M., Rexford, J., Wang, J., Katz, R.H.: Towards an accurate AS-level traceroute tool. In: ACM SIGCOMM. (2003)
21. Andersen, D., Feamster, N., Bauer, S., Balakrishnan, H.: Topology Inference from BGP Routing Dynamics. In: Internet Measurement Workshop. (2002)
22. Labovitz, C., Malan, G.R., Jahanian, F.: Internet routing instability. IEEE/ACM Transactions on Networking **6** (1998) 515–528
23. Chen, Q., Chang, H., Govindan, R., Jamin, S., Shenker, S., Willinger, W.: The origin of power laws in Internet topologies revisited. In: IEEE INFOCOM. (2002)

24. Siganos, G., Faloutsos, M., Faloutsos, P., Faloutsos, C.: Power-laws and the AS-level Internet topology. IEEE Transactions on Networking (2003)
25. Maennel, O., Feldmann, A.: Realistic BGP Traffic for Test Labs. In: ACM SIG-COMM. (2002)

Appendix

Detection of Session Resets

The problem of detection of BGP session resets has also been addressed by others. In [25] Maennel *et al.* propose a heuristic to detect session resets on AS–links in arbitrary Internet locations by monitoring BGP updates in RV. We are concerned with a seemingly less demanding task: detection of session resets with immediate neighbors of RV. Our algorithm is composed of two components. The first detects surges in the BGP updates received from the same peer over a short time window of s seconds. If the number of unique prefixes updated in s are more than a significant percent p of the previously known unique prefixes from the same peer, then a session reset is inferred. The second component detects periods of significant inactivity when a threshold t is passed from otherwise active peers. We combine both approaches and set low thresholds ($t = 4mins$, $p = 80\%$, $s = 4secs$) to yield an aggressive session reset detection algorithm. Then, we calculate NP and NL over a period of a month with and without aggressive session reset detection enabled. We find that the calculated statistics are virtually the same with less then 0.1% variation. Implying that the short time scale of the lifetime of session resets does not affect the span of the NP and NL statistics. Hence, we leave out the detection of session resets in the remaining NP and NL measurements.

Application, Network and Link Layer Measurements of Streaming Video over a Wireless Campus Network

Feng Li, Jae Chung, Mingzhe Li, Huahui Wu, Mark Claypool,
and Robert Kinicki

CS Department at Worcester Polytechnic Institute,
Worcester, MA, 01609, USA
{lif, goos, lmz, flashine, claypool, rek}@cs.wpi.edu

Abstract. The growth of wireless LANs has brought the expectation for high-bitrate streaming video to wireless PCs. However, it remains unknown how to best adapt video to wireless channel characteristics as they degrade. This paper presents results from experiments that stream commercial video over a wireless campus network and analyze performance across application, network and wireless link layers. Some of the key findings include: 1) Wireless LANs make it difficult for streaming video to gracefully degrade as network performance decreases; 2) Video streams with multiple encoding levels can more readily adapt to degraded wireless network conditions than can clips with a single encoding level; 3) Under degraded wireless network conditions, TCP streaming can provide higher video frame rates than can UDP streaming, but TCP streaming will often result in significantly longer playout durations than will UDP streaming; 4) Current techniques used by streaming media systems to determine effective capacity over wireless LAN are inadequate, resulting in streaming target bitrates significantly higher than can be effectively supported by the wireless network.

1 Introduction

The combination of the decrease in price of wireless LAN access points (APs) and the increase in wireless link capacities has prompted a significant increase in the number of wireless networks in homes, corporate enterprise networks, and academic campus networks. The promise of up to 54 Mbps capacity[1] from a wireless AP means that users now expect to see applications such as streaming video that require high bitrates running seamlessly from wired media servers to wireless media clients.

Although much is already known about wireless LANs and the individual components of the wireless LAN environment that make the delivery of high-demand applications over wireless a challenge, there has been little effort to

---

[1] IEEE 802.11g.

C. Dovrolis (Ed.): PAM 2005, LNCS 3431, pp. 189–202, 2005.

integrate measures of wireless link layer performance with streaming media application layer choices. Such knowledge can facilitate the redesign of streaming media systems to account for the trend towards a wireless last hop to clients. Moreover, a better understanding of the impact of wireless LAN transmission characteristics on streaming media is valuable to network practitioners concerned with providing adequate wireless LAN coverage and discovering trouble spots in network performance.

Previous work [8] has shown that streaming products such as RealNetworks and Windows Streaming Media make important decisions concerning the characteristics of the video stream prior to streaming the video to a client. These decisions are based on estimates of the underlying network characteristics obtained from network probes. However, it remains unclear which wireless channel characteristics, such as frame loss rate, signal strength, or link layer bitrate, are the most useful for streaming media strategies that improve the performance of a streaming video by adapting video transmission choices to current wireless network conditions.

A primary goal of this investigation is to correlate wireless link layer behavior and network layer performance with streaming media application layer performance. Application layer measurement tools [6] were combined with commercially available network layer measurement tools and publicly available IEEE 802.11 measurement tools to conduct wireless experiments and integrate the measurement results. Seeking to characterize the impact of wireless network conditions on streamed video performance, this active measurement study considers four hypothesis:

1. *Wireless LANs make it difficult for streaming video to gracefully adapt when network conditions degrade.* This investigation attempts to uncover specific characteristics of streams to poor locations that could trigger streaming server adjustments to improve video transmission quality. Increasing performance in poor locations is critical since a streaming wireless client with bad performance can negatively impact other wireless clients connected to the same AP [1].

2. *Videos encoded with multiple levels can stream better than videos encoded with only a single level when wireless LAN conditions are poor.* Commercial media encoders allow videos to be encoded with one or more target bitrate levels. When streaming, the server determines which encoding level to use based on feedback from the client regarding the client end-host network conditions. A video with multiple levels of encoding should make better use of a wireless LAN with limited capacity than a video with a single level of encoding.

3. *TCP is more effective than UDP for streaming video over wireless LANs.* Commercial media players typically let the client select the streaming transport protocol. UDP is often selected due to lower overhead and jitter. However, recent work [4, 5, 11] suggests TCP and TCP-like protocols can be at least as effective and potentially more effective at providing higher quality video to clients under poor network conditions.

4. *Current techniques used by streaming media systems to estimate available capacity to a wireless LAN client are inadequate for providing the best video performance.* Some commercial media players use packet-pair techniques [10] to estimate the capacity along the flow path prior to starting the streaming of the video to the client [8]. However, packet-pair was not designed for wireless networks where changes in transmission conditions cause mid-stream wireless capacity changes. By measuring frame errors and signal strength at the data link layer during wireless streaming experiments, changes in the wireless environment can be correlated with changes in video performance, and facilitate the development of better wireless capacity estimators.

2 Methodology

2.1 Tools

The unique aspect of this investigation is the concurrent use of measurement tools at multiple levels in the network protocol stack to evaluate streaming media performance over wireless LANs. This section discusses the tools employed in this study. For reference, the layer corresponding to each tool and examples of some of the performance measurements available from each tool are listed in Table 1.

Table 1. Measurement Tools

Layer	Tools	Performance Measures
Application	Media Tracker	Frame rate, Frames lost, Encoded bitrate
Network	UDP Ping, Wget	Round-trip time, Packet loss rate, Throughput
Wireless	Typeperf, WRAPI	Signal strength, Frame retries, Capacity

At the application layer, the WPI Wireless Multimedia Streaming Lab has experience measuring video client and server performance [4, 6, 8, 12]. An internally developed measurement tool, called *Media Tracker* [6], streams video from a Windows Media Server, collecting application layer data specific to streaming video including: encoding data rate, playout bitrate, time spent buffering, video frame rate, video frames lost, video frames skipped, packets lost and packets recovered.

For network layer performance measures such as round-trip time and packet loss rate along the stream flow path, *UDP ping*, an internally developed tool, was used. Preliminary experiments revealed that because the standard ICMP *ping* provided by Windows XP waits for the previous ping reply or a timeout before sending out the next ping packet, a constant ping rate could not be maintained in some poor wireless conditions where 10 second and longer round-trip times were recorded. Thus, a customized ping tool using application-layer UDP packets was built to provide constant ping rates, ping intervals configurable in milliseconds, and configurable ping packet sizes.

At the wireless data link layer, a publicly-available library, called *WRAPI* [2] was enhanced to collect information at the wireless streaming client that includes: signal strength, frame retransmission counts and failures, and information about the specific wireless AP that handles the wireless last hop to the client. Additionally, *typeperf*, a performance monitoring tool built-in to Windows XP, collected processor utilization and network data including data received bitrate and the current wireless target capacity.

Although the above four tools were deployed concurrently on the wireless streaming client, baseline measurements indicated these tools consume only about 3% of the processor time. Given that streaming downloads consumed about 35% of processor time, the assumption is the measurement tools do not significantly effect the performance of the streaming downloads to the wireless clients.

2.2   Experiment Setup

This investigation conducts a series of experiments where video clips are streamed from a Windows Media Server over a wired campus network to a wireless streaming client at pre-determined locations in the WPI Computer Science Department building. As Figure 1 shows, the wireless portion of the WPI campus network is partitioned from the wired infrastructure. Thus, the assumption is that all video streams traverse the same network path except for the last two hops from a common exit off the wired campus LAN to a wireless AP and from the AP to the streaming client. The media server runs Windows Media Service v9.0 as part of the Windows Server 2003 Standard Edition, and the wireless client resides on a Dell laptop with a Centrio mobile CPU running Windows XP sp1 and an IEEE 802.11g wireless network adaptor based on the Broadcom[2] chipset. The WPI wireless LAN uses Airespace[3] APs and provides IEEE 802.11 a/b/g wireless service for all the experiments.

Two distinct video clips, known as Coast Guard and Paris, were used in this study. Both clips were encoded to run at 352×288 resolution and 30 frames per second. Both clips run for approximately two minutes.[4] Coast Guard is a high-motion video clip (5.4% skipped macro blocks) with a camera panning scene of a moving Coast Guard cutter. Paris is a low-motion video clip (41.2% skipped macro blocks) with two people sitting and talking with some rapid motion from two small objects in the scene.

Windows Media Server selects the streaming rate based upon the encoded bitrate of the layers in the video clip and an estimate of available capacity for the bottleneck link along the flow path. During this investigation, two distinct versions of each video were streamed to every client location: a single level version of the video encoded at 2.5 Mbps to stress the wireless link; and a multiple level version that includes eleven encoding layers such that the streaming server has

---

[2] http://www.broadcom.com/

[3] http://www.airespace.com/

[4] The median duration of video clips stored on the Internet [7] is 2 minutes.

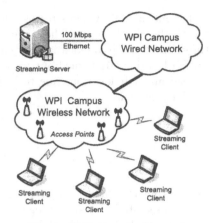

Fig. 1. WPI Campus Network

the opportunity to do media scaling to dynamically choose the encoded clip to stream based on perceived network capacity. To compare the performance of standard streaming protocol choices, each of the four videos instances was streamed using TCP and repeated using UDP.

2.3   Experiment Design

At the beginning and the end of each experimental instance, the client downloaded a large file using *wget*, a publicly-available HTTP/FTP download application,[5] to estimate the effective throughput of a TCP bulk transfer. Thus, each experiment consisted of an initial bulk download, eight different video downloads (2 clips (Paris and Coast Guard) × 2 versions (Single Level and Multiple Level) × 2 transport protocols (UDP and TCP)) and a final bulk download. While each video was streamed, the client initiated UDP ping requests to determine round-trip time and packet losses. The UDP ping requests were 200 milliseconds apart, with 1350-byte packets for the single level video and 978-byte packets for the multiple level video. The choice of packet sizes came from the observation that 90% of the packets are 1350 bytes and 978 bytes for single level and multiple level video, respectively. While streaming, measurement data was also collected by WRAPI, typeperf and Media Tracker at the client side on a stationary laptop.

Clearly, wireless networking transmission performance is dependent on current network conditions. To reduce the variability in the network conditions, all the experiments were conducted during the Winter Break (December 23-25, 2004 and December 29-30, 2004) in the Computer Science Department on the WPI campus. During these testing periods, there was only occasional network activity and virtually no other wireless users in the Computer Science department. Each experiment was repeated five times at three distinct locations on three different floors in the Computer Science department. Thus the results come from a total

---

[5] http://www.gnu.org/software/wget/wget.html

of 45 experimental runs that include 360 video streams. On each floor, an AP was selected to interact with the client laptop. Then, preliminary experiments were conducted to find three laptop reception locations for each AP, representing good, fair, and bad reception locations. It turned out to be difficult to make a clear distinction between bad and fair locations due to high variability in the signal strength at fair and bad locations.

3   Results

3.1   Data Collected

Ten data sets were removed from the 360 video streaming runs due to wireless connection failures that caused abnormal streaming terminations. Thus, 350 video instances (see Table 2) are included in the analysis of the results.

Table 2. Data Collected

	TCP Streaming	UDP Streaming	Total
Multiple Level Video	86	85	171
Single Level Video	89	90	179
Subtotal	175	175	350

Comparison of the two clips, Paris and Coast Guard, with analysis similar to the other experimental factors presented in Section 3.2-3.4, produced no statistically significant differences in performance. This suggests that the differences in motion between the low-motion Paris video and the high-motion Coast Guard video did not impact performance over a wireless network. Thus, all subsequent analysis combines the data obtained for both clips for each of the categories in Table 2.

3.2   Categorization

Figure 2 depicts the throughput obtained versus signal strength for all the streaming and bulk download instances. The streaming data and the bulk download data are separately fit with logarithmic functions. The root mean square value of the deviation of the data from the fitted function[6] are 0.49 Mbps and 1.47 Mbps for streaming throughput and bulk downloading throughput, respectively. Note, there is a "cliff" where throughput degrades suddenly when the signal strength is between -70 dBm and -80 dBm.

To provide a clearer picture of streaming video behavior, the experiments were classified by the average signal strength recorded for a download from the server to the instrumented video client. For the remainder of the analysis, the experiments are categorized in one of three distinct regions: "Bad" locations (less

---

[6] The *stdfit* reported from the *gnuplot* fit function.

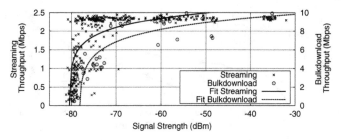

Fig. 2. Average Wireless Signal Strength versus Average Throughput

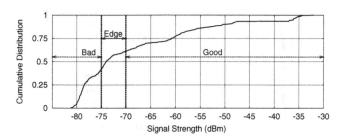

Fig. 3. CDF of Wireless Signal Strengths

than -75 dBm); "Edge" locations (between -75 and -70 dBm); and "Good" locations (greater -70 dBm).[7] This classification facilitates focusing on understanding the performance differences between the Good and the Bad locations. Figure 3 shows a cumulative distribution function (CDF) of the average signal strengths gathered and depicts the Good, Edge and Bad regions, with approximately 1/3 of the data points in each region.

3.3   Single Level Encoding Versus Multi-level Encoding

As described in Section 2.1, both video clips were encoded twice, once at a single, high-bandwidth encoded level and again with multiple encoded levels. Figure 4 and Figure 5 provide CDFs to compare the impact of the server having multiple encoding levels versus only a single encoding level for wireless streaming. These figures indicate that when the client is at a Good location, the number of encoded levels has little effect on the average video frame rate and the coefficient of variation of the video frame rate. Since a Good wireless connection can generally support both the single level and the highest level in the multiple level clip, the stream does not need to be scaled to a lower bitrate.

However, at Bad locations, multiple level encoding provides better streaming performance than single level encoding. More than 2/3 of the time, the multiple level clip has a higher frame rate than the single level clip, and the multiple level

---

[7] The variance in signal strength is about the same for both Good and Bad locations.

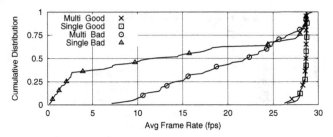

Fig. 4. Average Application Frame Rate for Multiple and Single Level Encoding

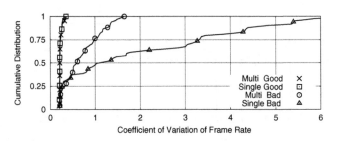

Fig. 5. Coefficient of Variation of Application Frame Rate for Multiple and Single Level Encoding

clip has a median frame rate of 22 frames per second compared to a median of 11 frames per second for the single level clip.

3.4   TCP Streaming Versus UDP Streaming

Figures 6-10 provide CDFs to compare the impact of choosing TCP versus UDP when streaming videos to clients at Good and Bad wireless locations. These figures show that at Good wireless locations, the choice of TCP or UDP has little effect on the average and coefficient of variation of frame rate. However, Figure 6 demonstrates that at Bad wireless locations, streams received by TCP streaming clients have a higher median frame rate (24 fps) than streams received by UDP streaming clients (15 fps). Moreover, the TCP streams have a higher frame rate about 2/3 of the time. Similarly, in Figure 7 the TCP streams have a lower median variation in frame rate than the UDP streams, and for 2/3 of the Bad locations TCP streams have a lower variation in frame rate than the UDP streams.

TCP video streams may be able to achieve better application frame rates under Bad conditions than UDP because when the wireless layer loses data, TCP retransmits the data and allows it to be played. However, without built-in retransmissions, UDP does not automatically recover lost data. The inter-frame dependencies in video can cause loss rates as low as 3% to result in up to 30% of application frames being unplayable [3]. Figure 8 graphs the CDF of wireless

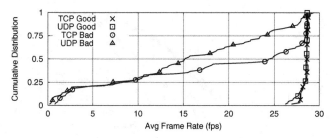

Fig. 6. Average Application Frame Rate for TCP and UDP Streams

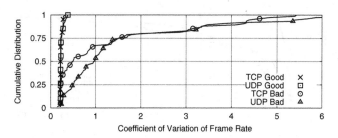

Fig. 7. Coefficient of Variation of Application Frame Rate for TCP and UDP Streams

layer retry fraction for upstream (from the client to the server) data and Figure 9 shows a CDF of network ping loss rates. Under Bad conditions, approximately 1/3 of all wireless layer frames need to be retransmitted and when the same wireless frame is retransmitted too many times, the wireless layer drops the frame and this yields network ping packet loss. Under Bad wireless conditions, nearly 1/3 of the time the network loss rate is about 15%.

The CDF for round-trip times in Figure 10 demonstrate that UDP packets suffer significant delays. Since the CDF of network ping packet loss rates measured for these UDP streams do not rise nearly as swiftly as the round-trip times in Figure 10, the conjecture is that the downstream wireless AP queues are large. Previous experience with Windows Streaming Media UDP streams [6, 8] suggests that excessively high average round-trip times occur when the initial UDP streaming stage uses a high data rate to fill the playout buffer. In Bad wireless situations, the downstream AP queue grows excessively long and the AP is never able to drain the queue since the actual wireless layer capacity is limited by degraded capacity and wireless layer retries.

In the presence of loss, the TCP stream may take longer to play out the same length video due to retransmissions. Severe loss causes TCP timeouts that delay video playout further. Figure 11 illustrates this behavior where total application playout duration (including buffering and playout) has been normalized by dividing it by the encoded (real-time) playout duration. In this figure, a

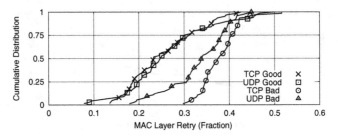

Fig. 8. Wireless Layer Retry Fraction for Upstream Traffic

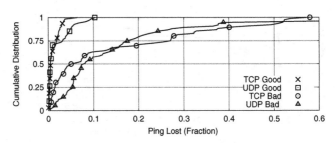

Fig. 9. Network Loss Rates for TCP and UDP Streams

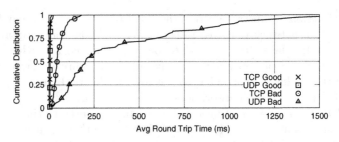

Fig. 10. Network Round-Trip Times for TCP and UDP Streams

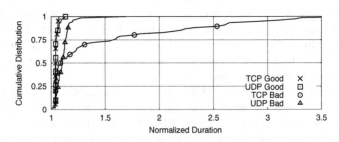

Fig. 11. Normalized Duration of Application Playout for TCP and UDP Streams

normalized duration of one[8] indicates that the clip playout was the same length as the encoded duration, while a 2 implies the clip took twice as long to play as the encoded duration. At Bad locations, TCP streaming can take significantly longer to playout than UDP streaming. For pre-recorded clips, it is not unreasonable to consider a stream duration extended by more than 10% to be unacceptable to users. Using this criteria, approximately 40% of the TCP Bad streams in Figure 11 are unacceptable.

4   The Challenges of Streaming over Wireless

Upon connection, video servers select an encoded send bitrate based on client feedback on network performance. Past work [8] indicates Windows Streaming Media uses a packet-pair technique [10] to estimate the bottleneck link capacity on the streamed path. Near the "cliff" in wireless performance, it is likely that a client will indicate an optimistically high average capacity that causes the video server to select a high encoding level. Figure 12 captures this phenomenon via a scatter-plot of the average encoding rate versus average wireless capacity both averaged over the duration of the video run. Points below the diagonal represent runs where the average encoding rate chosen for streaming is below the average capacity reported by the wireless network.

A conservative measure of effective capacity is the *TCP-Friendly* rate, namely, the data rate does not exceed the maximum rate of a conformant TCP connection under the same network conditions. The TCP-Friendly rate, T Bps, for a connection can be computed by [9]:

$$T = \frac{s}{R\sqrt{\frac{2p}{3}} + t_{rto}(3\sqrt{\frac{3p}{8}})p(1 + 32p^2)} \tag{1}$$

with packet size s, round-trip time R and packet drop rate p. TCP retransmission timeout t_{rto} is set to four times round-trip time by default. For each video clip for each run, Equation (1) is used to compute the TCP-Friendly rate (T), using a packet size (s) of 1350 bytes for the single level video and 978 bytes for the multiple level video, and the loss rate (p) and round-trip time (R) obtained from the corresponding ping samples.

Figure 13 shows a scatter-plot of the average encoding rate and average wireless network capacity both averaged over the video duration. Points above the diagonal line represent video runs in which the average encoding rate chosen for streaming are above the average effective capacity that can be supported by the wireless network. The preponderance of points above the diagonal line suggest the video streaming rate chosen is quite often higher than the capacity that the wireless network can effectively support. This results in the application streaming rate being too high to be supported by the network. Under such cases, when

---

[8] Note, the data points are all above one since the playout invariably includes at least one, initial buffering stage of about 10 seconds.

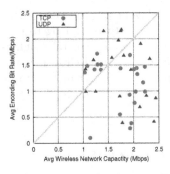

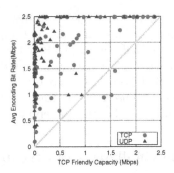

Fig. 12. Average Application Encoding Rate versus Wireless Capacity for TCP and UDP Streams

Fig. 13. Average Application Encoding Rate versus TCP-Friendly Capacity for TCP and UDP Streams

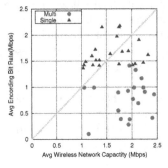

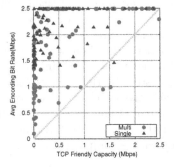

Fig. 14. Average Application Encoding Rate versus Wireless Capacity for Multiple and Single Level Stream

Fig. 15. Average Application Encoding Rate versus TCP-Friendly Capacity for Multiple and Single Level Stream

the video is streamed over UDP, the result is a reduced frame rate and when the video is streamed over TCP, the result is a longer playout duration.

Videos encoded with multiple levels provide modest performance improvement by enabling the video streaming rate to more easily adapt to the effective network capacity after streaming has commenced. This is depicted in Figure 14 and Figure 15 which show scatter-plots similar to Figure 12 and Figure 13, respectively, but broken down by multiple and single encoding levels. In Figure 15, the cluster of points in the bottom left corner of the graph are cases where the multiple level clips are able to stream at an average encoded rate closer to the capacity that the wireless network can effectively support.

This data suggests a need for more effective techniques to estimate the effective capacity for wireless networks to facilitate better choices for video encoding and streaming rates.

5   Conclusions and Future Work

This investigation reinforces the notion that IEEE 802.11 wireless networks can support streaming of high-quality video to wireless clients at high signal strength reception locations. Under such good conditions, nearly all video clips in this study played out at high frame rates. Moreover, server choices of multiple versus single encoding levels and TCP versus UDP streaming did not significantly impact performance at good locations. However, these experiments produce a noticeable cliff such that throughput drops off suddenly when signal strength degrades below -75dBm. The bad wireless environment for those experiments at the bottom of this cliff can be characterized by nearly 50% more retries for wireless MAC layer frames and median packet loss rates over 5%. Under such bad wireless environments, multiple level videos adapt better to volatile wireless conditions than videos encoded only at a single level. Under bad conditions, multi-level videos consistently had higher frame rates with a median of 24 application frame per second, approximately double the median application frame rate of single level videos.

At bad client locations, TCP streamed videos usually recorded higher frame rates than UDP streamed videos. For the TCP streams, the median of 24 application frames per second, was approximately 50% higher than the UDP median. The conjecture is that TCP retransmissions reduce network packet loss rates that yield more playable frames than UDP when wireless conditions are bad. Unfortunately, this higher TCP frame rate comes at a price, significantly longer video playout durations. Nearly 20% of the TCP streamed videos to bad client locations had the two-minute video clip produce four minute playout durations. Approximately 40% of these TCP videos had playout durations considered to be intolerable. While UDP streams also experienced extensions in playout durations under bad conditions, only 25% of the UDP durations were intolerable and no UDP playout reached a doubled duration in this investigation.

The effective capacities reported by the wireless MAC layer are significantly below the capacity the wireless MAC layer is expected to support, and the measured encoding rate for the streaming video, while lower than the wireless capacity, is higher than the effective capacity. The use of multiple encoding levels in a video clip partially alleviates this problem, but significant improvements to streaming performance under bad wireless conditions may require new techniques that identify and adapt to challenging wireless transmission situations.

Understanding packet and frame burst loss behavior is also critical to improving multimedia streaming encoding mechanisms designed to protect, correct or conceal video frame errors. Unfortunately, our tool set was unable to capture error bursts across layers. Developing measurement techniques to capture error bursts during real streaming events remains an important item for future research. Another missing component to improving the strategies used by video servers to adjust to volatile wireless network conditions is a better understanding of when and how a video server decides to do media scaling. Ongoing research is to measure the media scaling reaction of media players to changes in wireless network conditions.

Finally, two other commercial applications, Real Media and QuickTime, are also major contributors to streaming Internet traffic. However these servers probably behave differently than Media Player and investigations with customized measurement tools for these two application suites are also possible future work.

References

1. G. Bai and C. Williamson. The Effects of Mobility on Wireless Media Streaming Performance. In *Proceedings of Wireless Networks and Emerging Technologies (WNET)*, pages 596–601, July 2004.
2. A. Balachandran and G. Voelker. WRAPI – Real-time Monitoring and Control of an 802.11 Wireless LAN. Technical report, CS at UCSD, 2004.
3. J. Boyce and R. Gaglianello. Packet Loss Effects on MPEG Video sent over the Public Internet. In *Proceedings of ACM Multimedia*, pages 181–190, Bristol, U.K., Sept. 1998.
4. J. Chung, M. Claypool, and Y. Zhu. Measurement of the Congestion Responsiveness of RealPlayer Streaming Video Over UDP. In *Proceedings of the Packet Video Workshop (PV)*, Nantes, France, Apr. 2003.
5. A. Goel, C. Krasic, K. Li, and J. Walpole. Supporting Low Latency TCP-Based Media Streams. In *Proceedings of the Tenth International Workshop on Quality of Service (IWQoS)*, May 2002.
6. M. Li, M. Claypool, and R. Kinicki. MediaPlayer versus RealPlayer – A Comparison of Network Turbulence. In *Proceedings of the ACM SIGCOMM Internet Measurement Workshop (IMW)*, pages 131 – 136, Marseille, France, Nov. 2002.
7. M. Li, M. Claypool, R. Kinicki, and J. Nichols. Characteristics of Streaming Media Stored on the Web. *ACM Transactions on Internet Technology (TOIT)*, 2004. (Accepted for publication).
8. J. Nichols, M. Claypool, R. Kinicki, and M. Li. Measurements of the Congestion Responsiveness of Windows Streaming Media. In *Proceedings of the 14th ACM International Workshop on Network and Operating Systems Support for Digital Audio and Video (NOSSDAV)*, June 2004.
9. J. Padhye, V. Firoiu, D. Towsley, and J. Kurose. Modeling TCP Throughput: A Simple Model and Its Empirical Validation. In *Proceedings of ACM SIGCOMM*, Vancouver, Brisish Columbia, Candada, 1998.
10. R. Prasad, M. Murray, C. Dovrolis, and K. Claffy. Bandwidth Estimation: Metrics, Measurement Techniques, and Tools. *IEEE Network*, November-December 2003.
11. B. Wang, J. Kurose, P. Shenoy, and D. Towsley. Streaming via TCP: An Analytic Performance Study. In *Proceedings of ACM Multimedia*, Oct. 2004.
12. Y. Wang, M. Claypool, and Z. Zuo. An Empirical Study of RealVideo Performance Across the Internet. In *Proceedings of the ACM SIGCOMM Internet Measurement Workshop (IMW)*, pages 295 – 309, San Francisco, California, USA, Nov. 2001.

Measurement Based Analysis of the Handover in a WLAN MIPv6 Scenario

Albert Cabellos-Aparicio[1,*], René Serral-Gracià[1],
Loránd Jakab[2], and Jordi Domingo-Pascual[1]

[1] Universitat Politècnica de Catalunya (UPC),
Departament d'Arquitectura de Computadors, Spain
{acabello, rserral, jordid}@ac.upc.edu
[2] Universitatea Tehnică din Cluj-Napoca,
Facultatea de Electronică şi Telecomunicaţii, Romania
moriarty@bel.utcluj.ro

Abstract. This paper studies the problems related to mobile connectivity on a wireless environment with Mobile IPv6, specially the handover, which is the most critical part. The main goal of this paper is to develop a structured methodology for analyzing 802.11/IPv6/MIPv6 handovers and their impact on application's level. This is accomplished by capturing traffic on a testbed and analyzing it with two applications developed for this purpose. The analysis covers passive and active measurements. This methodology is applicable for measuring improvements on handover (such as Fast Handovers for Mobile IPv6, Hierarchical Mobile IPv6 or 802.11 handover).

1 Introduction

A great interest exists among users, in being on-line, permanently and without wires. On the last years, wireless technologies have improved and made cheaper. With WLAN (IEEE 802.11) [1] as one of the most used, it is possible to provide connectivity and bandwidth in a cheap and easy way.

This technology is able to provide "nomadism" to the Internet, in other words, an user can be connected to the Internet using WLAN, but he can't move, change his point of attachment and maintain his network connections. For that reason, IETF has designed Mobile IP, which, jointly with WLAN, provides this capability to the Internet (this is commonly known as mobility). In this paper, we focus on active and passive measurements using Mobile IPv6 with 802.11b.

The most critical part of these technologies is the handover. It is important to note that during this phase, the mobile node (MN) is not able to send or receive data, and some packets may be lost or delayed (due to intermediate buffers).

* This work was partially funded by the MCyT (Spanish Ministry of Science and Technology) under contract FEDER-TIC2002-04531-C04-02 and the CIRIT (Catalan Research Council) under contract 2001-SGR00226.

C. Dovrolis (Ed.): PAM 2005, LNCS 3431, pp. 203–214, 2005.

This lack of connectivity can affect some applications, especially streaming or real-time, which do not have retransmission mechanisms.

This paper focuses on measurements (active and passive) of the WLAN/ IPv6/ MIPv6 handover. Our goal is to study the handover in a real testbed using two different approaches. First, using passive measurements, analyzing the handover latency (the time where the mobile node is not able to send and receive data). Our aim is to compare layer 2, layer 3 and MIPv6 handover, and to find bottlenecks. Secondly, with active measurement; our goal is to study the effects of the handover on traffic sent or received by applications, studying differences depending on flow directions, packet losses, one-way delays and IPDV (IP Delay Variation).

Several papers focus on the same topic, [5] uses a mathematical model to study the handover latency but it does not take into account the wireless handover, [4] studies the Mobile IPv6 (and others) handover with a simulator, [2] makes an empirical analysis of the 802.11 handover, and, finally, [3] studies the WLAN/Mobile IPv6 handover in a real testbed proposing a new algorithm to improve the handover latency. Our paper goes further, analyzing bottlenecks, comparing the layer 2 and layer 3 handover and studying the effects suffered by the applications.

The reminder of this paper is organized as follows: section 2 and 3 are a summary of IEEE 802.11 and Mobile IPv6. Our measurement scenario is presented in section 4. In section 5 we propose an active and passive measurement methodology for handovers, in section 6 we present the results obtained in our handover analysis and finally, section 7 is devoted to the conclusions of the paper.

2   IEEE 802.11

This protocol is based on a cellular architecture, where the system is divided into cells. Each cell (Base Service Set or BSS) is managed by a Base Station (commonly known as Access Point or AP). WLAN can be formed by a single cell (or even by none, in "ad-hoc" mode) but, usually is formed by a set of cells, where AP's can communicate trough a backbone (Distribution System or DS). All this entities, including different cells, are viewed as a single 802.11 LAN from upper layers (in the OSI stack).

AP's announce their presence using "Beacon Frames" that are sent periodically. When a STA desires to associate to an AP, it has to search for one (scan). Scan can be performed using two different methods, either passive scanning, where STA "listens" for a "Beacon Frame" (which includes all related information to get associated), or active, where STA sends "Probe Requests" frames, expecting to receive "Probe Response" sent by AP's.

Once a STA has found an AP, and decided to join it, it will go through the "Authentication Process", which is the interchange of security information between the AP and the STA. When the STA is authenticated, it will start the

"Association Process", AP and STA will exchange information about capabilities and allow the DS to know about the current position of the station. Only after the association process is completed, the STA is able to transmit and receive data frames.

If the signal received by the STA degrades (possibly because has moved away from the AP) the handover procedure starts. First, STA must find a new AP; this is accomplished using "scan" (described previously). When a new AP is found and the STA decides to join it, it must "Reauthenticate" and "Reassociate".

3   Mobile IP

Mobile IP was designed by IETF in two versions Mobile IPv4 [8] and Mobile IPv6 (MIPv6) [9]. The main goal of the protocol is to allow mobile nodes to change its point of attachment to the Internet while maintaining its network connections. This is accomplished by keeping a fixed IP address on the mobile node (Home Address or HAd). This address is unique, and, when the mobile node is connected to a foreign network (not its usual network) it uses a temporal address (Care-of Address or CoA) to communicate, however, it is still reachable through it's HAd (using tunnels or with special options in the IPv6 header). In this paper, we focus in MIPv6, although the tools developed can be easily migrated to MIPv4, FastHandovers for Mobile IPv6 [10], or other handover improvements.

MIPv6 has three functional entities, the Mobile Node (MN), a mobile device with a wireless interface, the Home Agent (HA), a router of the home network that manages localization of the MN, and, finally the Correspondent Node (CN), a fixed or mobile node that communicates with the MN.

The protocol has four phases:

1. *Agent Discovery:* The MN has to discover if it is connected to the home network or to a foreign one. For this purpose uses "Router Advertisements" [11], those messages are sent periodically by all IPv6 routers and include information for client autoconfiguration. Using this information, the MN obtains a CoA.
2. *Registration:* The MN must register its CoA to the HA and to CN's. This way, they know "who" is the MN (HAd) and "where" it is (CoA). Some messages related to this phase are "Binding Update" (BU) and "Binding Acknowledgment" (BA).
3. *Routing and Tunnelling:* MN establishes a tunnel with the HA (if it is necessary), and it is able to receive and send data packets (using the tunnel, or directly).
4. *Handover:* MN changes its point of attachment. It must discover in which network it is connected (phase 1) and register its new CoA (phase 2). During this phase, some data packets (sent or received by the MN) can be lost or delayed due to incorrect MN location.

4   Measurement Scenario

This section describes the practical part surrounding the setup of a measurement scenario for networking tests. Also describes the different hardware and software used for all the tests shown in this paper.

4.1   Network Topology

Testbed's main reason is to compute Mobile Node handover latencies, the testbed in detail can be seen in Figure 1, there are all the different parts of the scenario. To avoid external interferences, this testbed is isolated from outside networks, all the input and output traffic on the testbed's network interfaces is controlled. Having this isolation, but without the lack of external access, the scenario has two parallel networks, the control network and the actual testing network, as highlighted on the figure. This is important because with uncontrolled sources of traffic all the delays will be miscalculated.

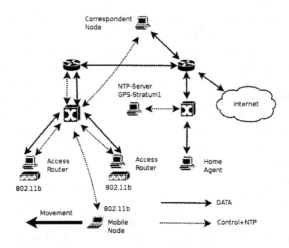

Fig. 1. Measurement scenario simplified structure

This testbed gives the tests all the privacy needed, this way, once the tests are prepared, no foreign agents are able to interfere with them. At the same time, the path followed by the packets is long enough to consider the possible clock skew too small to have any negative impact on the results.

Regarding synchronization, the testbed is configured to use four NTP (Network Time Protocol) sources [12], two of them belonging to a private network, Stratum 1 servers connected to a GPS source each. The other two sources are on the outside network and are as far as 3 hops away from the testbed. All the NTP traffic is routed through a parallel network (with the local NTP servers) where there isn't any other traffic. It is possible to access those remote NTP servers through the control network which can use external time sources. The NTP statistics shows that, with this setup, we obtain 1ms of measurement accuracy.

In order to confirm our synchronization accuracy, we made a simple test; we sent several ARP broadcast packets in our measurement network, those packets were captured on all the machines involved in our tests, and the timestamps were compared. The maximum difference among those timestamps agreed with the threshold stated by NTP.

4.2   Hardware and Software Equipment

Depending on the testbed description, all the machines involved on the tests are using the GNU/Linux Debian Sid distribution. Depending on the role of each computer, the hardware and the kernel varies accordingly:

- *Access Points/Routers (AP):* this testbed has two access points, each one with two wireless cards, one for communicating with the MN and the other one to monitor (capture frames). Those cards have the Atheros Chipset (802.11g) in 802.11b compatibility mode. The configured kernel is the 2.4.26.
- *Mobile Node (MN):* this mobile node uses a Cisco Aironet 350 card (802.11b) for wireless connectivity, here the kernel is 2.4.26 with MIPv6 1.1 [14] patch for Mobile IPv6 support.
- *Home Agent (HA)/Correspondent Node (CN):* the last two important hosts on the scenario have similar configuration with the 2.4.26 kernel patched the same way as the Mobile Node for Mobile IPv6 capabilities.

5   Methodology

This section is devoted to the description of the methodology developed for this paper. As our goal is to analyze the handover, we chose several tools which permit to measure the desired network parameters. Those tools are:

1. *MGen/DRec, NetMeter [15]:* for the active measurement part.
2. *Ethereal [13] and PHM Tool (Passive Handover Measurement) [6]:* for passive measurements.

Both applications depicted here: NetMeter's handover analysis module and PHM Tool are developed under the same code base. Their main goal is to analyze the Ethereal files and obtain for PHM Tool the handover latencies and for the NetMeter's part the packet losses and delays at application level.

The same capture is used for both solutions, the monitoring infrastructure is set up on the Access Points, given that is the only way of detecting all the handover latencies. Both captures (each on one access point) are merged (as they really represent the same traffic flow) and the data is analyzed.

The following subsection enumerates the set of tests prepared for this paper, following with the description of the *passive* analysis and later the paper focuses on the study of the *active* part.

5.1   Tests

For a good analysis of the handover, is necessary to build up a good set of tests. In this paper we ran a set of 16 tests, each 5 minutes long, from where extracted a set of 63 valid handovers.

Half the tests had the generated traffic from the Correspondent Node to the Mobile Node, while the other half was on the opposite direction.

Moreover, each direction of the tests where split as follows:

- *64Kbps Traffic*: this flow simulates with UDP the properties of VoIP traffic under IPv6, there are sent 34 packets per second with 252 bytes of payload as stated on [7].
- *1Mbps Traffic*: due to the low rate needed for VoIP the other tests are done on a higher packet rate, so the impact of a different bandwidth can be studied. This time there were 94 packets per second with a payload size of 1300 bytes per packet.

The VoIP simulation was chosen because all the traffic constraints of such technology are well known and will be easy to determine the user impact of the handover on such traffic regarding delays and packet losses.

5.2   Passive Handover Measurements (PHM)

Our main goal is to measure the handover latency or, in other words, the amount of the time where a MN is not able to send or receive data. This duration has several components, the amount of time spent by layer 2 (802.11b in our case) in scanning for a new AP, authenticate and re-associate to it, time used by IPv6 on connecting to the new network and, finally, the amount of time spent by MIPv6 in registering it's new CoA to HA and CN's.

The developed application "PHM Tool" monitors the signaling messages in both AP of our testbed. We capture all packets sent or received by their wireless interface. Handovers are "forced" attenuating the signal sent by the AP. The MN realizes this (it detects that the signal quality is poor) and tries to search for a new AP. In our testbed we do not have external interferences, and thus, the MN changes to the other AP.

When a set of handovers have been carried out, the captured packets are processed off line using "PHM", which analyzes the signaling messages providing results.

5.3   Active Handover Measurements

Usually, pure Active Measurements, have an end-to-end approach. The basis of such tests is to generate a synthetic flow travelling through the network under test. This paper proposes a new method for enhancing the Active Measurement framework. Our approach is based on a mixed use of Passive and Active Measurement systems. The whole point is to generate the Active flow and measure the typical end-to-end parameters. This flow is captured at its destination, but also at the Access Point (using typical capture software such as Ethereal [13]).

Once the tests are finished, the captured data is converted to the standard XML language for network packets (PDML [16]), this file is processed by our analysis tool, which will convert the data to a standard MGen file. This approach permits to calculate partial data delays, that's because the stored timestamps passed to the MGen file are taken from the monitoring machine (which is the actual Access Point on the testbed). Our solution can be used on a wide variety of scenarios on general network measurement systems.

Focussing now on the paper's tests, our method is to isolate the parameters computation on the wireless data flow from the wired one. This way, is possible to isolate all the handover incidences without taking into account the other parts of the tests. Besides, another possible use of the testbed on such conditions is to model the impact on the user's perception of the packet losses and bigger delays caused by the handover, this time with the end to end results.

6   Results

This section describes the results obtained from the tests discussed on the previous section. First the discussion focuses on the passive set of handovers for analysing its duration and all the parts throughout the process, later the analysis of the active results and the user level performance are shown.

6.1   Passive Handover Measurements

The whole system was tested doing a set of handovers, capturing all the signaling messages and processing them off line using PHM.

Table 1. Numerical results obtained using PHM (ms)

	Mean	Std. Dev.
Scan	257.224	108.007
Authentication	2.733	1.183
Association	1.268	0.311
IPv6	1836.413	430.196
Registration (HA)	3.914	1.017
Registration (CN)	9.262	4.881
Total time	2107.82	450.619

The table 1 (results in milliseconds) show the results obtained with our application and are a detailed version of the handover latency, and reveal time between two consecutive signaling messages. Wireless handover is detailed and we can see that the scan phase is the longest one; the MN uses in average 257ms to find a new AP. The whole 802.11 handover (Scan, Authentication and Association) represents 12% of the total handover latency.

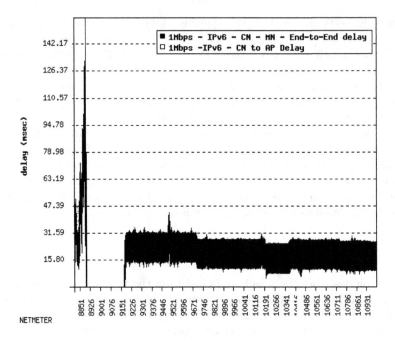

Fig. 2. Handover from CN to MN with 1Mbps traffic

The second phase is the time spent by IPv6 to realize that it is attached to a new network, and obtain a new CoA. IPv6 uses more than one second because it has to perform DAD (Duplicate Address Detection) and to realize that its old default router (the previous AR) it is unreachable [11]. For this last action, the MN has a timeout, if this was set to a very low value (less than one second) the MN, while communicating, would be sending "Router Solicitation" messages constantly trying to know if its default router is present or not. In our testbed, this timeout was set to its minimum value. This part is 87% of the total time.

The Mobile IPv6 handover is also detailed, the first part (Registration HA) is the time spent by the MN to indicate to the HA its new location, the second part (Registration CN) is the time used by the MN to announce its new point of attachment to the CN. It takes more time just because authentication between MN and CN includes more messages. This part is 1% of the total handover latency. This time is related to the round-trip time to the HA and to the CN.

We can conclude that in an 802.11/IPv6/MIPv6 handover, most part of the time is due to IPv6 (87%), and specifically, due to "Neighbor Unreachability Detection", the algorithm used to detect if the default router is present or not.

6.2   Active Handover Measurements

Once the analysis of the low level handover is finished, the next step on our study is to analyze the traffic's impact on user level, here the most important parameters are: packet losses and delays. As will be discussed, another important

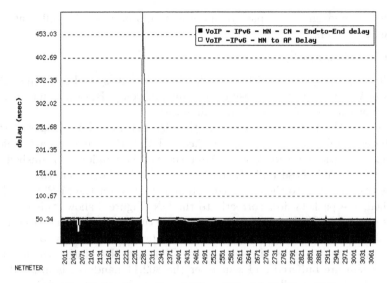

Fig. 3. Handover from MN to CN with VoIP traffic

result depicted here is the differences on the results regarding the traffic direction, from the CN to the MN or the other way around.

A - Packet Losses

To compute packet losses is a straight forward problem having the first and the last packet involved on the handover, which are provided by the capture on the access point.

Either Figure 2 and 3 have the packet's sequence number on the x-axis and the delay (in milliseconds) on the y-axis.

Figures 2 and 3 show two different handovers, the first one represents the flow going from the Correspondent Node to the Mobile Node, and the other one the opposite flow direction. Both handovers have similar duration, but, different packet losses (due to different rates) as shown in the table 2.

Table 2 shows the summary of mean and standard deviation of packet losses per handover on the whole set of tests done. As can be seen the higher is the packet rate higher is the packet's loss. The mean duration for the handovers treated on this part (at application's level) is *1,98 sec*, which is slightly different

Table 2. Packet Losses

	Mean		Std. Dev.	
	64Kbps	**1Mbps**	**64Kbps**	**1Mbps**
MN -> CN	65.80	162	9.78	16.97
CN -> MN	61.71	207.21	17.54	65.90

from the value given on the passive analysis, that's because of a different set of analyzed handovers.

B - Flow Directions

Regarding the effects suffered by a traffic flow when it is send from the MN to the CN or vice versa table 2 doesn't show any difference. However, figures 2 and 3 show that the handover has slight differences.

When the MN is sending packets to the CN and a handover starts, it stops immediately while searching for a new Access Point, those packets are buffered at Layer 2 by the wireless drivers. After the 802.11 handover is finished, the MN starts to send those buffered packets as fast as it can, but they are lost because they are sent to the old Access Router. Only after the MIPv6 handover is finished, the packets flow correctly to the CN. Figure 3 shows this behavior, packets marked as "End-to-End delay" (those are the packets that arrive correctly to the CN) reveal the handover gap while packets marked as "MN to AP delay" (not all those packet arrive correctly to the CN) show that during the handover some of them are buffered and sent after the 802.11 handover is finished with a higher delay, specifically the maximum delay is the duration of the wireless handover.

The MN's buffer may seem useless, but, in fact, is very effective in case of an 802.11 handover. In this case, the MN is not changing its default router, it doesn't need to send a Binding Update indicating a new location, it is just changing it's AP. The buffer will store packets that otherwise would be lost, those packets will be sent correctly (but with a higher delay) when the MN regains Layer 2 connectivity.

In the case that the CN is the source (figure 2) of the traffic flow the handover behaves differently. The CN sends packet constantly (it is connected to an Ethernet), when the handover starts, all those packets are lost because they are sent to the incorrect Access Router. Only after the MIPv6 handover part is finished, the CN realizes of the new location of the MN and sends the packets to the correct address.

C - QoS Parameters Consideration

Under a QoS environment, as stated above, there are other important parameters which highlight the level of provided QoS. Those parameters are the **One Way Delay** (OWD) and the **Inter Packet Delay Variation** (IPDV). Much discussion is possible on this subject, but only for the sake of simplicity, the study will be limited to the handovers studied on the previous figures, the statistically representative study is left as future work. The goal is to see if the QoS parameters are kept under those handovers. This is accomplished by taking near three seconds worth of packets before the handover and calculate the OWD, the IPDV and the same after it. With this is possible to see if there are grave variation of the above parameters when the Access Point signal's quality decreases just before the handover, or instead, if when associated to the new Access Point the system's convergence time to the new configuration causes any more problems.

Table 3. OWD and IPDV

	OWD (ms)		IPDV (ms)	
	Before	After	Before	After
VoIP	54.08	53.25	0.0125	0.0096
1Mbps	63.58	28.51	7.2566	-0.0011

The overall results for the displayed handovers can be seen on table 3, the results are very clear, when there is low traffic on the wireless link (VoIP), the loss of connectivity before the handover, hardly affects the packet delays, the same holds true for the system recovery once the handover is finished.

Another result, though, is the case when the link is more overloaded (1Mbps), where is easy to see the increment on the delivery delays of the flow, the reason is the loss of link quality on the wireless link, although the system's recovery is pretty fast and reliable. The same results can be seen on Figures 2 and 3.

7   Conclusions

This paper analysis focuses on all levels involved in the handover process, from 802.11 handover until application's level. That's why we designed a testbed and developed two applications to make active and passive handover measurements.

Passive measurements are intended to compute the handover latency in order to find bottlenecks and to compare layer 2, layer 3 and MIPv6. In the other hand, we expect to compute important parameters such as delay, IPDV and packet losses with active measurements in order to analyze the impact at application's level forced by such handovers.

Passive results show that an 802.11/IPv6/MIPv6 handover takes 2.107 seconds in average. The 802.11 part is 12% of the total time; most of this time is spent searching for a new AP. The IPv6 part is the longest one, takes 87% of the total time, the MN has to realize that its previous default router is no longer reachable and switch to the new one. Finally, the MIPv6 part is 1% of the total time.

Summarizing all the obtained results for the active measurement, the handover as is doesn't forbids the QoS on low bandwidths in terms of one way delay. The problem, though, is uncovered by the packet losses (which is proportional to the handover latency), where its value, depending on the packet's rate is about 63 losses per handover (VoIP), which is unbearable for a proper quality voice transmission. The only solution for this matter is to improve the handover times, that is, to improve Mobile IPv6, or change it to better protocols such as Fast Handovers.

Mixing both worlds (passive and active measurements paradigm) opens up a new set of possibilities for analyzing all the different aspects of the handover. We plan to extend this handover analysis, using the same methodology, to other protocols such as Mobile IPv4, Fast Handovers for Mobile IPv6, Hierarchical Mobile IPv6 or some IEEE 802.11 handover improvements. We also want to extend the methodology in order to know, exactly, how many packets, and which packets have been lost or delayed in a given handover phase, this will be useful for protocols such as Fast Handovers that uses intensively buffering.

References

1. IEEE: 802.11: Wireless LAN Medium Access Control (MAC) and Physical Layer (PHY) Specification Arch. Rat. Mech. Anal. (1997)
2. A. Mishra, M. Shin and W. Arbaugh: An Empirical Analysis of the IEEE 802.11 MAC Layer Handoff Process **Volume 33** *ACM SIGCOMM Computer Communications Review* (2003)
3. N. Montavont and T. Noel: Handover Management for Mobile Nodes in IPv6 Networks *IEEE Communications Magazine* (2002)
4. Xavier Pérez Costa and Hannes Hartenstein: A simulation study on the performance of mobile IPv6 in a WLAN-based cellular network *40 Issue 1 Computer Networks: The International Journal of Computer and Telecommunications Networking* (2002)
5. Marco Liebesch, Xavier Pérez Costa and Ralph Schmitz: A MIPv6, FMIPv6 and HMIPv6 Handover latency study: Analytical Approach *IST Mobile and Wireless Telecommunications Summit* (2002)
6. Loránd Jakab, Albert Cabellos-Aparicio, René Serral-Gracià, Jordi Domingo-Pascual: Software Tool for Time Duration Measurements of Handovers in IPv6 Wireless Networks *UPC-DAC-2004-25* (2004)
7. John Q. Walker, NetIQ Corporation: A Handbook for Successful VoIP Deployment: Network Testing, QoS, and More (2002)
8. C. Perkins: IP Mobility Support for IPv4 *RFC 3344* (2002)
9. D. Johnson, C. Perkins and J. Arkko: IP Mobility Support for IPv6 *RFC 3775* (2004)
10. Rajeev Koodl: Fast Handovers for Mobile IPv6 *draft-ietf-mipshop-fast-mipv6-03.txt* (2004)
11. T. Narten, E. Nordmark and W. Simpson: Neighbor Discovery for IP version 6 (IPv6) *RFC 2461* (1998)
12. Internet2 Consortium: OWAMP - NTP Configuration *http://e2epi.internet2.edu/owamp/details.html#NTP* (2004)
13. Gerald Combs: Ethereal: The world's most popular network protocol analyzer *http://www.ethereal.com* (2004)
14. Helsinki University of Technology: MIPL Mobile IPv6 for Linux *http://www.mobile-ipv6.org/* (2004)
15. René Serral, Roberto Borgione: NetMeter a NETwork performance METER *http://www.ccaba.upc.es/netmeter* (2002)
16. PDML Specification: *http://analyzer.polito.it/30alpha/docs/dissectors/PDMLSpec.htm* (2002)

A Distributed Passive Measurement Infrastructure

Patrik Arlos, Markus Fiedler, and Arne A. Nilsson

Blekinge Institute of Technology, School of Engineering,
Karlskrona, Sweden
{patrik.arlos, markus.fiedler, arne.nilsson}@bth.se

Abstract. In this paper we describe a distributed passive measurement infrastructure. Its goals are to reduce the cost and configuration effort per measurement. The infrastructure is scalable with regards to link speeds and measurement locations. A prototype is currently deployed at our university and a demo is online at http://inga.its.bth.se/projects/dpmi. The infrastructure differentiates between measurements and the analysis of measurements, this way the actual measurement equipment can focus on the practical issues of packet measurements. By using a modular approach the infrastructure can handle many different capturing devices. The infrastructure can also deal with the security and privacy aspects that might arise during measurements.

1 Introduction

Having access to relevant and up-to-date measurement data is a key issue for network analysis in order to allow for efficient Internet performance monitoring, evaluation and management. New applications keep appearing; user and protocol behaviour keep evolving; traffic mixes and characteristics are continuously changing, which implies that traffic traces may have a short span of relevance and new traces have to be collected quite regularly.

In order to give a holistic view of what is going on in the network, passive measurements have to be carried out at different places simultaneously. On this background, this paper proposes a passive measurement infrastructure, consisting of coordinated measurement points, arranged in measurement areas.

This structure allows for a efficient use of passive monitoring equipment in order to supply researchers and network managers with up-to-date and relevant data. The infrastructure is generic with regards to the capturing equipment, ranging from simple PCAP-based devices to high-end DAG cards and dedicated ASICs, in order to promote a large-scale deployment of measurement points.

The infrastructure, which currently is under deployment at our university, was designed with the following requirements in mind:

1. *Cost.* Access to measurement equipment should be shared among users, primarily for two reasons: First, as measurements get longer (for instance for detecting long-range dependent behaviour) a single measurement can tie

C. Dovrolis (Ed.): PAM 2005, LNCS 3431, pp. 215–227, 2005.

up a resource for days (possibly weeks). Second, high quality measurement equipment is expensive and should hence have a high rate of utilization.

2. *Ease of use.* The setup and control of measurements should be easy from the user's point of view. As the complexity of measurements grows, we should hide this complexity from the users as far as possible.

3. *Modularity.* The system should be modular, this to allow independent development of separate modules. With separate modules handling security, privacy and scalability (w.r.t. different link speeds as well as locations). Since we cannot predict all possible uses of the system, the system should be flexible to support different measurements as well as different measurement equipment.

4. *Safety and Security.* Measurement data should be distributed in a safe and secure manner, i.e. loss of measurement data should be avoided and access to the data restricted.

To solve these requirements we came up with an infrastructure consisting of three main components, *Measurement Point* (MP), *Consumer* and *Measurement Area* (MAr). The task of the MP is to do packet capturing, packet filtering, and distribute measurement data. The approach to the second design requirement was to use a system with a web interface. Through this interface users can add and remove their desired measurements. The MAr then handles the communication with the MPs. The cost for implementing this architecture is not very high, compared to a normal measurement setup you need two additional computers and an Ethernet switch of suitable speed, and this basic setup can grow as the requirements change.

There are several other monitoring and capturing systems available, here we describe only a few.

CoralReef [1] is a set of software components for passive network monitoring, it is available for many network technologies and computer architectures. The major difference between CoralReef and our infrastructure is that CoralReef does not separate the packet capturing and analysis as we do. Furthermore, the CoralReef trace format does not include location information as our does.

IPMON [2] is a general purpose measurement system for IP networks. IPMON is implemented and deployed by Sprint. IPMON separates capturing from analysis, similar to our infrastructure. On the other hand, the IPMONs store traces locally and transfer them over a dedicated link to a common data repository. The repository is then accessed by analyzers.

Gigascope [3] uses a similar approach as IPMON, by storing captured data locally at the capturer. This data is then copied, either in real time or during off-peak hour, to a data warehouse for analysis. It uses GSQL as an interface to access the data.

The IETF has (at least) two work groups that are relevant for this work; Packet Sampling (PSAMP) [4] and IP Flow Information Export (IPFIX) [5]. PSAMP works on defining a standard set of capabilities for network elements to sample subsets of packets by statistical and other methods. Recently an Internet draft was published [6], which describes a system at a higher level than our infrastructure, but they are very similar and our system could benefit by adjusting somewhat to the PSAMP notation. The IPFIX group is interesting since they deal with how to export measurement data from A to B, thus it is interesting with regards to consumers.

In Section 2 we will discuss the components and how they interact. This is followed by Section 3 where we describe how the system handles rules and filters. In Section 4 we discuss privacy and security related to the infrastructure. In Section 5 we describe two cases where the system has been deployed. In Section 6 we describe some of the ongoing and future work. And in Section 7 we conclude the paper.

2   Components

The three main components in the infrastructure will be described in the following subsections.

2.1   Measurement Point

In Figure 1 the components of a schematic MP are shown. This is the device that does the actual packet capturing. It is managed from a Measurement Area Controller (MArC) and transfers the captured data to consumers attached to the Measurement Area Network (MArN). The MP can either be a logical or a physical device. A logical MP is simply a program running on a host, whereas a physical MP could either use a dedicated computer or custom hardware in order to create high-speed high-performance MPs.

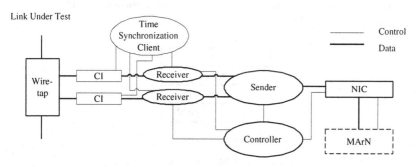

Fig. 1. Schematic overview of a MP

A MP can tap one or more links; each link is tapped via a wiretap. For full-duplex Ethernets, a wiretap has two outputs, one for each direction. These are connected to separate capture interfaces (CI). A receiver listens to a CI and filters the packets according to the filter rules stated by the MArC. If the CI hasn't timestamped the packet the receiver will do so. The packets are then delivered to the sender, which is responsible for sending the captured packets to the appropriate consumers. Such a measurement frame can contain several packets, where the number of packets is controlled by the maximum transfer unit (MTU) of the MArN. Each MP also has a controller that is responsible for the configuration of the MP and the communication with the MArC. A time synchronization client (TSC) is used to keep all the MPs with in a MAr synchronized, which can be done using a dedicated device or a simple NTP server.

The filter rules used by the receiver specify, in addition to packet properties, a consumer and the amount of the packet to be captured (currently the upper limit is 96 bytes). For each frame that passes the filter, the MP attaches a capture header (Figure 2). In this header, we store a CI identifier, a MP identifier, a timestamp when the packet was captured (supporting an accuracy of picoseconds), the packet length, and the number of bytes that actually were captured. The filters are supplied to the MP from the MArC, and they will be discussed in Section 3. Once a packet matches a filter, it is stored in a buffer pending transmission. Once the buffer contents reaches a certain threshold the buffer is transmitted using Ethernet multicast. This way, it is simple to distribute frames to several consumers in one transmission. The duplication of data is done by the MArN. This approach will also reduce the probability of overloading the MArN, and hence preventing loss of measurement frames as far as possible. However, in order to detect frame loss each measurement frame is equipped with a sequence number that is checked by the consumer upon reception. If a measurement frame is lost it is up to the consumer to handle this particular loss and notify the MArC. Given this information the MArC can take actions to prevent future losses. Actions can be to alter filters as well as requesting additional switching resources inbetween the MPs and the Consumers. The current implementation only notifies the consumer "user", who has to take appropriate actions.

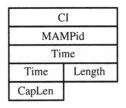

Fig. 2. Capture Header

The capture header enables us to exactly pinpoint by which MP and on what link the frame was captured, which is vital information when trying to obtain spatial information about the network's behaviour. This also enables us to use several MPs to measure a single link, which is interesting when the measurement task of a link speed becomes too great for a single MP to handle. This would require a device that is capable of distributing the packets such that the wiretap feeds different MPs in a round robin approach.

2.2   Measurement Area

In Figure 3 an example of a MAr is shown. The MAr provides a common point of control for one or more MPs. It uses a dedicated network in between MPs and the MAr subsystems for reasons of performance and security. A MAr consists of the following subsystems: a MArC, a time synchronization device (TSD), a MArN and at least one consumer and one MP. The MArC is the central subsystem in a MA. It supplies the users with a GUI for setting up and controlling their measurements. It also manages the MPs by supplying filters and by keeping track of their status. The TSD supplies all the MPs in the MA with a common time and synchronization signal. It can utilize the existing Ethernet structure to the MPs, or it can utilize some other network to distribute the time signal.

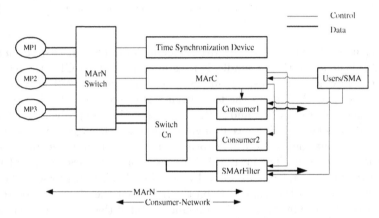

Fig. 3. Simple overview of a MA with three MPs, four consumers, one MArC and a time synchronization unit

The capacity of the MArN should be such that it can handle the peak rate of the measured traffic. Assume that a MP monitors a 10Base-T link, with a frame rate of 800 fps where each frame is 1500 bytes long ($\approx$ 9.6 Mbps). From each frame we collect 96 bytes, add a capture header of 36 bytes and store the data in a measurement frame, see Figure 4. Given a MArN MTU of 1500, a measurement frame can contain 1480 bytes of measurement data, consisting of capture headers and frames, the remaining 20 bytes are used by a measurement header (MH). In the current example we can store 11 frames in each measurement frame $(11 * (36 + 96) = 1452 \leq 1480$ bytes), causing the MP to send only $800/11 \approx 72$ fps into the MArN, see Figure 5. If the monitored link would have a frame rate of 14000 fps, each frame would only be 85 bytes long ($\approx$ 9.6 Mbps), the measurement frame would contain 12 frames $(12*(36+85) = 1452 \leq 1480$ bytes), yielding a frame rate of $14000/12 \approx 1167$ fps. However, if the MArN MTU was 9000, the measurement frame could contain 74 frames, yielding a frame rate of 189 fps.

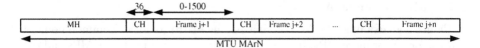

Fig. 4. Measurement frame encapsulation

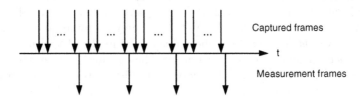

Fig. 5. After capturing N frames one measurement frame is sent from the MP

A consumer that attaches to the MArN should not request more data than the link that it is attached to can handle. For instance a consumer C1 is the recipient of two measurement streams, S1 and S2, each generating 1272 measurement frames per second. As long as the total frame rate of S1 and S2 is less or equal to the capacity offered by link and switch there should be no problems, but if the consumer desires to get full frames it might run into problems quite fast, since the MP adds a capture header to each captured frame potentially generating more traffic than it captures. The current implementation addresses this problem by having a maximum capture size of 96 bytes. The MArC also provides the user with an estimation of the frame rate on the links that the MPs are monitoring, giving the user an indication of the amount of traffic that his consumer might receive.

The example in Figure 3 contains a consumer network (CN). It is placed on a separate switch to minimize processing required by the MArN, thus enabling additional consumers to be easily connected to the MArN, for instance new probes, analyzers etc. to be evaluated in parallel. If the number of consumers is low, the MArN switch might handle them directly, and no CN switch is necessary. This would be the normal setup, see Figure 6. In Figure 7 a minimal MAr is shown. In both cases the MPs are using a separate network for the time signal distribution.

2.3 Consumer

A consumer is a user-controlled device that accepts packets according to the format specified by the system. A consumer should filter the content of the measurement frame that it receives, since the MP merges multiple user requests some filters will capture packets that match several requests. Such a joint filter might not perfectly match the desired frame description; this is discussed in the following section.

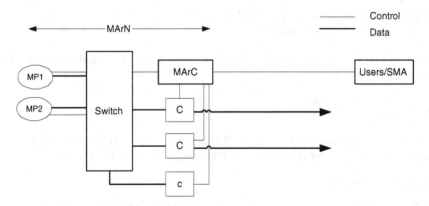

Fig. 6. Normal MAr

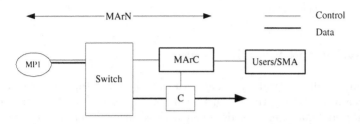

Fig. 7. Minimal MAr

3 Filters and Rules

A user supplies rules to the MArC. These rules describe *what* data the user desires to collect, *where* the data should be collected, *when* the data should be collected and *where to send* the data. The MArC uses this information to create filters that the MPs understand. The filters that the MP uses are a combination of all the user supplied rules, combined in such a manner that all requests are met in a best effort style. The MArC keeps track of the MPs and their capabilities, thus it knows how many filters a MP can handle before it runs into performance problems. The MArC also monitors the performance of the MArN and reject user rules that could cause performance problems within the MArN. If a MP is to obtain a filter list that would push it into a region of potential performance problems, the MArC will alter the filters in order to minimize the number of filters. By doing this the load on the MP is kept at a reasonable level, but this approach requires the consumers to do some filtering of their own. Hence, it is up to the user to supply the desired Consumer with a filter. The filters within a MP are arranged in such a manner that no packet is reported twice by the MP.

Let's give a simple example, we have one MP and two consumers C1 and C2. Initially we have two rules (using BPF syntax):

R1 $\{tcp\ host\ A.a\}$ which sends its data to C1.
R2 $\{ip\ net\ A\}$ which targets C2

Here two approaches are possible; the first during low load would have the following filters sent to the MP:

F1 $\{tcp\ host\ A.a\} \to$ M1
F2 $\{ip\ net\ A\} \to$ C2

Here M1 is a multicast address that C1 and C2 listens to. If the load on the MP approaches a high level then only one filter would be sent to the MP

F1 $\{ip\ net\ A\} \to$ M1

In this case the C1 consumer would need to perform filtering in order to select the TCP segments of host A.a. By default a consumer should always filter the measurement data that it receives, ensuring that it passes a correct stream to the analysis/storage entity.

4   Privacy and Security Issues

A MP will see all the traffic passing on a link that it is tapping, which can be viewed as a intrusion of privacy. Furthermore, since the majority of the network protocols used today were not designed with security in mind, user credentials might pass on the link and be clearly visible to the MP. This can be an intrusion of privacy and should require special care on behalf of the measurement system and its users. If the data collected from the system is only intended for internal use, it might be enough that all users and the network-owner have agreed to that their traffic can be monitored to allow for measurements. However, if the data is to be shared with researchers in other organizations, the data should be deprivatized. Deprivatization [7] can be done on various levels, from the removal of parts in the application data to the removal of all network data. We believe that the system should minimize the alternation of the captured data and leave the anonymization to the consumers. If the MP would anonymize the data, e.g. through scrambling of addresses [8], some consumers such as intrusion detection systems or charging systems might not be able to operate anymore. However, if the system does deprivatization by default, this should be done in the MPs. If address scrambling is utilized, this causes problems when the user specifies the measurement rules. If the unscrambled address was used, the user will obtain scrambled addresses matching his requirement and then it is possible to reverse-engineer the scrambling system. If the scrambled address was used, the user would need to know how to create that scrambled address. Probably, the first method should be chosen. In that case, the only person that is capable of reverse-engineering the packet trace is the user requesting the trace, since he knows both scrambled and unscrambled address. Now, if the packet trace is stolen, the thief cannot match packets to individual hosts/users unless he has access to a descrambler and the scrambling key.

Privacy issues will probably have to be addressed by specialized consumers. For instance, we have two consumers, a intrusion detection system (IDS) and a link utilization estimator (LUE). The IDS needs undistorted information. The LUE could on the other hand use deprivatized data, but since the MP will not send two copies of the same packet there is a problem. It is probable that a network owner would like to have control of the information that leaves his network, so it would be easier for the network owner to supply an export consumer that deprivatizes the data according to his own policies, which might not meet the particular desires of the user. For our own measurements, the agreement we made with the system owner was the following: The MPs are only allowed to capture headers, not user payload. Furthermore, the data leaving a consumer may only be in statistical form, or deprivatized in such a manner that it is impossible to reverse-engineer the data to obtain information that allows you to identify a particular individual.

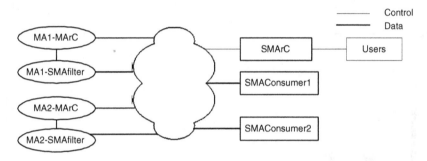

Fig. 8. Example of a SMAr

From a security point of view, all components in the system should be protected from unauthorized access. The simplest way to do this is to have the system operating on a separate network, with no connection to any other networks. This would however be expensive and unpractical in measurements distributed over a wide area. The solution to this it to utilize Super Measurement Areas (SMAr), see Figure 8. SMAr's are used to connect to MAr's at different locations using existing infrastructure. A SMAr can be seen as a MAr at a higher level, the MAr's MP becomes SMArFilters (specialized consumers that attach to the MArN), the MArs consumers are called SMArConsumers. Between the SMArFilters and SMArConsumers TCP is used to provide reliable communication. The MPs and the MArN need to be protected from unauthorized access, both physical and logically. Physical protection of the MAr subsystems is the first requirement in giving logical protection; the consumers and the MArC need to be protected from intrusions via their connection to the users.

5   Examples of Use

As of writing two MAr have been implemented and used. One is available online via http://inga.its.bth.se/projects/dpmi and is mainly used in a controlled environment. The second MAr consisted of two measurement points each monitoring a gigabit link on a campus network. In both cases only one physical consumer was used, but it was sufficient to handle up to eight logical consumers. Examples of consumers are: estimation of traffic distribution (at link, network, transport and application level); link utilization; packet inter arrival time; communication identification; and bottleneck identification [9]. At the time of writing we are preparing a third MAr to be deployed in an ISP network, where it will initially be used for bottleneck identification. In Figure 9 we visualize the result from a analyzer that identifies bottlenecks. It uses two consumers to estimate the link bit rate over a given time intervall, these are then transferred to a database which is accessed by the visualizer that estimates the bottleneck.

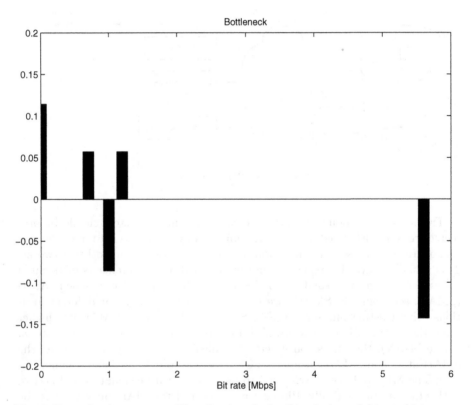

Fig. 9. Example of a consumer: Visualization of a bottleneck through bitrate histogram difference plots (c.f. [9])

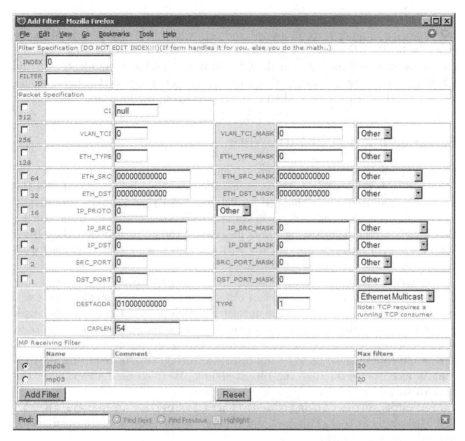

Fig. 10. User interface for adding rules

In Figure 10 the MArC (prototype) interface for adding a rule is shown. In this implementation all tasks are done manually, the goal was to develop the MP not the MArC. The following filtering options are availible, the MASK fields are used to mask the packet value.

- CI: Physical interface identifier.
- VLAN_TCI: VLAN number and priority.
- ETH_TYPE: Ethernet type.
- ETH_SRC/DST: Ethernet source/destination address.
- IP_PROTO: IP payload type.
- IP_SRC/DST: IP source/destination address.
- SRC/DST_PORT: Transport protocol source/destination port numbers (if applicable).
- DESTADDR: What Ethernet address should receive the measurement data?
- TYPE: Which type of transport should the MP use? Ethernet, UDP or TCP.
- CAPLEN: How much of each captured frame should we store?

FilterID is a number that specifies in which order the MP should check its filters, starting with number zero. Index will indicate which fields that are used in the rule specification. For instance if we wish to collect all packets caught on a specific CI the index would be 512, and the CI field would hold a string identifying the CI. If we would like to capture IP packets caught on a specific CI, index would be 640, ETH_TYPE=2048 and CI a string specifying the interface.

6   Ongoing and Future Work

Initial experiences with the system are encouraging, and development of consumers is currently ongoing. The experience of the demo has indicated that the MP's software needs to be changed in such a manner that the MPs periodically flush their measurement buffers, in order to prevent consumers from waiting long times. We are considering a modification of the system so that the MArC supplies the consumers automatically with the information that they need with regards to filters and multicast addresses.

To handle the increased link speeds, new devices with better timestamping accuracy are needed. Even if we can obtain this accuracy, a single device will probably run into problems when measuring such a link. Hence another task would be to investigate how to distribute the measurement task of a link onto several MPs. Compression of frame data is also considered to be implemented, this would could enable us to do full frame capturing without requiring a MArN that is more powerful that the observed link. We also need to evaluate the performance of a MArN.

The infrastructure is being considered as a part of the EuroNGI WP.JRA.4.3 [10] Measurement tool. This tool will support traffic generation, measurement, analysis and visualization.

7   Conclusions

In this paper we have presented a distributed passive measurement infrastructure, which has separate components for packet capturing, control and analysis. We discussed how the system deals with multiple users and their request for data. Since the infrastructure is passive we addressed the security and privacy issues associated with this. Furthermore, we gave examples of current usage and future work.

References

1. CAIDA: CoralReef. (2005) http://www.caida.org/tools/measurement/coralreef (Verfied in January 2005).
2. Sprint: IPMON (2005) http://ipmon.sprint.com (Verified in January 2005).
3. AT&T: Gigascope (2005) http://www.research.att.com/info/Projects/Gigascope (Verified in January 2005).

4. IETF: PSAMP Workgroup. (2005) http://www.ietf.org/html.charters/psamp-charter.html (Verfied in January 2005).
5. IETF: IPFIX Workgroup. (2005) http://www.ietf.org/html.charters/ipfix-charter.html (Verfied in January 2005).
6. IETF: A Framework for Packet Selection and Reporting. (2005) http://www.ietf.org/internet-drafts/draft-ietf-psamp-framework-10.txt (Verfied in January 2005).
7. Pang, R., Paxson, V.: A high-level programming environment for packet trace anonymization and transformation. In: SIGCOMM '03: Proceedings of the 2003 conference on Applications, technologies, architectures, and protocols for computer communications, ACM Press (2003) 339–351
8. Xu, J., Fan, J., Ammar, M., Moon, S.B.: On the design and performance of prefix-preserving ip traffic trace anonymization. In: IMW '01: Proceedings of the 1st ACM SIGCOMM Workshop on Internet Measurement, ACM Press (2001) 263–266
9. Fiedler, M., Tutschku, K., Carlsson, P., Nilsson, A.A.: Identification of performance degradation in ip networks using throughput statistic. In: Proceedings of the 18th nternational Teletraffic Congress (ITC-18), ELSEVIER (2003) 399–408
10. EuroNGI: Homepage (2005) http://www.eurongi.org (Verified in January 2005).
11. TCPDUMP Public Repository: Homepage. (2005) http://www.tcpdump.org (Verfied in January 2005).
12. Endace Measurement Systems: Homepage. (2005) http://www.endace.com (Verified in January 2005).

lambdaMON – A Passive Monitoring Facility for DWDM Optical Networks

Jörg B. Micheel

NLANR/MNA[1], San Diego Supercomputer Center/UCSD,
10100 John Hopkins Dr, 92092-0505 La Jolla, CA, USA
joerg@nlanr.net
http://pma.nlanr.net/~joerg/

Abstract. This paper presents lambdaMON - a novel approach to passive monitoring of very high performance optical networks based on dense wavelength division multiplexing (DWDM). The approach offers very attractive cost/benefit scaling properties, which are further refined by introducing state-of-the-art transparent fiber switching equipment. The rapid pace at which we intend to implement lambdaMONs opens new opportunities to apply passive monitoring facilities for debugging, troubleshooting and performance analysis of novel protocols and applications. To the best of our knowledge, this is the first attempt at designing a passive monitoring facility for optical networks. We report detailed architectural parameters, measurements and experience from laboratory tests and initial field deployment.

1 Introduction

Optical networking, based upon dense wavelength division multiplexing (DWDM), is rapidly becoming the technology of choice for the high-performance networking community, as well as major national and international commercial network providers.

Optical networking uses individual light rays (also referred to as colors, wavelengths, carriers, or channels) to carry very high-performance point-to-point connections over long distances. The capabilities offered by optical networking fundamentally challenge traditional means of time division multiplexed IP services, which have dominated the development of the Internet for the past two decades.

Traditional circuit-based provisioning, such as OC3/OC12 leased lines, ATM, MPLS, or virtual private networks (VPN) are rapidly being replaced by fiber optic communication channels, or lambdas. The opportunities for improving connectivity and performance between regions, countries, or very demanding end user communities, such as the high energy physics research community, are tremendous. At present, it is difficult to imagine the full impact that DWDM technology might have for end-to-end networking, especially when used to support very sensitive and demanding 10-Gigabits/second and quality-of-service aware user applications.

---

[1] This work was supported by the U.S. National Science Foundation Collaborative Agreement ANI-0129677 (NLANR/MNA, 2002) with subcontract to the University of Waikato in New Zealand.

C. Dovrolis (Ed.): PAM 2005, LNCS 3431, pp. 228–235, 2005.

With networks being rolled out at a rapid pace, new protocols being developed and field tested, and new applications emerging, the stakes for success are high and the risks that have to be taken by the research community are substantial. A readily available monitoring facility is highly desirable as a means to locate, debug, troubleshoot and resolve potential problems. In this regard, **passive systems have a unique advantage: they explore the network as is, without interfering with the actual traffic data pattern as generated by end systems or as modified through intermediate devices, such as switches and routers.** In addition, passive monitoring systems present an excellent means to understand and address issues at network layers two to seven.

In this paper we present lambdaMON – a new passive monitoring technology that enables the debugging and troubleshooting of applications and protocols that operate over long-distance DWDM networks. While the lambdaMON is fundamentally a new passive network monitoring technology, it preserves the traditional means to precisely collect and analyze workload and performance characteristics of edge, access, and backbone network links.

The rest of the paper is organized as follows. In the next section we provide a brief overview of traditional passive monitoring systems, their advantages and shortcomings. In section 3 we look at the specifics of DWDM technology as deployed in the field today and determine the point of instrumentation for lambdaMONs. In section 4 we outline the constraints for the lambdaMON architecture. Section 5 looks at implementation challenges. Section 6 summarizes the achievements in architecting and designing lambdaMONs.

2   Traditional Passive Monitoring Systems

Passive network monitors interoperate with the live network at the link level (see Figure 1). They are considered a vendor independent means of gathering data as they do not depend on features that would otherwise have to be provided by active network equipment operating as routers, switches, hubs or end systems [1]. Operating at the medium dependent physical layer, these OCxMON systems are equipped with link layer specific network measurement cards (NMCs), which reimplement all layer one and two functions, such as deserialization, packetization and various packet encapsulations, and then pass the data on to analysis-specific functions, such as arrival time stamping, selective filtering and payload discard or flow state analysis. Such functions are executed by reconstructing and accessing information that is specific to network layers three to seven. If real-time analysis applications are used, the monitor may deliver a complete solution by means of a graphical user interface.

The biggest strength of passive monitoring technology (i.e., link layer dependency) is at the same time, also one of its biggest weaknesses. Every emerging technology advance demands that a new set of PC cards (NMCs) be developed in order to support a compatible interface to the network link. Passive network monitors are, by design, lower-cost devices, and their implementation is hence based on off-the-shelf components. Therefore NMCs are typically designed and implemented after a given link layer technology has been rolled out in the field, and as a result they are generally

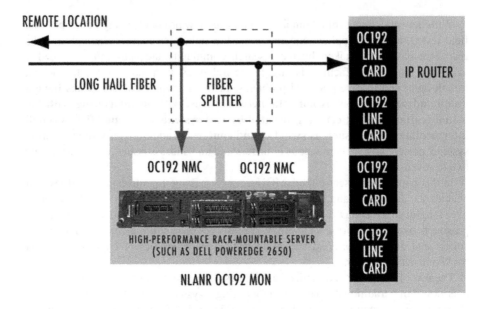

Fig. 1. Classic OCxMON monitoring setup

available only for the second half of the life cycle of any high-performance (backbone) link layer technology. This means that the use of passive monitors, so far, has been restricted to traffic and workload characterizations of mature networks.

While passive monitors are extremely powerful once deployed in the field, their widespread use faces some steep resistance for both cost and technology reasons. Being a per-link facility limits their deployment to dedicated research environments for cost reasons, thus preventing them from becoming a more general-purpose operational network facility. Examples of larger scale deployment include the infrastructures operated by NLANR/MNA and CAIDA. Perhaps the largest known infrastructure in a commercial setting is operated by Sprint ATL's IPMON research group. In addition to the financial obstacles, the process for installing fiber optic splitters still presents a technical hurdle for most users new to passive monitoring, and misconfigurations are frequent. At best, the monitor will be unable to collect data on one or both of its interfaces. At worst, the network link itself will be unable to operate, or remain intermittent and unreliable – unacceptable for any type of network operations.

As a result of these obstacles, passive monitors have never been used for locating, debugging, troubleshooting and eliminating end-user application, protocol, and performance problems. For a random end-user or problem the chances of one (or more) passive monitors being present, available, and accessible along an end-to-end networking path are very small.

3 DWDM Optical Networks

DWDM harnesses a spectrum of lambdas within the third optical window (1520 to 1620 nanometers) for long-distance high-performance data transmission. This spectrum has been chosen for its low attenuation and for its ability to amplify signals at the optical level without the need for regeneration, which would otherwise force a technically expensive optical-electrical-optical conversion.

Major amplification technologies in use today are erbium doped fiber amplifiers (EDFA) and Raman pump lasers. Unlike EDFAs, which can be purchased and integrated as modules into the fiber path, Raman amplifiers use the entire long-distance fiber span as a medium, with the pump laser located at the receiver of the optical transmission line. Due to the way these amplifiers work, the third optical window has been subdivided into the C (1520 to 1560 nanometers) and L (1565 to 1620 nanometers) bands. With the use of EDFAs, fibers spans of up to 600 kilometers (375 miles) can be achieved. EDFAs combined with Raman amplifiers will reach up to 2000 kilometers (1250 miles) for production use.

The use of lambdas within the C and L bands has been standardized by the International Telecommunication Union – Telecommunication Standardization Sector (ITU-T). ITU-T Recommendation G.694 defines a grid of wavelengths rooted at 1552.52 nanometers (193.1 THz). The spacing between carriers is implementation dependent and includes 200 GHz, 100 GHz, 50 GHz, and 25 GHz options. The choice of grid spacing controls the number of channels that can be supported by any single channel. For instance, with a 50 GHz grid, up to 80 channels can be supported in the C band. Due to the modulation noise band and necessary isolation between channels, a 50 GHz grid limits the digital carrier signaling frequency to about 20 GHz. For technical reasons it is unlikely (but not impossible – see [6]) that carriers beyond OC192/10-Gigabit-Ethernet will be deployed on a 50 GHz grid any time soon.

DWDM networks are presently built in a static setup as single-vendor single-product implementations. The simplest configuration will involve a pair of DWDM

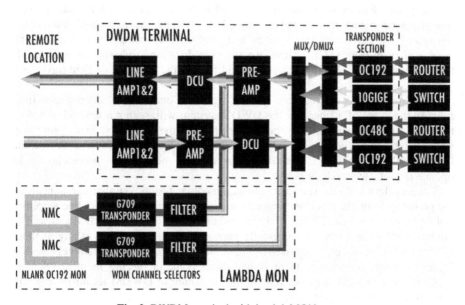

Fig. 2. DWDM terminal with lambdaMON setup

terminals operating at either end of a bidirectional long distance fiber (see Figure 2). A DWDM terminal supports access to the individual carrier wavelengths by means of a transponder, which supports the connection of traditional carrier class equipment (SONET OC12, OC48, OC192) or local area networking gear (such as 1-Gigabit and 10-Gigabit Ethernet devices).

The transponder converts a traditional SONET/SDH or Ethernet LAN PHY signal via a G.709/G.975 encoder by employing a Reed-Solomon (RS[239,255]) forward error correction (FEC) code to lower the expected bit error rate (BER) on the transmission link. The use of FEC supports the operation of longer fiber spans without the need for regeneration, which in turn makes the entire system significantly more cost effective. The G.709 encoded signal operates at an approximately 7 percent higher rate relative to the original 10 Gigabit carrier and is launched via a laser at a specified wavelength into the first stage multiplexer (MUX). It is this signal that the lambdaMON will pick up once the signal has passed through additional, but optional, stages of multiplexers and amplifiers (AMP), but before entering the dispersion compensation unit (DCU).

For the inbound direction, all lambdas will pass through one or more stages of the demultiplexer (DEMUX), with a single channel (still G.709 encoded) eventually reaching the decoding section of the transponder, where it will be converted back into the original SONET/SDH or Ethernet LAN PHY signal.

All of the passive modules forming a DWDM terminal, such as MUX, DEMUX and DCU, introduce a device dependent attenuation, typically between 5 dB and 10 dB. However, the operating range of transmitters, receivers and other parts of the system is typically limited to between +5 dBm and -20 dBm. Therefore, signals will have to be strengthened at least once within the terminal, which is the role of preamplifiers (PRE AMP).

The point of instrumentation at a DWDM terminal requires a fully multiplexed, non-dispersed signal that is strong enough to tolerate the additional attenuation introduced by the fiber optic splitter and lambdaMON components, such as the tunable channel filter (TCL). Therefore, the exact location to instrument is the right-hand side of any DCU and past any preamplifiers in the direction of light travel (see Figure 2).

DWDM transmission lines can be built from just one, or two independent, fiber runs. In the event of a fiber cut, the DWDM system will, within a defined period of time, automatically switch from the primary fiber to the secondary. Since protection switching is handled at the line amplifier section of a DWDM terminal or device, the lambdaMON architecture is not affected by whether the network owner chooses to operate in protected mode or not.

Since modern DWDM sites are built in a modular fashion, similar components as are presently found in DWDM terminals will also be present at optical line amplifier (OLA), regenerator (REGEN) or optical add/drop multiplexer (OADM) sites. The lambdaMON architecture fundamentally permits deploying monitors at any of those DWDM sites, however, since passive monitors are best operated remotely with a legacy network connection, there is not much point in placing them at OLA or REGEN sites. OADM sites are very similar in nature to DWDM terminals and will support the proposed architecture.

4 lambdaMONs

A lambdaMON is a bidirectional 10 Gigabits/second device capable of collecting and analyzing packet header data at line rate from any one (at a time) of the active wavelength carriers at a single DWDM link.

A multi-feed lambdaMON, or lambdaMON node, is an advanced configuration permitting a lambdaMON to monitor any one (at a time) of a number of DWDM links at a given location.

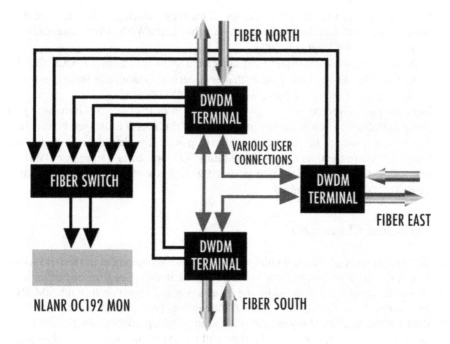

Fig. 3. Multi-feed lambdaMON or lambdaMON node

Our design for lambdaMONs and lambdaMON nodes is driven by the following considerations:

1. We are keen to break the technology dependency cycle that lets passive monitors miss the first 18 months of the life cycle of any high performance network link layer technology.
2. We want to demonstrate the utility of passive monitoring systems for troubleshooting and debugging of networks, applications and protocols. Such a facility is critical within the first year or two after network rollout, much less once the network is mature.
3. We want to retain the ability to collect and analyze workload profiles of mature networks.

4. We believe that there is no strict need for a static association between network links and monitors. It appears that the community would rather have a passive monitoring facility available on demand, and possibly on short notice, if and when such a requirement arises.
5. We believe that the success of passive monitoring technologies has been hampered by cost considerations as well as technical obstacles. We can break those constraints by designing a readily available facility central to the network.

Based on those constraints, we derive our lambdaMON architecture as follows:

1. Make best use of existing OC192MON technology, which permits loss free data capture and real-time analysis at 10 Gigabit LAN/WAN PHY and OC192c Packet-over-SONET network links.
2. Address the per-link constraint (and associated costs) of traditional OC192MON deployment by designing a system that will permit dynamically tuning into any one (at a time) of the active lambdas at a given DWDM link.
3. Increase the coverage on a given optical infrastructure by introducing a transparent fiber switch to tap into multiple DWDM links at a site (see Figure 3).
4. Retain the option to increase analysis power by introducing additional lambdaMONs to the fiber switch in the future. As opposed to a static per-DWDM-link setup, this arrangement will permit the analysis of multiple lambdas at a single DWDM link, if required.

5 Technical Challenges

DWDM is still an emerging technology and some of the equipment in the field today may not easily support passive fiber splitters, and the resulting engineering effort may turn out to be prohibitive to consider the installation and operation of lambdaMONs. As an alternative, existing one percent monitoring ports, which appear to be a standard feature with most equipment, can be used in replacement for the fiber tap. The advantage is that equipment in the field will not experience any service disruption when installing lambdaMONs; however, on the downside, the signal levels are too low for the operation of fiber optic transceivers and other passive optical equipment. Therefore, the signal levels will need to be amplified by an EDFA first, which causes an overall increase in costs. First field tests have shown that signal levels at one percent monitoring ports vary, and a single stage amplifier may or may not be sufficient to increase the signal power level in such a way that a commercially available G.709 transponder can be operated reliably at the expected BER.

Tunable channel filters (TCL) are just becoming commercially available for field deployment, and are restricted to operating in either the C or the L band. In addition, the selection of filters that support stable operation on a 50 GHz grid is presently very small. Field tests have shown that TCLs come at an affordable price and work reliably to the full satisfaction of the project.

It appears to be attractive, from a cost point of view, to integrate G.709 transponders into OC192MONs. Even though there is some loss of flexibility, the cost reductions are significant, specifically in a large scale rollout, like in a distributed infrastructure.

6 Conclusion

We have successfully carried out laboratory tests using a C band setup simulating a total of 600km of long distance fiber terminated by a pair of CISCO 15808 terminals. We amplified signals from the one percent monitoring ports in both the transmit and receive directions via a CISCO 15501 EDFA and successfully transferred 1.4 Terabytes of data over the course of several hours with varying packet sizes and data loads without incurring any bit errors or packet drops.

Our next step, at the time of publication, is to stage a phase 2 prototype of the lambdaMON, which will involve a full setup of the lambdaMON node in Los Angeles, with the active support of CENIC. Such a node may include OC48MON monitoring equipment right from the start, as there are a number of links that are of particular interest to us.

We are also looking at expanding the passive measurement node concept to traditional SONET-based networking. It is expected that we will continue to publish our progress via the project's Web site [7].

Acknowledgement

The author expresses his thanks for the tremendous support received by key staff at the Corporation for Network Research Initiatives in California (CENIC), National LambdaRail (NLR), the Internet2 HOPI team, the Indiana GlobalNOC and TransPAC. This project would have been impossible without the help of CISCO Systems. We also appreciate the support received from Iolon Corporation. Jim Hale of the NLANR/MNA team at SDSC contributed significantly in preparation of the field trials and assisted in the performance tests. Thank you, Jim!

References

1. OC3MON: Flexible, Affordable, High-Performance Statistics Collection. Joel Apisdorf, kc claffy, Kevin Thompson and Rick Wilder. Proceedings of INET'97, 24-27 June 1997, Kuala Lumpur, Malaysia. http://www.isoc.org/isoc/whatis/conferences/inet/97/proceedings/F1/F1_2.HTM
2. Precision Timestamping of Network Packets. Joerg Micheel, Stephen Donnelly, Ian Graham. ACM SIGCOMM Internet Measurement Workshop, San Francisco, CA, USA. November 1-2, 2001. http://www.icir.org/vern/imw-2001/imw2001-papers/23.pdf
3. ITU-T Recommendation G.694.1: Spectral grids for WDM applications: DWDM frequency grid. February 27th, 2003. http://www.itu.int/itudoc/itut/aap/sg15aap/history/g.694.1/g6941.html
4. ITU-T Recommendation G.709/Y.1331: Interfaces for the optical transport network. February 11th, 2003. http://www.itu.int/itudoc/itut/aap/sg15aap/history/g709/g709.html
5. Understanding Fiber Optics: Jeff Hecht. 4th Edition. Prentice Hall. ISBN 0-13-027828-9
6. An ISP's view of an LSR run. Peter Lothberg, Svensk TeleUtveckling & ProduktInnovation AB, ESCC/Internet2 Joint Techs Workshop, July 18th – 21st, Columbus, OH. http://events.internet2.edu/2004/JointTechs/Columbus/sessionDetails.cfm?session=1505&event=218
7. lambdaMON prototyping Web page http://pma.nlanr.net/lambdamon.html

Internet Routing Policies and Round-Trip-Times

Han Zheng, Eng Keong Lua, Marcelo Pias, and Timothy G. Griffin

University of Cambridge Computer Laboratory
{han.zheng, eng.lua, marcelo.pias, timothy.griffin}@cl.cam.ac.uk

Abstract. Round trip times (RTTs) play an important role in Internet measurements. In this paper, we explore some of the ways in which routing policies impact RTTs. In particular, we investigate how routing policies for both intra- and inter-domain routing can naturally give rise to violations of the triangle inequality with respect to RTTs. Triangle Inequality Violations (TIVs) might be exploited by overlay routing if an end-to-end forwarding path can be stitched together with paths routed at layer 3. However, TIVs pose a problem for Internet Coordinate Systems that attempt to associate Internet hosts with points in Euclidean space so that RTTs between hosts are accurately captured by distances between their associated points. Three points having RTTs that violate the triangle inequality cannot be embedded into Euclidean space without some level of inaccuracy. We argue that TIVs should not be treated as measurement artifacts, but rather as natural features of the Internet's structure. In addition to explaining routing policies that give rise to TIVs, we present illustrating examples from the current Internet.

1   Motivation

Since round trip times (RTTs) play an important role in Internet measurements, it is important to have a good understanding of the underlying mechanisms that give rise to observed values. Measured RTTs are the result of many factors — "physical wire" distance, traffic load, link layer technologies, and so on. In this paper, we explore a class of factors that are often ignored — the ways in which routing policies can impact minimum RTTs.

In particular, we investigate how routing policies for both intra- and inter-domain routing can naturally give rise to violations of the triangle inequality with respect to RTTs. The existence of Triangle Inequality Violations (TIVs) impact two areas of current research, one positively, and the other negatively. For overlay routing [1], TIVs represent an opportunity that might be exploited if the layer 3 routed path can be replaced with one of lower latency using a sequence of routed paths that are somehow stitched together in the overlay. On the other hand, TIVs pose a problem for any Internet Coordinate System (ICS) [2,3,4,5,6,7,8] that attempts to associate Internet hosts with points in Euclidean space so that RTTs between hosts are accurately captured by distances between their associated points. The problem is simply that any three points having RTTs that violate the triangle inequality cannot be embedded into Euclidean space without some level of inaccuracy, since their distances in Euclidean space must obey this inequality. We feel that current work on Internet Coordinates too often treats TIVs as measurement artifacts, either ignoring them entirely or arguing that they are not important. We have come to

C. Dovrolis (Ed.): PAM 2005, LNCS 3431, pp. 236–250, 2005.

the opposite conclusion — we feel that TIVs are natural and persistent features of the Internet's "RTT geometry" and must somehow be accommodated. We illustrate how TIVs can arise from routing policies and present illustrating examples from research networks in the Internet. Our measurement results are consistent with those reported in PAM 2004 [9], and indicate that the commercial Internet is even more likely to exhibit such policy-induced TIVs.

2   A Bit of Notation

A *metric space* is a pair $M = (X, d)$ where X is a set equipped with the distance function $d : X \rightarrow \mathbb{R}^+$. For each $a, b \in X$ the *distance between a and b* is $d(a, b)$, which satisfies the properties, for all $a, b, c \in X$,

(anti-reflexivity) $d(a, b) = 0$ if and only if $a = b$,
(symmetry) $d(a, b) = d(b, a)$,
(triangle inequality) $d(a, b) \leq d(a, c) + d(c, b)$.

A *quasi-metric space* (X, d) satisfies the first two requirements of a metric space, but the triangle inequality is not required to hold. This paper argues that Internet RTTs naturally form a quasi-metric space, with routing policies being an important, but not sole, factor in the violation of the triangle inequality.

A Triangle Inequality Violation (TIV) is simply a triple (a, b, c) that violates the triangle inequality. It is not hard to see that for any TIV, there must be one edge that is longer than the sum of the other two edges.

Suppose that $M_1 = (X_1, d_1)$ is a quasi-metric space, and $M_2 = (X_2, d_2)$ is a metric space. Every one-to-one function ϕ from X_1 to X_2 naturally defines a metric space called *the embedding of M_1 in M_2 under* ϕ, defined as $\phi(M1) = (\phi(X_1), d_2)$. We normally abuse terminology and simply say that ϕ embeds X_1 into X_2, We will be interested in the case where X_1 is a finite set, X_2 is $\mathbb{R}^n$ with the standard notion of Euclidean distance, $d_2(x, y) = \|x - y\| = \sqrt{\sum_{1 \leq i \leq n} (x_i - y_i)^2}$.

The number of possible embeddings is quite large. In addition, the *accuracy* of an embedding can be measured in various ways, as is outlined in [10]. In this paper, we will not focus on any particular embedding, nor on any particular notion of accuracy. We simply note that any TIV embedded into Euclidean space must involve some "distortion" since the triangle inequality will hold on the embedded points. Although one might attempt to embed RTT distances into a non-Euclidean space (such as [11]), by far the most common techniques are Euclidean.

3   Routed Paths Versus Round Trip Time

Most data paths are determined by dynamic routing protocols that automatically update forwarding tables to reflect changes in the network. Dynamic routing never happens without some kind of manual *configuration*, and we will refer to routing protocol configuration as implementing *routing policy*. The Internet routing architecture is generally described as having two levels [12] — Interior Gateway Protocols (IGPs) are designed to route within an autonomously administered routing domain, while Exterior Gateway Protocols (EGPs) route between such domains. In this section we explore the ways in which routing policies can give rise to data paths that violate the triangle inequality with respect to delay.

Intra-domain routing is typically based on finding shortest paths with respect to configured link weights. The protocols normally used to implement shortest path routing are RIP, OSPF, or IS-IS. We note that Cisco's EIGRP [13] presents a slightly more complex routing model, and in addition some networks actually use BGP for intra-domain routing. Nevertheless, for simplicity we will investigate here only how shortest path routing can give rise to TIVs.

In order for there to be no TIVs in shortest path routing, the link weights must be consistent with the actual link delays. However, delay is just one of the many competing demands in the design of intra-domain routing. So the disagreement between the link weight assignment and the actual link delay will cause structural TIVs in the intra-domain case.

We now consider how inter-domain routing can introduce triangle inequality violations (TIVs).

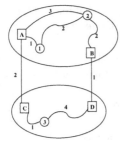

Fig. 1. Nodes 1, 2 and 3 form a TIV due to *Hot Potato Routing*. The numbers on the edges represent link propagation delay

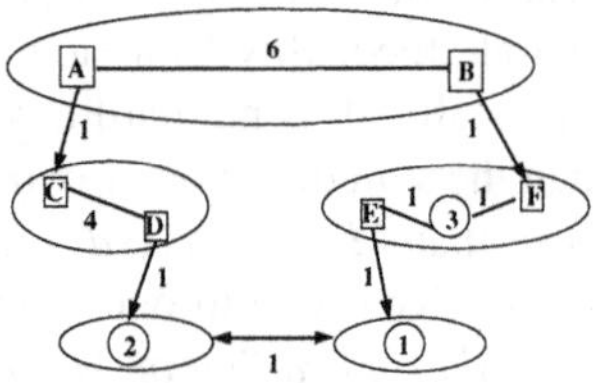

Fig. 2. Nodes 1 and 2 share a "private peering shortcut" that cannot be used to transit traffic between nodes 2 and 3

Hot potato routing [14] refers to the common practice of sending inter-domain traffic towards the closest egress point. Consider two ASes presented as large ovals in Fig. 1. We inspect the "triangle" formed by nodes 1, 2, and 3. The upper AS has two egress points, A and B, the first of which is closer to node 1, while the second is closer to node 2. Node 3 is in the lower AS and is closer to egress C. Note that hot potato routing will result in asymmetric routing between nodes 2 and 3. Traffic from 3 to 1 and 2 will always exit the lower AS at egress point C, whereas traffic from 2 to 3 will exit the upper AS at egress point B. The distance matrix for the nodes 1, 2, and 3, all calculated as "round trip" distance, is

d	1	2	3
1	0	4	8
2	4	0	13
3	8	13	0

Here we see that $13 = d(2,3) > d(2,1) + d(1,3) = 12$, and so this represents a TIV.

Lest the reader think that the problem is asymmetric routing alone, we now show how economic relations between networks can give rise to TIVs even when routing is symmetric. Private peering links are common routing shortcuts used to connect ISPs. Fig. 2 presents five ASes, the upper AS representing a large transit provider, the middle two ASes representing smaller providers, and the lower two ASes representing customers. The directed arrows represent customer-provider relationships pointing from a provider to a customer. The bi-directional arrow between the lower ASes represents a *private peering* [15] link. This link transits only traffic between the lower two ASes, and is not visible to the providers above. (This type of peering is very common on the Internet.) Traffic between nodes 1 and 2 uses this link for a round trip path cost of 2. Traffic between nodes 3 and 1 goes through border router E for a round trip path cost of 4. However, traffic between nodes 2 and 3 must go up and down the provider hierarchy for a round trip path cost of 28!. Here we see that $28 = d(2,3) > d(2,1) + d(1,3) = 6$, so this represents a TIV.

TIV can also be caused by traffic flowing through three independent AS paths between a triple of nodes, where at least some AS along one path is not in the other two paths. This usually happens due to multi-homing [16] and because peering relationship is a bilateral agreement and typically not transitive [17]. In this case, there is absolutely no reason to believe that triangle inequality must hold. We will see some examples of this type of TIV in section 4.

In fact, the interaction between inter-domain and intra-domain routing can also introduce TIVs. This type of TIV applies to the majority of systems that use end-to-end measurement results. This interaction often makes it very difficult to classify the root cause of an observed TIV.

The current inter-domain routing protocol, BGP, conveys only AS-level paths information. Nothing is learned about the actual router-level path within an AS. Therefore, when BGP makes a decision based on the shortest AS path, nothing can be inferred about the actual router-level path. An example of this is shown in Fig. 3.

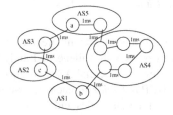

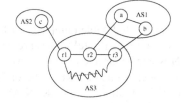

Fig. 3. When choosing the AS-level path between nodes a and b, BGP prefers AS $\underline{4\ 1}$ to AS $\underline{3\ 2\ 1}$, although the router-level path along AS4 is much longer

Fig. 4. A TIV caused by the interaction between hot-potato routing and intra-domain TIV

More complicated interactions are also possible. Fig. 4 shows an example of a TIV caused by both hot-potato routing and intra-domain TIV. Routers $r1$, $r2$ and $r3$ form an intra-domain TIV, and AS1 uses hot-potato routing between egress points a and $r2$, and b and $r3$. So the path between b and c (b $r3$ $r1$ c) exhibits a much longer RTT than the paths between a and b, and a and c (a $r2$ $r1$ c). We can see that sometimes the intra-domain behavior of the intermediate AS may change the existence of TIV through interaction with inter-domain routing.

4   Case Study: The Global Research and Education Network (GREN)

We define GREN (c.f. [9]) to be all the Autonomous Systems reachable from the Abilene network (AS11537), because we can reach almost all the research and education networks in the world from Abilene, and Abilene has no direct upstream commercial provider [18]. Fig. 5 illustrates the connectivity of most component networks of GREN. We study GREN instead of the commodity Internet because GREN is a relatively more open and transparent network, and we can understand its global structure more easily. In addition, a large percentage of PlanetLab nodes are hosted in GREN networks.

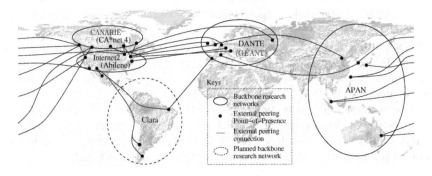

Fig. 5. The connectivity of most component networks of GREN

We take a BGP table dump (on 19 July 2004 at 12:00 GMT) from Oregon Internet Exchange (OIX), a RouteViews server which directly peers with Abilene, and study how GREN is inter-connected. We observe that quite a few commercial ASes are involved to glue together bits of GREN. According to the BGP *behavior* of each AS, we were able to classify all the 1203 GREN ASes into 30 commercial ASes and 1173 research ASes (details omitted). The particular reasons for 'leaking' these commercial ASes into GREN are still under further investigation, although we have seen that some are leaked in for very legitimate reasons.

4.1   Examples of Internal TIVs

We obtained the router-level topology [19] and IS-IS weights from a monitor box inside the GEANT backbone (the multi-gigabit pan-European research network managed by DANTE), as shown in Fig. 6. We also measured the minimum RTT values between all pairs of GEANT backbone routers using their looking glass interface [20]. The RTT

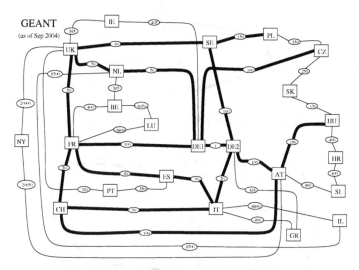

Fig. 6. The topology of GEANT backbone. Thick lines represent OC-192 10Gbps fast backbone links, and other lines represent slower links of various speeds. The numbers in the circles represent the IS-IS weights assigned to the links

measurements are taken 100 times for each pair (50 times starting from each end of the pair), and the minimum is used in our calculations. Our goal in this experimental work is to study the *structural* causes of TIV, therefore we take the minimum measurement to avoid biases in the results due to high variations in RTTs. The measurements are spread out into an 8 hour period, both to smooth out the variations in RTT caused by network conditions, and for rate limiting purposes. The experiment is repeated three times a day for a week from 12 August 2004 to 18 August 2004, so we had 21 RTT matrices. There are 23 backbone routers in the GEANT AS, so each matrix is 23x23 in size. We then obtained the final RTT matrix by taking the minimum measurement of each pair. Out of all the 1771 distinct triangles formed by triples of backbone routers, we observed 244 TIVs . This represents a significant 13.8% TIV inside the GEANT network.

When examined closely, it is observed that the TIVs in the GEANT network are mainly caused by the link weights disproportional to the link delay. For example (see Fig. 7), Slovakia has two OC-48 links to Czech Republic and Hungary, respectively. But their purpose is just to provide access for the Slovakia SANET, not to transit traffic between Czech and Hungary. So the weights on these two access links are intentionally set quite high, so that the traffic from Czech to Hungary would go via an alternative path through Germany and Austria, where the links are backbone OC-192 links with lower weights. When we look at the RTTs between the three nodes, however, the RTT between Czech and Hungary is much larger than the sum of the other two RTTs, causing a TIV.

We then looked at the Abilene network. Abilene publishes their router configuration files online [21], so we obtained their router-level topology and IS-IS weights from their website, as shown in Fig. 8. To verify that their published configuration file is up-to-date, we ran traceroute between directly connected nodes to see that every configured link is actually operational. The configuration data matches very well with the verification. We

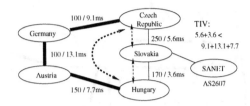

Fig. 7. An example TIV inside the GEANT network

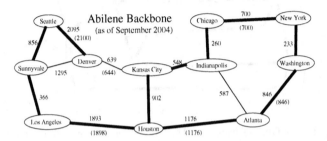

Fig. 8. The topology of Abilene, an Internet2 high-performance backbone network. Thick lines represent OC-192 10Gbps fast backbone links, and thin lines represent slower OC-48 links. The numbers on the links represent the IS-IS weights assigned to the links. Where there is a secondary backup link between two nodes, the IS-IS weight of the backup link is shown in brackets

then run the same measurements to collect the minimum RTT data between all pairs of Abilene backbone routers for the same whole week. Each measurement run takes around 8 hours, so we ran the experiments 3 times per day from 12 August 2004 to 18 August 2004, and obtained 21 matrices. There are only 11 backbone routers in Abilene, so each matrix is 11x11 in size. The minimum RTT between each pair of nodes is then taken to compute TIVs. We observed 5 TIVs out of all the 165 triangles. This represents 3.03% TIV inside the Abilene network.

We learned from the Abilene operators that the link weights are assigned according to geographic distance. As geographic distance is in a Euclidean space, we should expect the triangle inequality to always hold. The reality is, however, that even in an ideally designed network like this, TIV can still occur. On close inspection, we can see that all the TIVs are caused by traffic flowing through independent paths between the triple of nodes (e.g. between Indianapolis, Atlanta and Washington). Although geographically the path is shorter, behavior of the intermediate routers (e.g. load, processing delay, priority of traffic, or queuing delay) can affect the end-to-end RTT measurement. However, compared with GEANT, the violations are much less significant in terms of the r metric (defined in [9] as $r = \frac{a}{b+c} * (1 + (a - (b + c)))$, where a, b and c are the three edges of the triangle and a is the longest edge, as shown in Fig. 9.

4.2   Examples of External TIVs

To illustrate the effect of hot-potato routing on TIV, we picked three PlanetLab nodes from JANET (UK), BELNET (Belgium) and NYSERNet (USA), respectively. We run

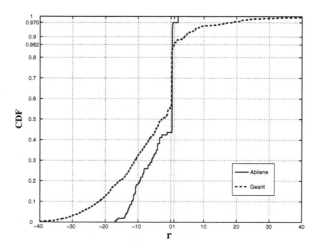

Fig. 9. CDF of the r metric for both Abilene and GEANT. TIVs are signified by $r > 1$, so it can be seen that GEANT exhibits a much higher percentage and magnitude of TIVs

traceroute from each node to the other two nodes to construct the exact path taken by data packets, as shown in Fig. 10. Here we use solid lines to represent a single direct link, and dotted lines to represent a few hops in the middle. We abstract out only the important routers along the paths. We can see that because of hot-potato routing, the traffic from node C to GEANT always goes through the New York router in Abilene. Similarly the GEANT network uses hot-potato routing as well for traffic going to Abilene. The primary link that is causing the problem in this case is the link between NL in GEANT and Chicago in Abilene. This link has a much longer RTT than the NY-to-NY peering link, but is preferred in hot-potato routing to route traffic from NL to Abilene. The end result is that the round-trip path between B and C is asymmetric and much longer than necessary, causing a TIV. We checked that the measurements we obtained with traceroute are within 0.25% error of the minimum RTT value taken during the first week of June 2004 (as mentioned in the dataset in [10]), so we use these RTT measurements in the figure for illustration.

To illustrate the effect of private peering shortcut on TIV, we picked three PlanetLab nodes from JANET (UK), DFN (Germany) and CERNET (China). We run traceroute from each node to the other two to construct the AS-level data path. We then use traceroute to collect RTT data once every 10 minutes between the triple for a 24 hour period on 18 August 2004, and the minimum RTT value is used to demonstrate the TIV. As shown in Fig. 11, the paths between pairs of nodes are symmetric, and there is a private peering shortcut between JANET and CERNET. DFN does not know about this private peering shortcut, so it has to go up the hierarchy tree to communicate with CERNET. This causes a TIV between the three nodes. This AS-level graph corresponds remarkably well with the theoretical analysis shown in Fig. 2.

To illustrate the effect of independent AS paths on TIV, we picked three PlanetLab nodes from Russia, Hong Kong and the UK. The AS-level paths between the three nodes are shown in Fig. 12. Here we ignore the router-level paths within individual ASes, as

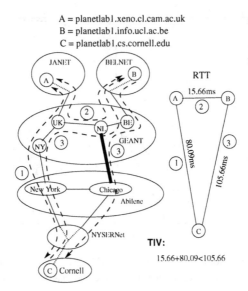

A = planetlab1.xeno.cl.cam.ac.uk
B = planetlab1.info.ucl.ac.be
C = planetlab1.cs.cornell.edu

Fig. 10. An example of TIV between PlanetLab sites introduced primarily by hot-potato routing in Abilene, GEANT and NYSERNet

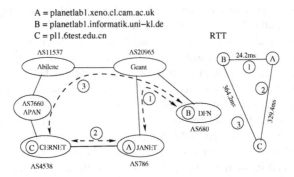

A = planetlab1.xeno.cl.cam.ac.uk
B = planetlab1.informatik.uni−kl.de
C = pl1.6test.edu.cn

Fig. 11. An example of TIV caused solely by private peering shortcut between ASes, i.e. in this case between JANET and CERNET

these are insignificant. What is of interest in this case is the complicated AS-level paths packets take between these three nodes. By *independent*, we mean that the intermediate ASes are independently engineered and that parts of the AS-paths do not overlap with any other. This is caused by the BGP import and export policies of the ASes involved, and can be explained by the economic incentives of inter-connecting networks [17]. As the minimum RTT measurements between the three nodes vary drastically from week to week in our earlier dataset [10], there are no representative values. However, the common observation is that they all show TIVs between these nodes. So we use the values of a particular measurement in the figure as illustration.

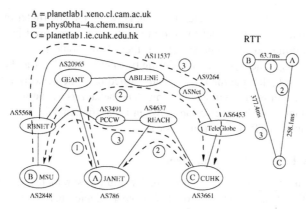

A = planetlab1.xeno.cl.cam.ac.uk
B = phys0bha–4a.chem.msu.ru
C = planetlab1.ie.cuhk.edu.hk

Fig. 12. An example of TIV between PlanetLab sites introduced by independent AS-paths

A = planetlab1.xeno.cl.cam.ac.uk
B = luna.hea.net
C = mad.so1–0–0.eb–iris4.red.rediris.es

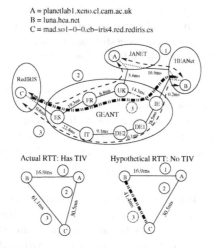

Fig. 13. An example of TIV caused by the interaction between intra-domain and inter-domain routing, or more specifically, between intra-domain TIV and private peering

4.3   Examples of End-to-End TIVs

To show an example of end-to-end TIV caused by the interaction between intra-domain and inter-domain routing, we picked three nodes from JANET (UK), HEANet (Ireland) and RedIRIS (Spain), respectively. As not all the domains have PlanetLab nodes in this case, we use Looking Glass nodes instead whenever necessary. We ran traceroute between the triple, and collected the data once every 10 minutes for a 24 hour period on 18 August 2004. The minimum RTT values between each pair of nodes are then chosen from the dataset. This time, however, we want to see how exactly the path affects TIV, so we use the minimum RTT observation to break the RTT down into segments with one RTT per link (or a set of links if the links are topologically unimportant). The detailed break-down of RTT values is shown in Fig. 13.

Here, we can see that there is actually a link between the UK router in GEANT and the IE router, but its weight is set to be quite high so it is not used in routing. If we used this link on the path from node C to A, then there would not be a TIV between the triple even though there is a private peering link between JANET and HEANet. This illustrates that just looking at the inter-domain structure of the network is not sufficient to determine whether a TIV will occur or not. The behaviors of the intermediate ASes are also important. It is the interaction between intra-domain TIV and private peering shortcut that causes a TIV to occur in this case.

4.4   TIVs in PlanetLab Measurements

To illustrate that TIV is not uncommon in real-world measurements, we took a week's PlanetLab RTT measurement trace from [22] from 13 September 2004 to 19 September 2004. The pair-wise RTT data were collected on consecutive 15-minute periods, and we take the median RTT value from each measurement. Thus for each day in this period there were 96 matrices of RTT measurements, and the size of each matrix is 399×399. Over the week we therefore had 672 such matrices. We then take the minimum RTT value of all the matrices for each pair, and construct a final RTT matrix. Some entries in the final matrix have no values due to unsuccessful measurements, so we denote those by 'NaN'. In calculating triangles, we discard any triangle that has 'NaN' as one of its edges.

We classify all the PlanetLab nodes into research and commercial nodes by looking at whether the IP address of the node matches any prefix in the GREN (i.e. Abilene) BGP table. To be on the safe side, we also manually check the list of nodes that are classified as being commercial nodes. In this way, of all the 399 PlanetLab nodes as of 16 September 2004, we identified 327 nodes as research nodes and 72 nodes as commercial ones. This means that 82% of the hosts are in the GREN, a slight decrease from the 85% we observed in [9].

Notice that the names of the PlanetLab nodes are not accurate indications of whether the node is research or commercial. For example, HP Labs (AS71) has a few PlanetLab nodes under the domain hpl.hp.com, but they are reachable from Abilene, and so are in fact research nodes. Conversely, the Computer Science and Artificial Intelligence Lab of MIT has a few PlanetLab nodes under the domain csail.mit.edu, but CSAIL (AS40) uses Cogent (AS174) as its upstream provider for connections, and so these nodes are in fact commercial nodes.

We classify all the triangles into 'R.R.R' with all research nodes, 'C.C.C' with all commercial nodes, and 'mixed' with a mixture of nodes. We use the r metric as defined in [9] to illustrate the amount of TIVs. Of all the 2537992 valid triangles formed by the 399 nodes, there were 467328 TIVs, so this represents 18.4% TIV in PlanetLab measurements. Table 1 shows the detailed break-down of TIVs by category for the node-by-node matrix. Fig. 14 shows the CDF distribution of r values for each category, when zoomed in to the area around $r = 1$. We can see that 'R.R.R' triangles behave the best. There are the fewest number of TIVs, and the TIVs are all quite small in magnitudes. 'C.C.C' and 'mixed' triangles behave very similarly, and both have a higher percentage of TIVs and much bigger r values.

Table 1. A detailed break-down of TIVs by category for the node-by-node matrix

Category	Total	TIVs	Percentage
$R.R.R$	1704809	282164	16.6%
$C.C.C$	9678	2306	23.8%
$Mixed$	823505	182858	22.2%
All	2537992	467328	18.4%

Table 2. A detailed break-down of TIVs by category for the site-by-site matrix

Category	Total	TIVs	Percentage
$R.R.R$	263062	52170	19.8%
$C.C.C$	966	227	23.5%
$Mixed$	119945	26084	21.8%
All	383973	78481	20.4%

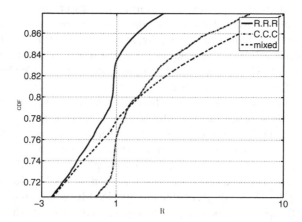

Fig. 14. The CDF distributions of r values for the three categories of triangles formed by PlanetLab *nodes*, when zoomed in to areas around $r = 1$

However, there are a few problems with using the node-by-node matrix. When we look at the r values close to 1, we can see that most of those corresponding triangles have two nodes physically located in the same site with a very small RTT value between them (typically <1ms). This makes the r value very vulnerable to measurement artifacts, as measurement error can often be much bigger than 1ms. So if we do not filter out those triangles, then our measured TIV percentage is not accurate. We have also seen some absurd RTT values (e.g. >3000ms), and triangles with one edge having those RTT values usually have huge r values. We suspect the high RTT values are due to a few nodes being overloaded, so we want to filter out those as well to get an accurate measure of TIV magnitude. For these reasons, we want to get a site-by-site matrix, where we pick a representative node for each site.

In our initial approach, we noted that all the nodes that belong to a physical site are located within the same '/24' prefix, so we picked one node from each '/24' prefix which has the most number of non-NaN measurements that are less than 1500ms. However, we also noticed that sometimes not all the nodes in the same '/24' prefix are physically located together. For example, the Internet2 PlanetLab nodes are physically scattered across the Abilene backbone and co-located with Abilene backbone routers, and so the RTTs between them are typically more than 10ms. Therefore, we decided to use physical closeness and the *behavior* of the nodes as the main criteria for grouping nodes into sites.

For each node (i.e. the *pivot* node), we first pick out all the nodes that have sub-millisecond RTT to that node. Within the group, we then check each pair-wise RTT to make sure it is either sub-millisecond or NaN. Then, we calculate the Correlation Coefficient (CC) between the *pivot* node and every other node, and throw away nodes that have $CC < 0.99$. In calculating CC, we use the standard definition on the two row vectors of the RTT matrix, but we modify it slightly so we treat NaN entries as perfect matches with the corresponding entries. Finally, for each remaining node in the group, we also search through the node list and add into the group any node that has the same '/24' prefix as this node but NaN in the RTT matrix. This accounts for nodes in the same physical location but was down during the measurement period. By using the above procedure, we were able to reduce the 399 nodes into 168 'sites', with 140 research 'sites' and 28 commercial 'sites'. Again we picked out the 'best' node from each site, and calculate TIVs on the reduced 168×168 matrix. This time, we observed 78481 TIVs out of all 383973 valid triangles. Table 2 shows the detailed break-down of TIVs by category for the site-by-site matrix.

We expected the 'R.R.R' triangles to behave the best, as it is very likely that traffic between them is only transited through GREN. In GREN, the networks are not driven purely by commercial relationships and are very cooperative in general, and the GREN inter-domain routing policies are often configured to use shortest path even if it violates economic provider-to-customer relationship. For example, RENATER2 (AS2200), the French research network, has a direct peering connection with APAN-Korea (AS9270), constructed under the TEIN2 project. Although RENATER2 is a customer of GEANT (AS20965), it still exports this private peering link to GEANT so the shortest path can be used. (Actually the European Commission is funding part of the TEIN2 project, so this private peering is logically peering with both RENATER2 and GEANT. GEANT can also reach APAN-Korea through Abilene and APAN-Japan, but the path is much longer.) In the commercial world, however, paths are often inflated due to economic reasons [23,24]. Thus, we would expect a higher percentage of TIVs between commercial PlanetLab nodes than between research ones. When there is a mixture of nodes, traffic between a research and a commercial node tends to go through the commercial Internet, so in a mixed triangle there are at least two paths between nodes through the commercial Internet. This makes the mixed case a lot like the 'C.C.C' case.

5   Conclusion

Although Internet routing policies play an important role in the global observed round-trip-times, we also want to emphasize that routing policy is not the only thing that contributes to TIVs. We have already seen that the private peering connection between JANET and CERNET is not giving too much savings on the RTT measurements. This relates to the Layer 2 technology used on this peering link. There is also the fact that the earth is a sphere, not a plane, so triangles on the surface of earth can go around the earth and do not have to satisfy the triangle inequality (although currently there are only very few fast links through continental Europe to Asia). Even as we are finishing this paper, we have heard that the South American research network is being re-structured and connected to GEANT directly, and the AMPATH network, which used to connect Brazil to Abilene, is being decommissioned. This would mean that temporarily all research traffic from Brazil to Abilene will need to go through GEANT, essentially traversing through the Atlantic twice. In particular, we have verified that traffic from Brazil to Mexico goes along this very much stretched path through GEANT and Abilene. This illustrates that the structural planning of the network can bring about unexpected TIVs as well.

In conclusion, TIVs are not just data collection artifacts, they can be structurally persistent as a result of routing policies (as well as many other factors). Both intra-domain routing policies and inter-domain routing policies, and the interactions between them, can cause structural TIVs. TIVs present an opportunity for overlay routing, but they make Internet Coordinate embeddings less accurate.

Acknowledgements

The bulk of the work done for this paper was carried out while the authors were at Intel Research, Cambridge. The authors would like to thank Derek McAuley of Intel Research for his support. In addition, Jon Crowcroft of the University of Cambridge provided many useful comments and suggestions.

References

1. Andersen, D.G., Balakrishnan, H., Kaashoek, M.F., Morris, R.: Resilient overlay networks. In: Proc. 18th ACM SOSP, Banff, Canada. (2001)
2. Ng, T.E., Zhang, H.: Predicting Internet Network Distance with Coordinates-Based Approaches. In: IEEE INFOCOM' 02, New York, USA (2002)
3. Pias, M., Crowcroft, J., Wilbur, S., Harris, T., Bhatti, S.: Lighthouses for Scalable Distributed Location. In: 2nd International Workshop on Peer-to-Peer Systems. (2003)
4. Tang, L., Crovella, M.: Virtual Landmarks for the Internet. In: ACM SIGCOMM Internet Measurement Conference (IMC'03), Miami (FL), USA (2003)
5. Lim, H., Hou, J., Choi, C.: Constructing Internet Coordinate System Based on Delay Measurement. In: ACM SIGCOMM Internet Measurement Conference (IMC'03), USA (2003)
6. Costa, M., Castro, M., Rowstron, A., Key, P.: PIC: Practical Internet Coordinates for Distance Estimation. In: 24th IEEE International Conference on Distributed Computing Systems (ICDCS' 04), Tokyo, Japan (2004)
7. Dabek, F., Cox, R., Kaashoek, F., Morris, R.: Vivaldi: A Decentralized Network Coordinate System. In: Proceedings of the ACM SIGCOMM 2004, Portland, Oregon (2004)
8. Shavitt, Y., Tankel, T.: Big-bang simulation for embedding network distances in Euclidean space. In: Proceedings of the IEEE INFOCOM 2003, San Francisco, California (2003)
9. Banerjee, S., Griffin, T.G., Pias, M.: The Interdomain Connectivity of PlanetLab Nodes. In: 5th International Workshop on Passive and Active Measurement, France (2004)
10. Pias, M., Lua, E.K., Zheng, H., Griffin, T.G.: On the Accuracy of Embeddings for Internet Coordinate Systems. In: Under submission. (2005)
11. Shavitt, Y., Tankel, T.: On the Curvature of the Internet and its usage for Overlay Construction and Distance Estimation. In: Proc. ACM INFOCOM 2004, Hong Kong (2004)
12. Perlman, R.: Interconnections: Bridges, Routers, Switches, and Internetworking Protocols. 2nd edn. Addison-Wesley (1999)
13. Pepelnjak, I.: EIGRP Network Design Solutions: The Definitive Resource for EIGRP Design, Deployment, and Operation. Cisco Press (2000)

14. Greenberg, A.G., Hajek, B.: Deflection routing in hypercube networks. IEEE Trans. Commun. **40** (1992) 1070–1081
15. Norton, W.B.: Internet Service Providers and Peering. In: Proceedings of NANOG 19, Albuquerque, New Mexico (2000)
16. Bu, T., Gao, L., Towsley, D.: On routing table growth. In: Proc. of Global Internet Symposium 2002. (2002)
17. Houston, G.I.: Interconnection, Peering and Settlements. Internet Protocol Journal (1999)
18. Abilene: Abilene Network Operations Center: Conditions of Use (COU) for the Abilene Network. http://abilene.internet2.edu/policies/cou.html (2004)
19. DANTE: GEANT Topology. http://www.dante.net/server/show/nav.007009007 (2004)
20. DANTE: GEANT Backbone Looking Glass. http://stats.geant.net/lg/lgform.cgi (2004)
21. Abilene: Abilene Network Operations Center: Abilene Backbone Network Router Configurations. http://loadrunner.uits.iu.edu/~gcbrowni/Abilene/vn/configs/configs.html (2004)
22. Jeremy Stribling: Pair-wise RTT data between all PlanetLab nodes by Jeremy Stribling (MIT). http://www.pdos.lcs.mit.edu/~strib/pl_app/ (2004)
23. Spring, N., Mahajan, R., Anderson, T.: Quantifying the Causes of Path Inflation. In: Proceedings of ACM SIGCOMM 2003. (2003)
24. Gao, L., Wang, F.: The Extent of AS Path Inflation by Routing Policies. In: Proceedings of Global Internet 2002. (2002)

Traffic Matrix Reloaded: Impact of Routing Changes

Renata Teixeira[1], Nick Duffield[2], Jennifer Rexford[2], and Matthew Roughan[3]

[1] U. California–San Diego
teixeira@cs.ucsd.edu
[2] AT&T Labs–Research
{duffield, jrex}@research.att.com
[3] University of Adelaide
matthew.roughan@adelaide.edu.au

Abstract. A *traffic matrix* represents the load from each ingress point to each egress point in an IP network. Although networks are engineered to tolerate some variation in the traffic matrix, large changes can lead to congested links and poor performance. The variations in the traffic matrix are caused by statistical fluctuations in the traffic *entering* the network and shifts in where the traffic *leaves* the network. For an accurate view of how the traffic matrix evolves over time, we combine fine-grained traffic measurements with a continuous view of routing, including changes in the egress points. Our approach is in sharp contrast to previous work that either inferred the traffic matrix from link-load statistics or computed it using periodic snapshots of routing tables. Analyzing seven months of data from eight vantage points in a large Internet Service Provider (ISP) network, we show that routing changes are responsible for the majority of the large traffic variations. In addition, we identify the shifts caused by *internal* routing changes and show that these events are responsible for the largest traffic shifts. We discuss the implications of our findings on the accuracy of previous work on traffic matrix estimation and analysis.

1 Introduction

The design and operation of IP networks depends on a good understanding of the offered traffic. Internet Service Providers (ISPs) usually represent the traffic as a matrix of load from each ingress point to each egress point over a particular time interval. Although well-provisioned networks are designed to tolerate some fluctuation in the traffic matrix, large variations break the assumptions used in most designs. In this paper, we investigate the *causes* of the traffic matrix variations. Identifying the reasons for these disruptions is an essential step toward predicting and planning for their occurrence, reacting to them more effectively, or avoiding them entirely.

The traffic matrix is the composition of the *traffic demands* and the *egress point selection*. We represent the traffic demands during a time interval t as a matrix $V(\cdot, \cdot, t)$, where each element $V(i, p, t)$ represents the volume of traffic entering at ingress router i and headed toward a destination prefix p. Each ingress router selects the egress point for each destination prefix using the Border Gateway Protocol (BGP). We represent the BGP routing choice as a mapping ε from a prefix to an egress point, where $\varepsilon(i, p, t)$ represents

C. Dovrolis (Ed.): PAM 2005, LNCS 3431, pp. 251–264, 2005.

the egress router chosen by ingress router i for sending traffic toward destination p. At time t each element of the traffic matrix TM is defined as:

$$TM(i, e, t) = \sum_{p \in P: \varepsilon(i, p, t) = e} V(i, p, t). \tag{1}$$

where P is the set of all destination prefixes.

Figure 1 presents a simple network with one ingress router i, two egress routers e and e', and two external destination prefixes p_1 and p_2. Given traffic demands $V(i, p_1, t)$ and $V(i, p_2, t)$ and a prefix-to-egress mapping $\varepsilon(i, p_1, t) = \varepsilon(i, p_2, t) = e$, the traffic matrix for this network is $TM(i, e, t) = V(i, p_1, t) + V(i, p_2, t)$ and $TM(i, e', t) = 0$.

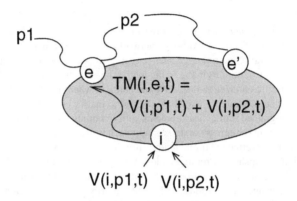

Fig. 1. Example of traffic matrix

Fluctuations in the traffic demands and changes in the prefix-to-egress mapping cause the traffic matrix to vary. This paper considers the natural question: *what are the causes of large variations in the traffic matrix?*

Most previous work on measuring [1–4] and analyzing traffic matrices [5, 6] has assumed that the prefix-to-egress mapping ε is stable. However, relying on periodic snapshots of routing data runs the risk of associating some traffic measurements with the wrong elements in the traffic matrix, obscuring real variations in the traffic. In this paper, we study how *changes* in ε impact the traffic matrix. A previous analysis of five traces of 6–22 hours in duration on the Sprint network [7] shows that most BGP routing changes do not lead to large traffic shifts. However, given that large traffic variations are infrequent (yet significant) events, we believe that longer traces are necessary to draw meaningful conclusions. Our previous work [10] shows that internal routing can cause ε to change for a large number of prefixes at the same time, which can potentially cause a large traffic shift. Neither [7] nor [10] study the significance of traffic shifts caused by routing relative to regular traffic fluctuations, which is the topic of this paper.

In this paper, we study the impact of routing changes on the traffic matrix over a *seven-month* period in a tier-1 ISP network. Using Cisco's Sampled Netflow feature [8] and feeds of internal BGP (iBGP) messages, we compute the traffic demands V and the prefix-to-egress mapping ε for eight ingress routers. Joining these two datasets allows us to construct a detailed view of the variation of the traffic matrix over time. We also collect measurements of the intradomain routing protocol [9] in order to identify the changes in ε that were caused by internal network events, using the algorithm described in [10]. Our analysis shows that:

1. **Although the likelihood of large traffic fluctuations is small, big changes do sometimes occur.** In any given ten-minute time interval, less than 0.02% of the traffic matrix elements studied have a traffic variation of more than 4 times the normal traffic variations. However, some elements vary by more than 4 times the normal variations several times a week.
2. **Most routing changes do not cause much variation in the traffic matrix.** Previous studies [7, 11] have shown that routing changes typically do not cause large traffic shifts; most BGP routing changes affect destination prefixes that receive very little traffic.
3. **Routing changes are responsible for many of the large traffic shifts:** 58.6% of instances where a traffic matrix element fluctuated by more than 10 times the normal variation for that element could be explained by a BGP routing change.

Although routing changes usually do not affect much of the traffic, many of the large traffic shifts are triggered by routing changes. Large traffic shifts caused by routing are rare, but important events. After introducing our measurement methodology in Section 2, we identify the causes of the big variations in Section 3. Section 4 discusses the implications of our results on other studies of traffic matrices. Section 5 concludes the paper.

2   Measuring Traffic Matrix Variation

Studying the variation of traffic matrix elements over time requires collecting fine-grained measurements of traffic and routing. We analyze data collected from a tier-1 ISP network for 173 days from March to September 2004. We collect data from eight aggregation routers that receive traffic from customers destined to peers and other customers. The eight routers are located in major Points of Presence (PoPs) that are spread throughout the United States.

We compute eight *rows* of the traffic matrix, considering all traffic from these eight ingress aggregation routers to all of the egress PoPs. This section describes how we compute the prefix-to-egress mapping $\varepsilon(i, p, t)$ from the BGP data and the traffic demands $V(i, p, t)$ from the Netflow data. Once we have computed ε and V, we use Equation 1 to compute the elements of the traffic matrix $TM(i, e, t)$. The BGP monitor and the Netflow collection servers are NTP-synchronized, allowing us to use the timestamps to join the two datasets.

2.1   Prefix-to-Egress Mapping

A BGP monitor collects internal BGP update messages directly from each vantage point. Configured as a route-reflector client of each vantage point, the BGP monitor receives updates reporting any change in the best BGP route at each router for each destination prefix. The monitor records each BGP update with a timestamp at the one-second granularity.

A single network event, such as a failure or policy change, can lead to a burst of BGP updates messages as the routers explore alternate paths. Rather than studying the details of routing convergence, our analysis focuses on the changes from one stable route to another. Similar to previous studies [10, 11], we group the BGP updates for the same destination prefix that have an interarrival time of 70 seconds or less. Our analysis considers the stable route that existed before the flurry of updates and the new stable route that exists at the end.

Based on an initial BGP table dump and a sequence of BGP updates, we generate the prefix-to-egress mapping $\varepsilon(i, p, t)$ for any given time. The egress point corresponds to a *PoP* rather than a specific router. We associate each egress router with a PoP based on the router name and configuration data.

2.2   Traffic Demands

Every vantage point has the Cisco's Sampled Netflow feature [8] enabled on all links that connect to access routers and exports flow records to a collection server at the same location. The collection server samples the flow records using the technique presented in [12] in order to reduce processing overhead, and computes 10-minute aggregated traffic volumes for each destination prefix. We use these aggregated reports to extract $V(i, p, t)$ for each vantage point i and destination prefix p at every 10-minute interval. Consequently, a reference to a time t indicates the end of a 10-minute interval.

Because of sampling, the volumes $V(i, p, t)$ are random quantities that depend on the sampling outcomes. Through a renormalization applied to the bytes reported in sampled flow records, the quantities $V(i, p, t)$ are actually unbiased estimators of the volumes of the original traffic from which they were sampled, i.e., their average over all possible sampling outcomes is the original volume. The standard error associated with an aggregate of size V is bounded above by $\sqrt{k/V}$ for some constant k that depends on the sampling parameters [12]. For the parameters employed in the current case, $k < 21$MB. Note that the standard error bound decreases as the size of the aggregate increases. This property aligns well with our focus on the largest changes in traffic rates: these are the most reliably estimated. As an example, for a 10-minute aggregate of traffic at a rate of 10 MB per second, the standard error due to sampling is no more than 6%.

Even though the traffic data is divided into 10-minute intervals, our 70-second grouping of BGP updates is important for cases when path exploration crosses the boundary between two ten-minute intervals. This ensures that we focus our analysis on stable changes of ε. If the mapping $\varepsilon(i, p, t)$ changes more than once in a 10-minute interval, then we cannot distinguish the volume of traffic affected by each of them individually. Therefore, we exclude those cases from our analysis by ignoring intervals with prefixes that have more than one stable routing changes in that bin; this excludes 0.05% of the (i, e, t) tuples from our study. We also exclude all traffic for the small number of flows that had no matching destination prefix in the BGP routing tables or update messages; we verified that these flows corresponded to an infinitesimal fraction of the traffic.

3   Causes of Large Traffic Variations

In this section, we explore the contributions of changes in the traffic demands V and prefix-to-egress mapping ε to the variations in the traffic matrix elements $\mathcal{TM}$. Our analysis shows that, although most changes in ε have a small effect on the traffic matrix, many of the large variations in the traffic matrix are caused by changes in ε. Also, we show that, while most changes in ε are caused by external routing events, the small number of internal routing events are more likely to cause larger shifts in traffic.

3.1   Definition of Traffic Variations

Figure 2 shows an example of how two traffic matrix elements (with the same ingress point i) change over the course of a day. The total traffic entering at the ingress point varies throughout the day, following a typical diurnal cycle. For the most part, the traffic $TM(i, e_1, t)$ has the same pattern, keeping the proportion of traffic destined to e_1 relatively constant. For most of the day, no traffic travels from ingress i to egress point e_2. The most significant change in the two traffic matrix elements occurs near the end of the graph. The traffic leaving via egress point e_1 suddenly decreases and, at the same time, traffic leaving via egress point e_2 increases. This shift occurred because a routing change caused most of the traffic with egress point e_1 to shift to egress point e_2. The egress point e_2 also starts receiving traffic that had previously used other egress points (not shown in the graph), resulting in an increase for e_2 that exceeds the decrease for e_1. In the meantime, the total traffic entering the network at ingress i remained nearly constant.

The traffic experiences other relatively large downward spikes (labeled as load variation). These spikes may very well be associated with a routing change in another AS in

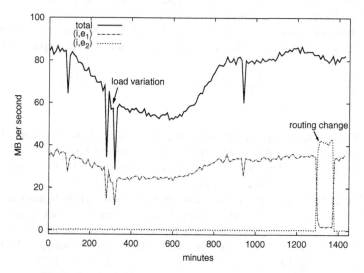

Fig. 2. Sample traffic volume from one ingress to two egresses

the Internet that caused traffic to enter at a different PoP (this kind of traffic variation was called an "ingress-shift anomaly" in [6]). In this paper, we analyze traffic shifts caused by routing changes experienced by our network. Finding a signature of routing-induced traffic variations for one network is an important first step to infer other traffic variations that are caused by routing changes in other networks.

To analyze these kinds of traffic fluctuations, we define the variation of a traffic matrix element at an interval t as:

$$\Delta TM(i, e, t) = TM(i, e, t) - TM(i, e, t - 1).$$

3.2   Changes in Traffic Demands Versus Egress Points

The variation of a traffic matrix element (ΔTM) is composed of the load variation (ΔL), which represents volume fluctuations on the traffic demands V, and the routing shifts (ΔR), which accounts for changes in the prefix-to-egress mapping ε:

$$\Delta TM(i, e, t) = \Delta L(i, e, t) + \Delta R(i, e, t)$$

$\Delta L(i, e, t)$ represents the change in the volume of traffic for all destination prefixes that did *not* change their egress point from the previous time interval (i.e., $\varepsilon(i, p, t) = \varepsilon(i, p, t - 1) = e$):

$$\Delta L(i, e, t) = \sum_{\substack{p \in P: \\ \varepsilon(i, p, t) = e \\ \varepsilon(i, p, t - 1) = e}} V(i, p, t) - V(i, p, t - 1)$$

Fluctuations in the traffic demands may occur for a variety of reasons, such as changes in user or application behavior, adaptations caused by end-to-end congestion control, or even routing changes in other domains.

The routing variation $\Delta R(i, e, t)$ considers the destination prefixes that shifted *to* egress point e during time interval t or shifted *from* e to another egress point in t:

$$\Delta R(i, e, t) = \sum_{\substack{p \in P: \\ \varepsilon(i, p, t) = e \\ \varepsilon(i, p, t - 1) \neq e}} V(i, p, t) - \sum_{\substack{p \in P: \\ \varepsilon(i, p, t) \neq e \\ \varepsilon(i, p, t - 1) = e}} V(i, p, t - 1)$$

Note that if a routing change occurs within the time interval t, we associate *all* of the traffic associated with that prefix in that time interval with the new egress point.

Not all traffic matrix elements carry the same volume of traffic, and the volume of traffic from an ingress to an egress PoP varies over time. How do we judge if a change in the traffic is "large"? There is no absolute standard: one approach might be to judge the size of the change in traffic matrix element relative to the average traffic for that element. However, this is not useful here, because the traffic process itself is non-stationary. It has daily and weekly cycles, as well as level shifts resulting from routing changes. The *relative* change $\Delta TM(i, e, t)/TM(i, e, t)$ (or $\Delta TM(i, e, t)/\max(TM(i, e, t), TM(i, e, t-1))$) seems appealing. However, this metric places too much emphasis on large relative changes to small values; for example, a traffic matrix element with 1 kbit/sec might easily experience a 50% relative change in traffic without having any significant effect on the network. An alternative metric would be the *absolute* change $\Delta TM(i, e, t)$. However, a shift of (say) 10 MB/sec may be significant for one ingress point but not for another. Another option would be to normalize the value of $\Delta TM(i, e, t)$ by the total traffic entering ingress point i at time t, which would capture changes in the *fraction* of the incoming traffic that uses a particular egress point. However, this metric depends on the "current" traffic demand at ingress i (which could be low at certain times) and may not accurately reflect the strain imposed on the network by the traffic change. Another extreme approach would be to consider the capacity of the network, and define as large any traffic shift that causes a link to be overloaded. Besides being difficult to compute, this metric is too closely tied to the current design of the network, and is not useful for most typical applications of the traffic matrix such as capacity planning or anomaly detection. Instead, we want a metric that captures properties of the traffic matrix itself, such as how large the traffic changes are relative to the normal variations of traffic matrix elements.

For that, we should consider what type of process we observe, namely, a difference process. Over short time periods, we can approximate the traffic with a linear process $y_t = \alpha + \beta t + x_t$, where x_t is a zero-mean stochastic process, with variance σ^2. We observe the differences $\Delta y_t = y_t - y_{t-1}$, which will form a *stationary* process, with mean β and variance $2\sigma^2$. Thus we can approximate the difference process by a stationary process, and measure deviations from the mean, relative to the standard deviation of this process. We measure $2\sigma(i, e)^2$ on the traffic variation process $\Delta L(i, e, \cdot)$ (using the standard statistical estimator), and use this to normalize the traffic variations, i.e. we then observe $\Delta \tilde{L}(i, e, t) = \Delta L(i, e, t)/\sqrt{2}\sigma(i, e)$, and $\Delta \tilde{R}(i, e, t) = \Delta R(i, e, t)/\sqrt{2}\sigma(i, e)$.

If the variance of the process x_t was time dependent, it might make sense to use a moving average to estimate the process variance at each point in time, i.e. $\sigma(i, e, t)^2$, and use this to normalize the traffic variations. We tried such an approach, but it made little difference to the results, and so we use the simpler approach described above.

Figure 3 presents a scatter plot of $\Delta \tilde{TM}(i, e, t)$ versus $\Delta \tilde{R}(i, e, t)$ for all the valid measurement intervals t. The high density of points close to zero shows that large traffic variations are not very frequent (99.88% of the traffic variations are in the $[-4, 4]$ range). Points along the horizontal line with $\Delta \tilde{R}(i, e, t) = 0$ correspond to traffic variations that are not caused by routing changes, whereas points along the diagonal line correspond to variations caused almost exclusively by routing changes. Points in the middle are caused by a mixture of routing changes and load variation. Figure 3 shows that both load and routing are responsible for some big variations. Routing changes, however, are responsible for the *largest* traffic shifts. Indeed, one egress-point change made a traffic matrix element vary more than 70 times the standard deviation.

3.3 Internal Versus External Routing Changes

The prefix-to-egress mapping ε may change because of either internal or external routing events. *External routing changes* represent any changes in the set of egress points that an AS uses to reach a destination prefix. For example, in Figure 1, the neighbor AS might withdraw the route for $p2$ from the router e, resulting in a change in ε. External

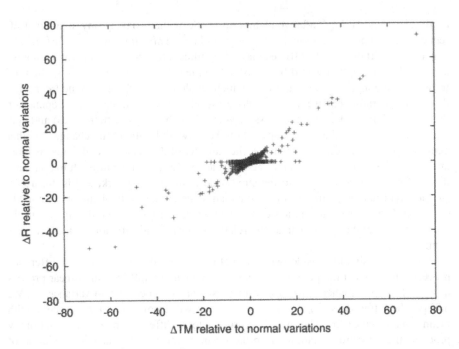

Fig. 3. Scatter plot of $\Delta\tilde{TM}$ versus $\Delta\tilde{R}$ for all traffic matrix elements over the seven-month period

routing changes may be caused by a variety of events, such as an internal routing change in another domain, a modification to the local BGP routing policy, or a failure at the edge of the network. In contrast, *internal routing changes* stem from changes in the routing inside the AS, due to equipment failures, planned maintenance, or traffic engineering. These events affect the prefix-to-egress mapping because the intradomain path costs play a role in the BGP decision process through the common practice of *hot-potato routing*.

When selecting a best BGP route, a router first considers BGP attributes such as local preference, AS path length, origin type, and the multiple exit discriminator. If multiple "equally good" routes remain, the router selects the route with the "closest" egress point, based on the intradomain path costs. Since large ISPs typically peer with each other in multiple locations, the hot-potato tie-breaking step almost always drives the final routing decision for destinations learned from peers, although this is much less common for destinations advertised by customers. In the example in Figure 1, an internal link failure might make router i's intradomain path cost to e suddenly *larger* than the path to e'. This would change the prefix-to-egress mapping for $p2$, causing a shift in traffic from egress point e to e'. Using the methodology described in [10], we identified which changes in ε were caused by internal events.

Figure 3 shows the cumulative distribution functions of $\Delta\tilde{R}$ caused by hot-potato routing and by external BGP changes. For comparison, we also present the cumulative distribution function (CDF) of a normal distribution, which is drawn from randomly generated Gaussian data with standard deviation equal 1, because $\Delta\tilde{R}$ has been normalized to have standard deviation equal 1. Although the routing events are rare (only 0.66%

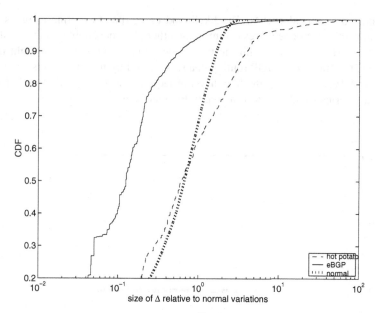

Fig. 4. Cumulative distribution function of $\Delta \tilde{R}$ caused by hot-potato routing and eBGP

of non-zero $\Delta \tilde{TM}$ are caused by eBGP changes and 0.1% by hot-potato changes), this result shows that there are significant cases where these events are big, to very big. In particular, approximately 5% of traffic shifts caused by hot-potato routing are at least one order of magnitude bigger than normal variations. A single internal change is more likely to affect a large number of destination prefixes [10], including the popular destinations receiving large amounts of traffic.

We analyzed the source of traffic variation for individual traffic matrix elements, and saw that the likelihood of changes in the prefix-to-egress mappings can vary significantly from one ingress router to another. Figures 6 and 5 present the same data as in Figure 3 for two sample traffic matrix elements (Note that the axis are different across the two graphs.). Some traffic matrix elements have no traffic variation caused by routing changes (Figure 5), whereas other have few very large egress shifts (Figure 6). We computed the percent of the traffic matrix elements (i, e) that have large to very large traffic shifts. We define *large* as more than 4 times the normal traffic variations for (i, e) and *very large* more than 10 times. Approximately 25% of ingress-egress pairs (i, e) in our study have no large traffic variation, and the vast majority of them (85.7%) have no very large traffic variation. The differences across the traffic matrix elements have two main explanations:

– **Size of traffic matrix element.** Some traffic matrix elements carry little traffic. Most of the traffic from an ingress router exits the network at few egress PoPs, because of hot-potato routing. For instance, most of the traffic entering in San Diego is likely to stay in the west cost. Therefore, the traffic element San Diego to New York carries very little traffic at any time.

– **Impact of internal events.** The likelihood of hot-potato routing changes varies significantly from one ingress point to the other [10], depending on the location in the network and the proximity to the various egress points. For our eight ingress points, the fraction of BGP routing changes caused by internal events varies from 1% to 40%. As a result, the likelihood of large traffic shifts caused by hot-potato routing varies significantly from one traffic matrix element to another.

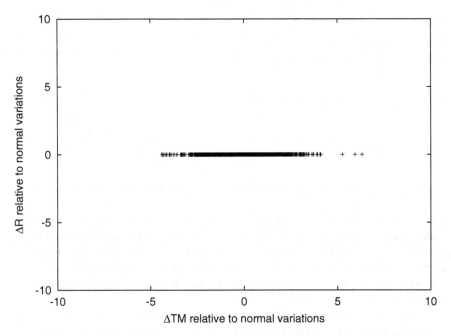

Fig. 5. Scatter plot of $\Delta\tilde{TM}$ versus $\Delta\tilde{R}$ for a traffic matrix element that have no routing-induced traffic variations over the seven-month period

Out of the traffic matrix elements that do experience large traffic variations 15% have an average of more than one large traffic variation per week. The small percentage of elements that experience large traffic variations combined with the low frequency large shifts per element may lead to the incorrect conclusion that these events are irrelevant. However, if we consider the network-wide frequency of large traffic shifts, these events happen fairly often. To show this, we have counted the number of 10-minute measurement intervals for which at least one of our eight vantage points experienced a large traffic variation. On average, the network experiences a large traffic variation every four and half hours. Large traffic variations caused by routing changes happen every 2.3 days, and very large routing-induced traffic variations happen every 5.9 days. If our analysis considered all of the PoPs in the network, the overall frequency of large traffic variations would be even higher.

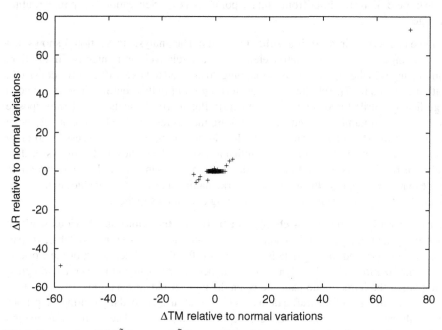

Fig. 6. Scatter plot of $\Delta \tilde{T}M$ versus $\Delta \tilde{R}$ for a traffic matrix element that has few very large routing-induced traffic shifts over the seven-month period. One traffic shift was over 70 times normal traffic variations!

4   Implication for Traffic Matrix Studies

Our analysis on traffic matrix variations has important implications for the results of previous measurement studies.

Differences across vantage points: The results in Section 3 show that the likelihood of changes in the prefix-to-egress mappings can vary significantly from one ingress router to another. In particular, some ingress points may be much more susceptible to hot-potato routing changes than others [10], making analysis of routing stability very dependent on where the data are collected. For example, the study in [11] showed that popular destination prefixes do not experience BGP routing changes for days or weeks at a time. In addition to studying RouteViews and RIPE BGP feeds, the analysis included iBGP data from two of the eight routers used in our current study. In our analysis, these two routers did not experience many hot-potato routing changes. Had the analysis in [11] analyzed a router that experiences several hot-potato routing changes a day, the conclusions might have been quite different. In fact, hot-potato routing changes can affect a large number of prefixes [10], both popular and not, so we might reasonably expect popular destinations to experience changes in their egress points. A preliminary analysis across all eight vantage points confirms that popular destination prefixes have more BGP instabilities from vantage points that experience more hot-potato routing changes.

Choice of metrics in studying unlikely events: The analysis in Section 3 shows that large changes in the traffic matrix elements occur relatively infrequently. In addition, most changes in the prefix-to-egress mapping do not lead to large traffic shifts, consistent with the results in [7]. Yet, these two results do *not* imply that routing changes are not a significant contributor to large changes in the traffic matrix elements. In fact, the opposite is true. A small number of routing changes are indeed responsible for a relatively large fraction of the (few) large traffic shifts. In addition, long traces are necessary to draw conclusions about infrequent (yet significant) events. The study in [7] draws on five traces of 6–22 hours in duration, outside of the maintenance periods where operators made planned changes to the internal network, making it difficult to conclude definitively if large traffic shifts occur and whether routing contributes to them.

Errors from ignoring egress changes in traffic matrix analysis: Previous work on measuring and analyzing traffic matrices has assumed that routing is stable, in part because fine-grained routing data is sometimes difficult to collect. Most of the work on traffic matrix *estimation* [1, 2, 4] assumes that there are no changes in the prefix-to-egress mapping or the intradomain paths between the ingress and egress points. Even work on direct *measurement* of the traffic demands [5, 13] has used only daily routing snapshots, although the work in [7] is a notable exception. Using out-of-date routing information runs the risk of associating some traffic measurements with the wrong elements in the traffic matrix. In some cases, the routing changes might lead to second-order effects on the traffic (e.g., by causing congestion or increasing the round-trip time) that may appear in the data, but the primary affect of the traffic *moving* to a different egress point is obscured—as is the *reason* for the variation in the traffic. In addition, changes in the prefix-to-egress mapping may cause large fluctuations in *multiple traffic matrix elements at the same time*, which would be obscured if the traffic matrix is computed or analyzed without regard for routing changes. In our ongoing work, we plan to quantify the errors in the traffic matrix computed using daily snapshots, similar to the approach in [7] but focusing specifically on routing changes that have a large affect on multiple traffic matrix elements.

Dependence on network design, traffic, and goals: The results of any traffic matrix analysis, including ours, depend on the details of the network under study. For example, large ISP networks handle high volumes of aggregated traffic, which may experience much smaller statistical fluctuations in the traffic. In addition, a large ISP network connects to its peers and many of its customers in multiple locations in the network, increasing the likelihood that destination prefixes are reachable via multiple egress points. This makes an ISP network much more likely to experience changes in the prefix-to-egress mapping over time. Together, these two factors tend to make routing changes have a larger relative influence on the traffic matrix in ISP networks than in other kinds of networks. Even within a single network, the fluctuations in the traffic matrix may vary from one ingress point to another, due to hot-potato routing changes or the particular senders and receivers connected to that router. Identifying metrics that isolate each of these effects would be very helpful in deepening our fundamental understanding of what causes fluctuations in traffic matrices.

5   Conclusion

Our study shows that large traffic variations, while unusual, do sometimes happen. Although most routing changes typically do not affect much traffic, routing is usually a major contributor to large traffic variations. This implies that network operators need to design the network to tolerate traffic variations that are much larger than typical statistical fluctuations in the incoming traffic. In addition, research on traffic engineering and anomaly detection should take into account the impact of routing on the traffic matrix. Since both the traffic demands V and the prefix-to-egress mapping ε are necessary to compute an accurate traffic matrix, we believe it is more accurate to operate on V and ε directly, rather than simply on $\mathcal{TM}$.

This work has implications for both the research and network operations communities. Researchers should consider the impact of changes in the prefix-to-egress mapping when analyzing the traffic matrix. Ignoring these changes might lead to wrong conclusions about traffic matrix stability. Operators need to provision for traffic variations that are much larger than normal traffic fluctuations. In addition, operators often need to diagnose the cause of a large surge in traffic. Our work shows that the routing system is one important place they should look for explanations.

As future work we plan to quantify the inaccuracies introduced in studies of routing and traffic stability when changes in ε are ignored. We are also studying the duration of the traffic shifts. If traffic shifts are short-lived, then network operators should just over-provision to tolerate them. If they are long-lived, however, adapting the routing protocol configuration may be a better approach for alleviating congestion.

References

1. J. Cao, D. Davis, S. V. Wiel, and B. Yu, "Time-varying network tomography," *J. American Statistical Association*, December 2000.
2. A. Medina, N. Taft, K. Salamatian, S. Bhattacharyya, and C. Diot, "Traffic matrix estimation: Existing techniques and new directions," in *Proc. ACM SIGCOMM*, August 2002.
3. Y. Zhang, M. Roughan, N. Duffield, and A. Greenberg, "Fast, accurate computation of large-scale IP traffic matrices from link loads," in *Proc. ACM SIGMETRICS*, June 2003.
4. Y. Zhang, M. Roughan, C. Lund, and D. Donoho, "An information-theoretic approach to traffic matrix estimation," in *Proc. ACM SIGCOMM*, August 2003.
5. A. Lakhina, K. Papagiannaki, M. Crovella, C. Diot, E. Kolaczyk, and N. Taft, "Structural analysis of network traffic flows," in *Proc. ACM SIGMETRICS*, June 2004.
6. A. Lakhina, M. Crovella, and C. Diot, "Characterization of Network-Wide Anomalies in Traffic Flows," in *Proc. Internet Measurement Conference*, October 2004.
7. S. Agarwal, C.-N. Chuah, S. Bhattacharyya, and C. Diot, "Impact of BGP dynamics on intra-domain traffic," in *Proc. ACM SIGMETRICS*, June 2004.
8. Sampled Netflow. http://www.cisco.com/univercd/cc/td/doc/product/ software/ios120/120newf%t/120limit/120s/120s11/12s_sanf.htm.
9. A. Shaikh and A. Greenberg, "OSPF monitoring: Architecture, design, and deployment experience," in *Proc. USENIX/ACM NSDI*, March 2004.
10. R. Teixeira, A. Shaikh, T. Griffin, and J. Rexford, "Dynamics of hot-potato routing in IP networks," in *Proc. ACM SIGMETRICS*, June 2004.
11. J. Rexford, J. Wang, Z. Xiao, and Y. Zhang, "BGP routing stability of popular destinations," in *Proc. Internet Measurement Workshop*, November 2002.

12. N. Duffield, C. Lund, and M. Thorup, "Estimating flow distributions from sampled flow statistics," in *Proc. ACM SIGCOMM*, August 2003.
13. A. Feldmann, A. Greenberg, C. Lund, N. Reingold, J. Rexford, and F. True, "Deriving traffic demands for operational IP networks: Methodology and experience," *IEEE/ACM Trans. on Networking*, June 2001.

Some Observations of Internet Stream Lifetimes

Nevil Brownlee

CAIDA, UC San Diego, and
The University of Auckland, New Zealand
nevil@auckland.ac.nz

Abstract. We present measurements of stream lifetimes for Internet traffic on a backbone link in California and a university link in Auckland. We investigate the consequences of sampling techniques such as ignoring streams with six or fewer packets, since they usually account for less than 10% of the total bytes. We find that we often observe large bursts of small 'attack' streams, which will diminish the integrity of strategies that 'focus on the elephants'. Our observations further demonstrate the danger of traffic engineering approaches based on incorrect assumptions about the nature of the traffic.

1 Introduction

Over the last few years there has been considerable interest in understanding the behaviour of large aggregates of Internet traffic flows. *Flows* are usually considered to be sequences of packets with a 5-tuple of common values (protocol, source and destination IP addresses and port numbers), and ending after a *fixed timeout* interval when no packets are observed. For example, Estan and Varghese [1] proposed a method of metering flows which ensures that all packets in *elephant* flows, i.e. those that account for the majority of bytes on a link, are counted, while packets in less significant flows may be ignored.

In contrast, *streams* are *bi-directional* 5-tuple flows, ending after a *dynamic timeout* interval of at least 10s and terminating after a quiet period of ten times their average packet inter-arrival time. Brownlee and Murray [2] investigated stream lifetimes, using a modified NeTraMet [3] meter. By using streams rather than flows, NeTraMet is able to measure various stream distributions at regular intervals (typically five or 10 minutes) over periods of hours or days. In [4] Brownlee and Claffy used this methodology to observe stream behaviour at UC San Diego and Auckland, where about 45% of the streams were *dragonflies* lasting less than two seconds. However, there were also many streams with lifetimes of hours to days, and those *tortoises* carried 50% to 60% of the link's total bytes.

At U Auckland, we use NeTraMet to measure Internet usage (bytes in and out for each user). In recent years the character of our Internet traffic has changed; the total volume has steadily grown, and we now see frequent network-borne attacks. Such attacks frequently appear as short time intervals during which we see large numbers of *dragonfly* streams. With our production NeTraMet rulesets (meter configuration files), attacks like address scans can give rise to tens of

C. Dovrolis (Ed.): PAM 2005, LNCS 3431, pp. 265–277, 2005.

thousands of flows. Such large bursts of flows tend to degrade the performance of our measurement system.

To minimise the effect of bursts of 'attack' streams, we investigated a strategy similar to that proposed by Estan and Varghese [1]. To do that we modified NeTraMet to ignore streams carrying K or fewer packets. That, however, posed the question of choosing a value for K.

In this paper we present some observations of stream lifetimes on a tier 1 backbone in California, which are consistent with earlier work by Brownlee and Claffy [4], and compare them with similar recent observations at Auckland.

We present measurements of the varying population of active streams at Auckland and compare that with the packet rate, using data gathered at one-second intervals over several days.

We investigate the proportion of the total bytes accounted for by streams with K or fewer packets, so as to help determine a suitable value for K. We often see measurement intervals when a high proportion of the total traffic is carried in *dragonfly* streams; for such intervals there are few *elephant* streams.

Lastly, we show that ignoring streams with six or fewer packets can provide effective usage monitoring for U Auckland.

2   Methodology, 'Overall' Traffic Observations

2.1   Understanding Flows and Streams

Traffic Flows were first defined in the seminal paper by Claffy, Polyzos and Braun [5]. A *CPB flow* is a set of packets with common values for the 5-tuple (IP protocol, Source and Destination IP Address and Port Number), together with a specified, fixed inactivity timeout, usually 60 seconds. Note that a CPB flow is unidirectional, with the 5-tuple specifying a direction for the flow's packets. CPB flows are widely used, providing a convenient way to summarise large volumes of Internet traffic data.

The IETF's RTFM architecture [6] provided a more general definition of a traffic flow. RTFM flows are bidirectional, with any set of packet attribute values being allowed to specify a flow. For example, an RTFM flow can be as simple as a CPB flow, or something more complex such as "all flows to or from network 192.168/16."

NeTraMet is an RTFM traffic measuring system that implements an extended version of RTFM flows. Streams were introduced to NeTraMet as a way of collecting data about subsets of a flow. For example, if we specify a flow as "all packets to/from a particular web server," then NeTraMet can recognise a stream for every TCP connection to that server, and build distributions of their sizes, lifetimes, etc.

NeTraMet's ability to handle streams in real time allows us to produce stream density distributions (e.g. lifetime and size in bytes or packets) over long periods of time – eight hours or more – while maintaining stream lifetime resolution down to microseconds. Furthermore, NeTraMet can collect such distributions at

5-minute intervals for days, without needing to collect, store and process huge packet trace files.

Although streams are bidirectional, that only means that NeTraMet maintains two sets of counters, one for each direction of the stream. If the meter can only see one direction of the stream, one set of counters will remain at zero. Bidirectional streams are, however, particularly useful for security analysis, where we need to know which attack streams elicited responses from within our network!

2.2   Streams in NeTraMet

From our earlier study of stream lifetimes [4] we know that a high proportion of traffic bytes are carried in *tortoise* streams. We modified the NeTraMet meter to use this fact to cache flow matches for each stream. The meter always maintains a table of active streams; when a new stream appears it is matched so as to determine which flow(s) it should be counted in. The set of matching flows is cached in the stream table, so that later packets can be counted in their proper flows without requiring further matching; we find that for most rulesets, average cache hit rates are usually well above 80%.

Since NeTraMet is now based on stream caching, it is straightforward to collect distributions of byte, packet and stream density, using a set of bins to build histograms for a range of stream lifetimes. We use 36 bins to produce distributions for lifetimes in a log scale from 6 ms to 10 minutes, and read these distributions every ten minutes.

Streams are only counted when they time out, so longer-running streams do not contribute to our distributions directly. Instead we create flows for them, so that they produce flow records giving the number of their packets and bytes every time the meter is read. From those 10-minute flow records we construct two more decades of logarithmic bins, producing lifetime distributions from 6 ms to 30,000 seconds (roughly 8 hours), i.e. nearly seven decades.

2.3   Tier-1 Backbone in California, December 2003

Fig. 1 gives an overview of traffic on a tier-1 OC48 backbone in California over Friday, 6 December 2003. Only one direction is shown, the other direction had about one-quarter the traffic volume. There is a clear diurnal variation from about 450 to 700 Mb/s. Most of the traffic is web (upper half of bars) or non-web TCP (lower half), plus a background level of about 50 Mb/s of UDP and other protocols.

Fig. 2 shows the stream density vs lifetime (upper left traces) for every 10-minute reading interval. There is little variation, and about 95% of all streams have lifetimes less than ten seconds. The lower right traces, however, show stream byte density vs lifetime. Again there is little variation, but only 60% of the bytes are carried by streams with lifetimes less than 1000 s. In other words, most streams are short but the bulk of the bytes are carried in long-running streams.

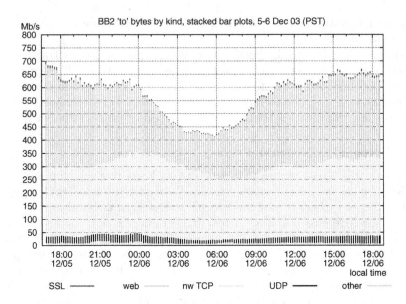

Fig. 1. Stacked-bar plot of traffic on an OC48 backbone in California

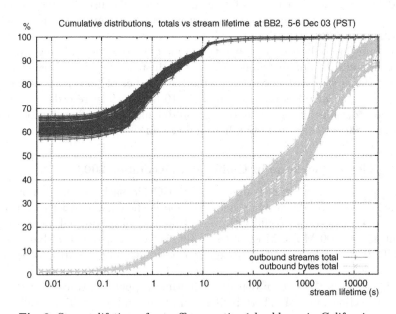

Fig. 2. Stream lifetimes for traffic on a tier-1 backbone in California

2.4   U Auckland Gateway, October 2004

Fig. 3 shows the traffic on U Auckland's 100 Mb/s Internet gateway for Friday-Saturday 1-2 October 2004. There is only around 15 Mb/s of traffic, and it

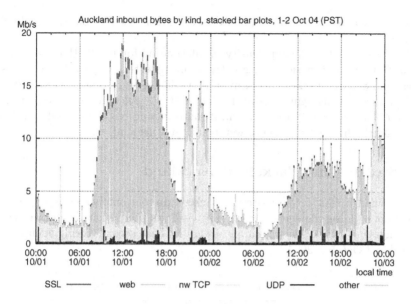

Fig. 3. Stacked-bar plot of traffic on the U Auckland (100 Mb/s) gateway

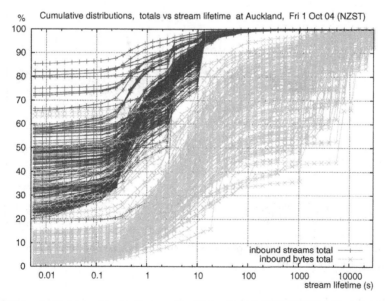

Fig. 4. Stream lifetimes for traffic on the U Auckland (100 Mb/s) gateway at ten-minute intervals for Friday, 1 October 2004

is rather bursty, probably because the total rate is low. During the day web traffic dominates, especially on Friday. In the evenings there are periods of high non-web TCP usage when we update local mirrors for databases outside New Zealand.

Fig. 4 shows the stream density vs lifetime as for fig. 2. Here the stream lifetime and byte densities vary greatly, again reflecting the low traffic levels at Auckland. Stream lifetimes are similar at Auckland and California, with 70% to 95% of the streams again lasting less than 10 seconds. However, at Auckland up to 60% of the bytes are carried in streams lasting only 10 seconds; probably reflecting the high proportion of web traffic at Auckland.

3   Streams and Packets at Auckland

We modified NeTraMet to write the packet rate and number of active streams and flows to a log file every second. Fig. 5 shows the packet rate (lower trace) and number of active streams (upper trace) for each second during Friday 1 and Saturday 2 October 2004.

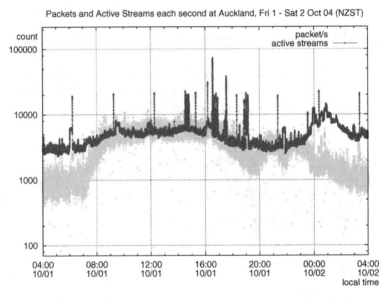

Fig. 5. Packet rate and number of active streams at one-second intervals at Auckland for Friday 1 October 2004

The diurnal variation in stream numbers generally follows the variation in packet rate, i.e. it rises from about 0600 to 0900, falls from about 1700 to 2000, then rises again in the evening. Unlike the packet rate, however, the number of streams rose while the traffic rate fell around midnight on Friday 1 Oct 04. That rise was not repeated over the weekend; it appears to have been a one-off event (e.g. a database replication job copying many tiny files) rather than part of the diurnal pattern.

At regular three-hour intervals we see a short, high step in the number of streams. Our network security team were well aware of this; they are investigating. We believe that such steps are caused by some sort of network attack. Similarly, every day at 1630 we see a bigger spike. We have also observed other, less regular, spikes taking the number of active streams as high as 140,000. Fig. 6 shows more detail for two of these spikes.

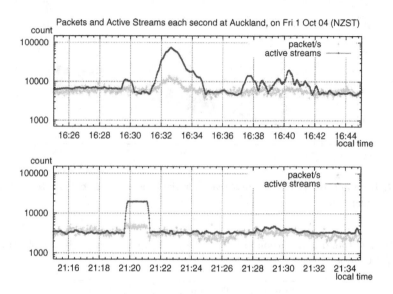

Fig. 6. Details of fig. 5 showing spike in streams at 1633, and step at 2120

4   Usage Metering at Auckland

For usage accounting at Auckland we want to ignore streams with K or fewer packets. To help select a K value, we plotted distributions of byte density vs stream size (packets). Fig. 7 shows distributions for inbound (lower traces) and outbound (upper traces) byte-percentage distributions for ten-minute sample intervals from three hours from 2100 on Friday 1 October 2004. For most of those intervals it seems that we could ignore streams with six or fewer packets in either direction. However, there is one *outbound* trace, for the interval ending at 2120, which has 29% of its bytes in streams with only one or two packet. Fig. 6 shows that at 2120 the number of active streams had risen sharply.

Table 1 shows that the interval ending at 2120 had two unusual features: a high inbound UDP traffic rate, and a low outbound non-web TCP traffic rate. We examined the ten intervals with their highest proportion of bytes in short streams. Few of those had low outbound non-TCP rates, but all had high UDP inbound rates. We hypothesise that the step in streams was caused by an inbound address or port scan, i.e. a flood of single-packet UDP streams.

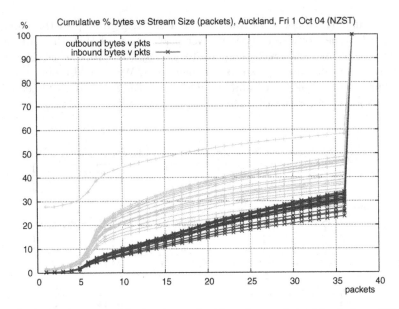

Fig. 7. Byte density vs packets in stream for three hours at Auckland, from 2100 on Friday, 1 Oct 2004

Table 1. Inbound and outbound traffic rates (Mb/s) for various Kinds of traffic on Friday 1 October 2004

Inbound rate	UDP non-web	web	SSL	other
2110	0.15	2.91	8.85	0.51	0.03
2120	*1.66*	2.23	10.15	0.52	0.04
2130	0.21	1.37	9.86	0.50	1.09

Outbound rate	UDP	nonweb	web	SSL	other
2110	0.10	1.47	3.31	0.73	0.03
2120	0.10	*0.92*	3.34	0.850	0.03
2130	0.10	3.71	3.54	0.859	0.07

Although few of those inbound UDP probe packets elicited any response, those that did increased the proportion of bytes in small streams enough to dominate the outbound traffic.

Since U Auckland has about five times as many inbound traffic bytes as it does outbound, we plotted the total (inbound+outbound) byte-percentage distributions for every ten-minute interval over 1-2 October 2004, producing fig. 8. We often see intervals when the short streams contribute a significant proportion of the total link bytes, suggesting that we should not simply "focus on the elephants" for our usage measurements.

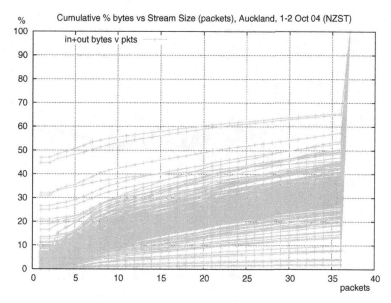

Fig. 8. Total (inbound+outbound) byte density vs packets in stream at Auckland, 1-2 October 2004

5   Ignoring Short Streams at Auckland

Our observations in section 4 suggest that on our link, intervals when traffic is dominated by short streams are caused by network attacks *(plagues of dragon-flies)*. Although we need to know about those for security monitoring, they are less important for usage accounting. We decided to try metering while ignoring streams with six or fewer packets (total in both directions).

We ran the meter with $K = 6$ for five days, using our normal 'usage accounting' ruleset. All five days were similar (including regular three-hourly spikes and a daily spike at 1640); fig. 9 shows the packet rate (lower trace), active streams (middle trace) and flows (upper trace) for every second of Thursday, 7 October 2004. The packet rate and streams traces are similar to those in fig. 5; the number of flows is stable and tracks the packet rate.

Fig. 10 shows the three hours from 2200 in more detail. The number of active flows rises steadily as new flows appear, and falls rapidly when flows are read every ten minutes, allowing the meter to recover flow table space for newly-inactive flows. The 'sawtooth' behaviour, clearly visible in the plot, is thus an artifact of the RTFM architecture. (The important point here is that when we ignore small flows, the average number of active flows remains stable over long periods, minimising the load on the flow data collection system.)

Fig. 10 also shows that when the number of active streams increases sharply (showing spikes and steps), the number of flows is not affected. That supports our hypothesis that such 'attack' increases are caused by bursts of short-lived streams.

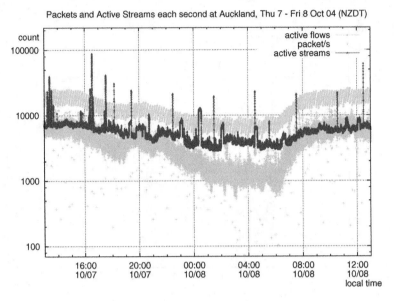

Fig. 9. Packet rate, active streams and active flows at one-second intervals at Auckland for Thursday, 7 October 2004. Our NeTraMet meter used $K = 6$, i.e. streams with six or fewer packets were not matched to flows

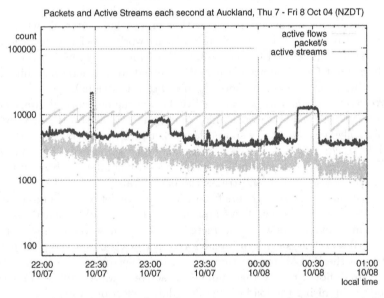

Fig. 10. Detail of fig. 9 showing spike and steps in streams, and sawtooth variation in flows

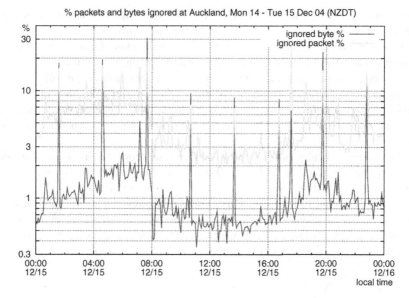

Fig. 11. Percent of bytes and packets ignored at five-minute intervals at Auckland, measured using $K = 6$, i.e. ignoring streams with six or fewer packets

To verify our estimate that ignoring streams with six or fewer packets would exclude between 5% and 10% of the total bytes, we modified NeTraMet so as to collect distributions of the ignored packets and bytes as a function of stream size. One day of typical 'ignored' data, collected at 5-minute sample intervals, is shown in fig. 11.

The 'ignored packet' percentage (upper trace) generally varies between about 2.5% and 8%. Furthermore, its average varies inversely with the average packet rate, suggesting that small *(dragonfly)* streams provide a more or less constant background all the time, with their 'ignored' percentage more obvious when the packet rate is low.

For about 95% of our sample intervals, between 0.5% and 2% of the bytes were ignored (lower trace). In the other 5% of the intervals we saw large spikes in the packet rate and in the number of active streams, as shown earlier in fig. 9. During those spikes, 10% to 30% of the bytes were ignored. Overall, the percentage of bytes ignored is acceptably low, with high 'ignored byte' percentages only occurring during attack events.

6   Conclusion

At Auckland we see frequent bursts of incoming 'attack' streams, which can dominate the traffic mix on our Internet gateway. We believe that traffic engineering and accounting approaches that ignore streams with six or fewer packets ($K = 6$) means that in the long term only about 2% of our total bytes are not measured as 'user' traffic. In return we achieve a significant reduction in the number of flows we have to create, read, store and process.

However, for some traffic mixes, this sampling bias against small flows can radically warp the inferences one makes about the aggregate traffic.

We are continuing our investigation of stream behaviour, especially that relating to the 'attack streams' *(plague of dragonflies)* events. We have not observed these on the California backbone link, where traffic levels are much higher and there is more statistical mixing, but such 'attack streams' probably do appear there.

An alternative approach is the adaptive one proposed in [7], which adapts its sampling parameters to the traffic in real time That approach avoids the bias against small flows and should give a true picture of the actual traffic load, within its sampling limitations. Our approach, however, may be more useful for accounting applications, since we are not sampling. Instead we preserve detail for all the larger streams (which we can bill to a user) while ignoring the small 'attack' streams (which are overhead, not billable to a user).

6.1   Future Work

We are continuing to investigate the *plague of dragonflies* events at Auckland. We would like to improve our network attack detection ability by recognising and reporting frequently-occurring attack patterns. The ability to summarise large groups of small streams would also reduce the number of packets we ignore in our traffic monitoring.

At this stage it is clear that NeTraMet can handle our network's data rate at 100 Mb/s. We are confident that this can be done – without having to use sampling techniques – at 1 Gb/s.

6.2   The Need for Ongoing Measurements

At U Auckland we use NeTraMet for usage accounting and traffic analysis, *Snort*[1] and *Argus*[2] for security monitoring, and *MRTG*[3] for traffic engineering. Each of these tools is specialised so as to perform its intended function well, but there is little overlap between the tools. Indeed, when an unusual event occurs on the network, it can be useful to have data from many tools, providing many different views of that event.

We believe, therefore, that *every large network should collect traffic data on an ongoing basis, using several different tools.* The work required to support such monitoring is well justified by the ability it provides to investigate incidents soon after they occur. In addition, the understanding gained about the network, its traffic, and the ways that traffic changes over time, provides a sound basis for long-term improvements in the network's performance and in service to it's users.

---

[1] Security Monitoring, http://www.snort.org/

[2] Network Auditing, http://www.qosient.com/argus/

[3] Traffic Rate Monitoring, http://people.ee.ethz.ch/ oetiker/webtools/mrtg/

Acknowledgement

Support for this work is provided by DARPA NMS Contract N66001-01-1-8909, NSF Award NCR-9711092 'CAIDA: Cooperative Association for Internet Data Analysis,' and The University of Auckland.

References

1. C. Estan and G. Varghese, *New directions in traffic measurement and accounting: Focusing on the Elephants, Ignoring the Mice,* ACM Transactions on Computer Systems, August 2003
2. N. Brownlee and M. Murray, *Streams, Flows and Torrents,* PAM2001, April 2001
3. N. Brownlee, *Using NeTraMet for Production Traffic Measurement,* Intelligent Management Conference, IM2001, May 2001
4. N. Brownlee and K. Claffy, *Understanding Internet Stream Lifetimes: Dragonflies and Tortoises,* IEEE Communications magazine, October 2002
5. K. Claffy, G. Polyzos and H-W. Braun, *A parameterisable methodology for Internet traffic flow profiling,* IEEE Journal on Selected Areas in Communications, 1995.
6. N. Brownlee, C. Mills and G. Ruth, *Traffic Flow Measurement: Architecture,* RFC 2722, October 1999.
7. C. Estan, K. Keys, D. Moore and G. Varghese, *Building a better NetFlow,* SIGCOMM, September 2004

Spectroscopy of Traceroute Delays

Andre Broido, Young Hyun, and kc claffy

Cooperative Association for Internet Data Analysis,
SDSC, University of California, San Diego
{broido, youngh, kc}@caida.org

Abstract. We analyze delays of traceroute probes, i.e. packets that elicit ICMP
TimeExceeded messages, for a full range of probe sizes up to 9000 bytes as observed on unloaded high-end routers. Our ultimate motivation is to use traceroute
RTTs for Internet mapping of router and PoP (ISP point-of-presence) level nodes,
including potentially gleaning information on equipment models, link technologies, capacities, latencies, and spatial positions. To our knowledge it is the first
study to examine in a reliable testbed setting the detailed statistics of ICMP response generation.

We find that two fundamental assumptions about ICMP often do not hold in
modern routers, namely that ICMP delays are a linear function of packet size and
that ICMP generation rate is equal to the capacity of the inteface on which probes
are received. The primary causes of these violations appear to be optimizations
that suppress size dependence, e.g. buffer carving, and rate-limiting of internal
ICMP packet and bit rates. Our results suggest that the linear model of packet
delay as a funcion of packet size merits revisiting for many situations, especially
for packets over 1500 bytes. Our findings also suggest possibilities of developing
new techniques for bandwidth estimation and router fingerprinting.

1   Introduction

Remote network mapping is usually done via active measurement. Generally a measurement host sends packets that trigger ICMP replies from routers, and the reply information
is integrated into a map. ICMP *time exceeded*, *echo reply* and *port unreachable* responses
are commonly elicited for this purpose.

An ICMP reply carries binary ("host is alive"), discrete ("9 hops away") and temporal
("replied in 15 ms") data. The last of these, per-hop delay (in the form of round trip
time or RTT), is potentially the richest source of information about a router. However,
extracting the useful components from a delay value is difficult, since not only are the
delay summands unavailable but even their statistics and their dependence on other
factors are unknown.

In the common linear model, packet delay is split into three summands, with one
being proportional to packet size. Specifically, the delay, d, is modeled as follows:

$$d = ax + b + \xi \qquad\qquad (*)$$

where a and b are positive real constants, x is the size of the packet or frame, and ξ
is a positive random variable ("residual delay") that can be arbitrarily close to 0. This

C. Dovrolis (Ed.): PAM 2005, LNCS 3431, pp. 278–291, 2005.

representation implies that $d = ax + b$ is a tight lower bound for all observed delays. Most network spectroscopy and bandwidth estimation experts assume that delay is a linear function of packet size, [1] [2] [3].

Our main goal in this study is to test the validity of this linear model, at least with respect to delays seen in ICMP responses (we do not cover forwarding delay in this study). Our underlying motivation is to find ways of using traceroute RTTs to:

- construct router and PoP-level Internet maps [4] [5]
- obtain metric maps with link latencies and capacities
- enable *user-level path diagnosis* [6]
- improve the integrity of variable-size bitrate estimation tools [7]; and
- fingerprint routers.

For example, one approach to identifying a PoP would be to look at traceroute paths that branch between backbone and access routers. Given that the routing to external destinations is common among all routers within a PoP, return paths to the monitor will be the same. One could thus use the topological closeness of forward paths together with the numeric closeness of RTTs to identify interfaces that belong to the same PoP. We recognize that this aggregation technique requires precise knowledge of typical latencies across a PoP, as well as how often and for how long ICMP TimeExceeded generation can be delayed.

A typical traceroute covers 14–20 hops [8], and during a traceroute all but the last hop responds with an ICMP TimeExceeded packet. The last hop responds with an ICMP EchoReply or ICMP PortUnreachable. We will discuss properties of delays obtained from TimeExceeded packets in detail. We hope to report on destination-based (EchoReply, PortUnreachable) ICMP delays in the future.

The rest of the paper is organized as follows. We review previous work in Sec.2. The description of our testbed and experiment design is in Sec.3. In Sec.4 we present our results, and Sec.5 contains discussion and conclusions.

2   Previous Work

Although the need for precise and detailed measurement of packet delays is recognized by the networking community, equipment constraints render it challenging, and the literature on this topic is scant. In particular, few researchers have access to high-precision (sub-microsecond precision) capture cards or to high performance routers representative of those deployed in Tier-1 ISP backbones.

Further, most previous work does not focus on ICMP delays, per se, but rather on separating *forwarding* (that is, router transit) delays from queueing delays [9] or delays caused by network distance [10]. Bovy, *et al.*, estimated the forwarding delay of three office-class routers to be 224 μs per 100-byte packet per hop [10]. A wide variety of work in bandwidth estimation, much of it surveyed in [11] and [12], also assumes that delays are amenable to linear modeling.

Researchers from Sprint's Advanced Technology Laboratory (ATL) did several studies of instrumented operational routers in a setup close to ours [13], [14], [9], and support the claim that queueing delay in a well-provisioned network is small enough to effectively allow VOIP deployment [15].

A *Light Reading* test of Cisco, Juniper and Foundry measured forwarding delays at line rate (100% load) [16].

Govindan and Paxson [17] and Anagnostakis *et al.*[18] also study ICMP generation times, concluding that ICMP-based RTTs do not tend to include excessive (slow path) delays. Timing jitter in the network around routers complicates the attribution of these delays, but their value (0.1–0.3 ms) is comparable to those in [10] and to ours.

The goal of [18] is to infer link latencies and queueing from ICMP timestamp differences at both ends of a link (see also [6])[1]. The authors found routers (5 in 20 studied) with 95th percentiles of ICMP Timestamp delay around 10 ms; 2 had 95th percentiles at 80 ms. Remote link estimation is quite daunting in the face of such high uncertainty. For comparison, more than 99.6% of our TimeExceeded delays up to 9000 bytes are under 1 ms, except a few (0.4%) that are rate-limited by Juniper routers to incur approximately 10 ms delays.

Donnelly [20] and Mochalski *et al.*[21] demonstrate a piecewise linear size dependence for router/switch transit times, which shows a noticeable rate change at 512 bytes. This phenomenon is similar to our ICMP delay rate discontinuities occurring around 1500 bytes.

To the best of our knowledge, precision timestamping matching modern router speeds is available only with Dag cards from the Waikato group [22] and Endace [23]. The latest models (4.xx) can reach sub-microsecond accuracy when synchronized to GPS or CDMA [20] [24].

Some of the available studies use the now older model (3.xx) of Dag cards, with 5–6 μs precision [13] and 53-byte uncertainty with respect to the portion of the packet that is timestamped. Despite these limitations, the results obtained in [9], [13], and [14] have served as inspiration for this work.

3   Data Collection

We collected our measurements in CAIDA's high-speed testbed [25] [12] which includes (Fig.1): two IBM eServers (running FreeBSD 4.8); a Dell Gigabit Ethernet switch; Juniper, Cisco and Foundry routers; an OC48 link between the Juniper and Cisco; and Gigabit Ethernet links between all other devices. The testbed's path MTU is 9000 bytes. We tap both links at the Cisco router (OC48 and gigE) using NetOptics splitters, and capture packets with Dag cards. The Foundry router doubles as a 16-port switch that connects all equipment in the lab to the Internet and to CAIDA's production network via 100 M Ethernet.

We perform traceroutes on herald or post, and use CoralReef [26] utilities to capture, process, and extract delays from packets. A command line on herald of:

```
traceroute -q 4 -M 2 -m 3 -w 2 -P udp -t 64 post 214
```

---

[1] [19] suggests using traceroute delays for both purposes.

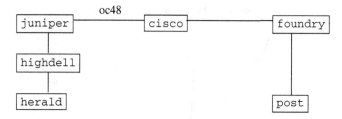

Fig. 1. Lab diagram. Equipment (clockwise): IBM eServer `herald`, Dell PowerConnect 5212 switch, Juniper M20 router, Cisco 12008 router, Foundry BigIron 8000 router/switch, IBM eServer `post`, Links: oc48 (Juniper to Cisco); GigabitEthernet (all other links). For details, see [12] (this volume)

specifies series of 4 probes (q) to hops 2 (M) through 3 (m), using a timeout of 2 sec (w), UDP[2] (P), TOS of 64 (t) and packet size 214 bytes. Its output looks like (numbers from real data):

```
2 cisco-oc48   0.221 ms  0.154 ms  0.254 ms  0.168 ms
3 foundry      0.217 ms  0.226 ms  0.230 ms  0.227 ms
```

Our experiments combine UDP and ICMP traceroutes with 9 TOS values (0, 1, 2, 4, 8, 16, 32, 64, 128), and sizes 64-9000 bytes, for a total of 160866 (2*9*8937) traceroutes, each probing 2 hops with 4 packets at each hop. The router configuration guarantees that the return path for an ICMP packet is symmetric with the forward path.

Traceroute dynamics determine the intervals between probes in our experiments (Fig.2). We call a time lag between two successive packets targeting the same interface an *interprobe gap* (IPG). When traceroute probes one hop, it sends the next packet immediately after receiving an ICMP TimeExceeded for the previous packet. These probes succeed each other within a few hundred microseconds (under 1 ms). The next traceroute command will probe the same hop after an OS scheduling quantum (10 ms) and after probing a subsequent hop (several milliseconds); in that case, the probes are separated by 10-20 ms. When a TimeExceeded is not generated or is lost before the source host receives it (the loss is in fact very rare in our experiments) the traceroute script waits for a 2-second timeout. This gap can affect the delay of the packet that follows, e.g. through route cache latency if the address has been flushed from the cache.

Parameter Scan. We walk the experiment design space (N_S packet sizes, N_P protocols, N_D destinations, N_T TOSes, etc.) using a pseudo-random scan. Scanning of other parameters (hop number, packets/hop) is a part of typical traceroute operation. We take the product of dimensions $m = N_S N_P N_D N_T \ldots$ and find a prime $p > m$. Then we find a primitive root $r \bmod p$ near $\sqrt{p}$, and try all combinations of parameter values as follows. For experiment k, $1 \leq k \leq m$, we use $a_k = r^k \bmod p$ in mixed-radix notation to get indices S (index for size), P (index for protocol), D (index for destination):

---

[2] Recall that traceroute sends UDP or ICMP packets, but always gets back ICMP. Our data contains half UDP and half ICMP probes. The analysis presented here does not distinguish between UDP and ICMP probes, or between TOS values.

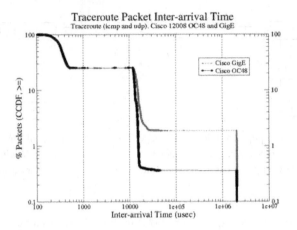

Fig. 2. Clustering of interprobe gaps for the Cisco router (OC48 and gigE): microsecond range, 10–20 ms, 2 sec. The higher fraction of 2-sec gaps on the Cisco gigE (upper curve) is caused by the Juniper not generating some ICMP messages

$$S = a_k \bmod N_S, \ P = [a_k/N_S]\bmod N_P, \ D = [a_k/(N_S N_P)]\bmod N_D, \text{ etc. } (a_k \le m)$$

Example. For two packet sizes ($N_S = 2$) and two protocols ($N_P = 2$), $m = N_S N_P = 4$ and $p = 5$; $r = 3$ is a possible choice of a primitive root. Combinations of packet size (e.g. (40, 1500) indexed by (0,1)) and protocol ((UDP, ICMP) indexed by (0,1)) follow each other in sequence[3] $(3^1, 3^2, 3^3, 3^4)\bmod 5 = (3, 4, 2, 1) = (11, 00, 10, 01)_2$, where 11 corresponds to (ICMP, 1500), and so on.

This approach, inspired by turbo codes [27] and Monte-Carlo integration techniques, is robust against outages, whether at the beginning (Dag cards warming up) or at the end (too small capture interval, disk space). All parameter values appear close to the start of experiment (as opposed to with a lexicographic scan), which allows us to debug problems with each dimension or value, e.g. too high chance of a timeout.

Table 1 presents a description of the data in terms of destinations, experiment duration, number of traceroutes and number of probes (packets). The second half of the table is a breakdown of the probes by interprobe gap (IPG). The longer duration of the second (PCJ) experiment is due to a higher level of ICMP non-generation on Juniper (12140 or 2% of all probes) which results in more occurrences of the 2-sec timeouts. This extra 10K (12140-2310) of timeouts increases the experiment duration by about 5.5 hours. In addition, Juniper's generation bitrate of TimeExceeded (at 8 ns/bit) is the slowest of all three routers (Table 2). ICMP bitrate limiting causes many packets in the 7000–9000 byte range (73K or 11%) to arrive more than 1 ms later than the previous probe. This lag applies to packets 2–4. Packet 1 is always delayed by an OS scheduling quantum of 10 ms, which explains the large number of packets (about 25% of the total) in the 10-100

---

[3] In this special case, one can read parameters from the two rightmost bits of $r^k \bmod p$.

Table 1. Experimental data and intereprobe gaps

Code	Source	Destination	Date	Start	End	Traceroutes	Packets sent
HCF	herald	Cisco, Foundry	2004-09-10	00:00	02:00	160866	1287 K
PCJ	post	Cisco, Juniper	2004-09-12	00:30	08:00	160866	1287 K

Code	Source	Dest.	i/face	IPG<1ms	1–10ms	10–100ms	0.1–1s	IPG>1s	Total
HCF	herald	Cisco	OC48	482546	20	158587	0	2310	643463
HCF	herald	Foundry	gigE	477557	539	160747	0	2310	641153
PCJ	post	Cisco	gigE	482570	19	148733	1	12140	643463
PCJ	post	Juniper	OC48	389211	72793	157178	1	12140	631323

ms bin. The drop rate (non-generation) for the Foundry is under 0.4%, and the Cisco returns all 643464 probes, i.e. has 0% drop rate.

4   Results

Table 2 provides a lower bound for size dependence parameters from equation $d = ax + b$: a (slope) and b (intercept) of TimeExceeded delay. We apply the $O(N)$ linear programming (LP) algorithm of [28] to delays observed at the Cisco and Juniper OC48 interfaces for all packet sizes, and to those at the Cisco and Foundry gigE interfaces separately for ranges ≤ 1500 and > 1500. This latter choice is based on the fact that these gigE interfaces have different ICMP generation rate for packet sizes under 1500 and over 1500 bytes.[4]

Each linear fit has a slope and an intercept. The slope is in ns/bit (not μs/byte), to match the gigE rate, 1 ns/bit. The intercept at 0 and the values of $ax + b$ at three packet sizes ($x =$40, 1500, and 9000 bytes) are the minimum delays including deserialization (but not serialization)[5] and ICMP generation.

The only router/probe type with ICMP generation rate equal to link rate is the Foundry TimeExceeded at over 1500 bytes; others have smaller or larger slopes. Note that small slopes a for packets under 1500 bytes can trick variable packet size (VPS) tools [7] into capacity overestimation, whereas slower-than-link rates (higher values of a) can result in underestimation. This situation is similar to the underestimation caused by extra serialization delays at Layer 2 switches that are invisible to traceroute [29].

Delays through the Juniper router are special in several respects (Fig.3). The minimum delay of the TimeExceeded packets grows stepwise by approximately 4.033 μs per 64-byte cell for sizes 64–320 bytes: $d = 4.033\lceil x/64 + 31\rceil\mu$s where $\lceil x\rceil$ is the smallest integer greater or equal to x. This formula is similar to that for ATM delays from [30], although the fixed cost (which for 64-byte packets is 128 μs, an equivalent of almost 40 KB at the OC-48 wire speed) is much higher than ATM's encapsulation cost. This

---

[4] We did not pinpoint a precise byte value for this boundary; it may be router-dependent. Also, the Cisco OC48 interface produces slightly different linear fits for packet size ≤ 1500 and > 1500, but since their relative difference is under 0.1%, we provide and discuss only the single fit of Cisco OC48 delays across the full range of packet sizes.

[5] Recall that Dag cards timestamp at the beginning of a packet [20].

Table 2. Linear fit of lower bound on TimeExceeded delay

Router	Slope (ns/bit)	Lower bound (μs)			
		0	40B	1500B	9000B
Cisco OC48, all	0.732	18.41	18.64	27.19	71.10
Juniper OC48, all	8.091	122.63	125.22	219.72	705.18
Cisco GE $\leq$ 1500	0.320	18.88	18.98	22.72	(41.93)
Cisco GE $>$ 1500	1.313	(6.66)	(7.08)	22.42	101.19
Foundry $\leq$ 1500	-0.075	29.87	29.84	28.97	(24.48)
Foundry $>$ 1500	0.996	(15.90)	(16.22)	27.85	87.59

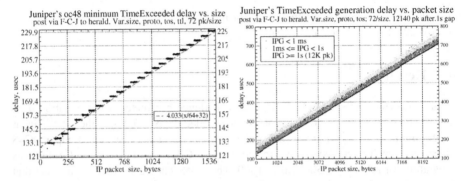

Fig. 3. (a) Minimum TimeExceeded delay from Juniper (left) with a staircase of 64-byte segments, 4 μs steps and an 8-μs jump at 320 bytes; (b) TimeExceeded delay from Juniper (right) showing about 30 μs of extra delay for an interprobe gap (IPG) of 2 sec. Three bands of delays result from the three ranges of interprobe gaps: light-colored band for IPG < 1ms, medium dark band for IPG between 1ms and 1s, and dark band for IPG $\geq$ 1s

cell rate would result in an average bitrate of 7.877 ns/bit, or 127 Mbps. However, the experimental curve jumps by an extra cell's worth of delay right after 320, 3712 and 7104 bytes (which are separated by 3392 bytes, or 53 64-byte cells). As a consequence, the slope in the linear programming-based lower bound is somewhat higher. That is, 54 cells worth of delay per 53 cells of size equals 8.026 ns/bit, but the LP estimate from Table 2 is 8.091 ns/bit, which may imply a 0.8% error in 4.033 μs cell time. Fig.3a shows a close-up for packets under 1600 bytes. The staircase of minimum delays starts under the line $y = 4.033x/64 + 32$ (recall that $\lceil x \rceil < x + 1$) but crosses over the line between 320 and 321 bytes. The 8-μs jump at 320 bytes and the accumulated discrepancy with the straight line (12 μs over 9000 bytes interval) can potentially be measured by traceroute-like tools, even though individual 4μs-steps may be hard to discern from network noise.

The Juniper router also delayed widely spaced (interprobe gap of 2s) packets by about 30 μs compared with closely spaced packets of the same size. This delay could be due to route cache flushing. The delay pattern in Fig.4 (bottom) which holds for 2400 (0.4%) probes rate-limited to 10 ms (packets 2–4 in some traceroutes) has a prominent negative trend that could potentially be used for fingerprinting.

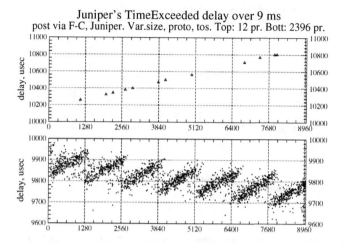

Fig. 4. TimeExceeded message delay from Juniper OC48. Values over 10 ms (top) and 9–10 ms (bottom). Values between 9–10 ms reveal unusual size dependence of ICMP TimeExceeded generation delay when ICMP is rate-limited to 100 packet per second (one packet in 10 ms)

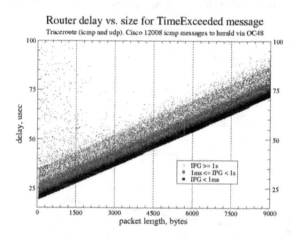

Fig. 5. TimeExceeded message delay from Cisco OC48. Compare with Fig.6 where each cluster of interprobe gaps is in its own panel

Figs. 5 and 6 show the dependence of ICMP delays on packet size for the Cisco OC48 interface, separated into three sets by interprobe gap (time between traceroute packets): under 1 ms, 1ms–1s, over 1 s. The actual distibution of the longer lulls clusters around 10 ms (kernel scheduling quantum) and 2 sec (traceroute timeout), both described in Sec.3 (Fig.2). Probes delayed by 10–20 ms span a wider range of ξ (reflected in the width of

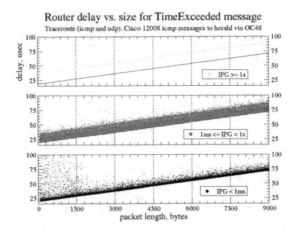

Fig. 6. TimeExceeded message delay from Cisco OC48. Panels from top to bottom: delay for interprobe gap of over 1 sec, 1 ms–1 sec and under 1 ms. The position of the curve in the top panel reflects about 20 μs of extra delay (presumably route cache warm-up) beyond the lower bound of all delays, which is indicated by the solid line. The bottom panel shows some additional scattering of delays (possibly from rate limiting) for closely-spaced packets under 3000 bytes

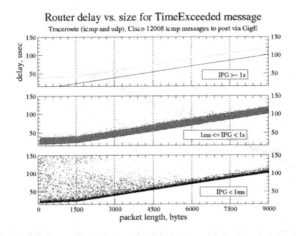

Fig. 7. TimeExceeded message delay from Cisco GigE. Panels from top to bottom: delay for interprobe gap of over 1 sec, 1 ms–1 sec, and under 1 ms. There is a rate change near 1500 bytes. However, unlike the Foundry data, the slope under 1500 bytes is positive, albeit smaller than the slope over 1500 bytes. The position of the curve in the top panel reflects about 20 μs of extra delay (presumably route cache warm-up) beyond the lower bound of all delays, which is indicated by the solid line. The bottom panel shows some additional scattering of delays (possibly from rate limiting) for closely-spaced packets under 4500 bytes

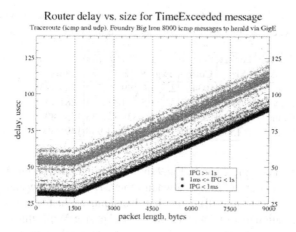

Fig. 8. TimeExceeded message delay from Foundry. The delay slope for packets under 1500 bytes is a small negative number (see Tab. 2)

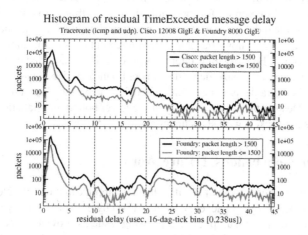

Fig. 9. Histogram of residual TimeExceeded delay. Positions of the maxima are similar for packets under and over 1500 bytes, suggesting that residual delay ξ is not strongly dependent on size

the middle strip in Fig. 6) than probes sent immediately after the previous probe, but at the same time many of them are close to the linear lower bound (dashed line at the bottom of the strip). On the other hand, probes sent after the 2-second timeout always encounter an extra delay of about 20 μs. The banding of ξ here and in Figs. 7, 8, and 9 may be due to route cache flushing and other state lost after certain time intervals. We observed similar dependence on the duration of the lull between the previous packet and the current probe for Juniper (Fig.3b) and Foundry routers (Fig.8). To give an idea of the

Table 3. Statistics of residual delay ξ for ICMP TimeExceeded (generation time in excess of delays attributed to packet size) on Cisco and Foundry's gigE line cards

Router	Packet Size	Packet Count	Delay (μs)			
			avg	95%	99%	max
Cisco	$\leq$ 1500	103463	2.598	6.199	20.504	296.593
Cisco	$>$ 1500	540000	2.252	5.484	18.835	281.096
Foundry	$\leq$ 1500	103073	4.364	3.338	31.233	1537.800
Foundry	$>$ 1500	538087	4.996	3.815	31.948	1492.500

average density of points in these bands, Fig.9 shows a histogram of residual delay ξ, i.e. the delay less the lower bound of delay shown in Table 2 for sizes below and above 1500 bytes (partial Radon transform [30]). Note that this summary histogram suggests (but does not prove) the stationarity of ξ with respect to packet size. While this stationarity is typically assumed, our preliminary results show that it at best only approximately holds.

A common assumption in network research is that an idle router processes packets with minimum possible delay [31]. Our experimental setup guarantees that no cross-traffic is present and that routers process probes one at a time. Table 3 presents statistics (average, 95%, 99% and maximum for the whole datasets without groupings by inter-probe gap) for the residual delay ξ, i.e. ICMP generation time in excess of linear lower bound $ax + b$ (where Table 2 shows the slope a and intercept b). We can summarize the data as follows: Cisco and Foundry gigE interfaces process TimeExceeded with no more that 6 μs of extra delay (over the size-dependent lower bound) in 95% of cases; however, for 1% of packets the extra delay is between 20 and 300 μs on the Cisco and 30–1500 μs on the Foundry. Despite piecewise linearity of the lower bound, the statistics of ξ are close to each other for packets with sizes under and over 1500 bytes.

5   Discussion, Conclusions, Future Work

We demonstrated that a linear model of ICMP delay is an approximation (like Newtonian mechanics) that breaks down for packet sizes over 1500 bytes. That so many measurement techniques rely on this assumption is a pressing issue as providers like Abilene, Geant, Switch are already supporting 9000-byte transparent paths, and the global Internet transition toward these larger packet sizes is only a matter of time. With a 1500-byte ICMP delay rate discontinuity at all three routers[6], and with packet forwarding (as opposed to ICMP message generation delay) having a similar break in linearity at 1500 bytes for at least one router (as our work in progress shows), we can safely say that there is commonly disparate treatment of packets under versus over 1500 bytes. Designers of bandwidth estimation and other measurement tools [12] must be aware of this reality.

We find that for all packet sizes (under and over 1500 bytes):

– delays above the minimum are not necessarily due to queueing. For example we observed that Juniper delays some closely spaced traceroute packets by 9–10 ms

---

[6] For Juniper, the delay *rate* discontinuity appears for EchoReply and PortUnreachable but not for TimeExceeded, for which there is no global rate change at 1500 bytes.

(Fig.4). However, our measurements of Cisco and Foundry's gigE interfaces (Table 3) show that for most (95%) probes the extra delay is within a few microseconds, and it is within 300 μs for Cisco (and 1.5 ms for Foundry) over the whole sample, which is negligible for many applications.

- buffer carving [32] can lead to "non-physical" size-delay dependence which can appear faster than link rate or decrease with packet size. Such buffering can also make loss rates size-independent [33].

The negative slope for the Foundry data in Fig. 8 could possibly be caused by the router zeroing out the rest of a 1500 byte buffer after a smaller packet arrives, when this operation is slower than the link's line rate. We emphasize that this is only speculation, and have not investigated this issue further.

Surprisingly, we found that the ICMP rate can differ by two orders of magnitude up or down from the link rate, depending on router and ICMP type. This ambiguity suggests that capacity estimates by ICMP-based tools [7] [34] [35] may need to make heavy use of router and even interface fingerprinting, rather than just filtering and fitting as if 'all RTT data are created equal'.

We found that Juniper's TimeExceeded processing is based on 64-byte cells (Fig.3a). We plan to investigate whether the 48-byte cell[7] granularity of the Cisco documented in [32] is present in our data.

Our analysis shows that ICMP delay can depend on packet size and header fields in various non-intuitive ways, including:

- different growth rates under and over 1500 bytes (piecewise linearity, Fig.7,8)
- jumps or drops (discontinuity, Fig.3)
- stepwise growth, e.g. each 64 bytes (Fig.3)
- negative (decreasing) slope with respect to packet size (Fig. 4, 8, Table 2)
- internal tasks can postpone packet scheduling by fixed delays (clustering in distinct "bands") on an absolutely empty device (Fig. 8, 9)
- warming up caches can cause significant (20-30 μs) extra latency for widely spaced probes, e.g., an interprobe gap of seconds (Fig. 6, 5, 7, 8)

Table 4 summarizes our main results and lists three cases of linearity of message generation delay with respect to packet size (fully linear, linear with a break, stepwise linear with jumps) observed for the three router types studied. In contrast with prevalent assumptions used by some rate estimation tools, none of our studied routers has a TimeExceeded generation rate equal to the line rate of the inbound link for packets under 1500 bytes. One router has an ICMP rate that is 20 times slower than its line rate (the ratio of generation rate to line rate is 0.05, Table 4). Other routers use optimizations that create an illusion of a faster ICMP rate at the expense of increasing minimal ICMP delay. These properties can facilitate remote device/link fingerprinting. Taken together, our results indicate surprisingly different attitudes of router vendors (from restrictive to receptive to acceptive) with regard to ICMP Time Exceeded messages. Our work in progress suggests that many of these attitudes apply to other ICMP messages too.

---

[7] "The Fabric Interface ASIC is set up to segment the packet into 48-byte cells." [32].

Areas for further investigation include confirming details on the phenomena mentioned above, as well as forwarding delays, payload dependent delays, cross-traffic effects, rate estimates based on optimization technique of [30], and independence tests.

Table 4. Observed behavior of routers responding with ICMP TimeExceeded messages

Property	Juniper OC48	Cisco OC48	Cisco GigE	Foundry GigE
Message generation linearity	steps w.jumps	linear	piecewise	piecewise
Min.latency, all packets $\geq$ 64B	128 μs	19.4 μs	19.4 μs	29.2 μs
Generation rate/Line rate, $\leq$ 1500B	0.05	1.37	3.1	negative
ICMP non-generation rate	2%	0%	0%	0.4%

Acknowledgements

Dan Andersen, Brendan White, Grant Duvall, Margaret Murray, and Kevin Walsh created the lab used in this study. Ken Keys helped with the Coral software. Yoshi Kohno provided the linear programming code. Thanks to the PAM reviewers for their comments, all of which we incorporated in the text. Thanks also to Allen Porter, Dave Berlin, Niheer Patel, Xin Cai, Andrey Shapiro, and Tin Tran for their useful feedback on this report.

References

1. Pasztor, A., Veitch, D.: The packet size dependence of packet pair like methods. In: IWQoS. (2002)
2. Jain, M., Dovrolis, C.: End-to-end available bandwidth: measurement methodology, dynamics, and relation with TCP throughput. In: Sigcomm. (2002)
3. Katti, S., Katabi, D., Blake, C., Kohler, E., Strauss, J.: Multiq: Automated detection of multiple bottlenecks along a path. In: IMC. (2004)
4. Spring, N., Mahajan, R., Wetherall, D.: Measuring ISP topologies with Rocketfuel. In: Sigcomm. (2002)
5. Spring, N., Wetherall, D., Anderson, T.: Reverse engineering the Internet. In: HotNets. (2003)
6. Mahajan, R., Spring, N., Wetherall, D., Anderson, T.: User-level Internet path diagnosis. In: SOSP. (2003)
7. Jacobson, V.: pathchar - a tool to infer characteristics of Internet paths (1997) ftp.ee.lbl.gov/pathchar.
8. Broido, A., claffy, k.: Internet Topology: connectivity of IP graphs. In: SPIE, vol.4526. (2001)
9. Hohn, N., Veitch, D., Papagiannaki, K., Diot, C.: Bridging router performance and queueing theory. In: Sigmetrics. (2004)
10. Bovy, C.J., Mertodimedjo, H.T., Hooghiemstra, G., Uijtervaal, H., van Mieghem, P.: Analysis of end-to-end delay measurements in Internet. In: PAM. (2002)
11. Prasad, R.S., Murray, M., Dovrolis, C., Claffy, K.: Bandwidth estimation: metrics, measurements, techniques and tools. In: IEEE Network. (2004)
12. Sriram, A., Murray, M., Hyun, Y., Brownlee, N., Broido, A., Fomenkov, M., Claffy, k.: Comparison of public end-to-end bandwidth estimation tools on high-speed links. In: PAM. (2005)
13. Papagiannaki, K., Moon, S., Fraleigh, C., Thiran, P., Tobagi, F., Diot, C.: Analysis of measured single-hop delay from an operational network. In: Infocom. (2002)

14. Choi, B.Y., Moon, S., Zhang, Z.L., Papagiannaki, K., Diot, C.: Analysis of point-to-point packet delay in an operational network. In: Infocom. (2004)

15. Fraleigh, C., Tobagi, F., Diot, C.: Provisioning IP backbone networks to support latency sensitive traffic. In: Infocom. (2003)

16. Newman, D., Chagnot, G., Perser, J.: The internet core routing test: Complete results (2001) www.lightreading.com/document.asp?doc_id=6411.

17. Govindan, R., Paxson, V.: Estimating router ICMP generation delays. In: PAM. (2002)

18. Anagnostakis, K., Greenwald, M., Ryger, R.: cing: Measuring network-internal delays. In: Infocom. (2003)

19. Akela, A., Seshan, S., Shaikh, A.: An empirical evaluation of wide-area internet bottlenecks. In: IMC. (2003)

20. Donnelly, S.: High precision timing in passive measurements of data networks (2002) Ph.D. thesis, University of Waikato, Hamilton, New Zealand.

21. Mochalski, K., Micheel, J., Donnelly, S.: Packet delay and loss at the Auckland Internet access path. In: PAM. (2002)

22. Graham, I.D., Pearson, M., Martens, J., Donnelly, S.: Dag - A cell capture board for ATM measurement systems (1997) www.cs.waikato.ac.nz /Pub/Html/ATMDag/dag.html.

23. Endace: Measurement Systems (2004) www.endace.com.

24. Micheel, J., Donnelly, S., Graham, I.: Precision timestamping of network packets. In: IMW. (2001)

25. CAIDA: Bandwidth estimation project (2004) www.caida.org/projects/bwest.

26. Keys, K., Moore, D., Koga, R., Lagache, E., Tesch, M., claffy, k.: The architecture of Coral-Reef: Internet Traffic monitoring software suite. In: PAM. (2001)

27. Berrou, C., Glavieux, A., Thitimajshima, P.: Near Shannon limit error-correcting coding and encoding: turbo codes. In: IEEE Int'l Conference on Conmmunications. (1993)

28. Moon, S.B., Skelly, P., Towsley, D.: Estimation and removal of clock skew from network delay measurements (1998) Tech.Rep.98-43, UMass Amherst (Infocom 1999).

29. Prasad, R., Dovrolis, C., Mah, B.: The effect of store-and-forward devices on per-hop capacity estimation. In: Infocom. (2003)

30. Broido, A., King, R., Nemeth, E., Claffy, k.: Radon spectroscopy of packet delay. In: ITC 18. (2003)

31. Ribeiro, V., R.Riedi, R.Baraniuk, J.Navratil, L.Cottrell: pathChirp: Efficient Available Bandwidth Estimation for Network Paths. In: PAM. (2003)

32. Cisco: How to read the output of the show controller frfab / tofab queue commands on a Cisco 12000 Series Internet Router. Document ID 18002 (2004) www.cisco.com.

33. Barford, P., Sommers, J.: Comparing probe and router-based packet loss measurement. In: Internet Computing. (2004)

34. Mah, B.:   pchar: a tool for measuring Internet path characteristics (1999) www.kitchenlab.org/www/bmah/Software/pchar.

35. Downey, A.B.: Using pathchar to estimate Internet link characteristics. In: Sigcomm. (1999)

Measuring Bandwidth Between PlanetLab Nodes

Sung-Ju Lee[1], Puneet Sharma[1], Sujata Banerjee[1], Sujoy Basu[1], and Rodrigo Fonseca[2]

[1] Hewlett-Packard Laboratories, Palo Alto, CA
{sjlee, puneet, sujata, basus}@hpl.hp.com
[2] University of California, Berkeley, CA
rfonseca@cs.berkeley.edu

Abstract. With the lack of end-to-end QoS guarantees on existing networks, applications that require certain performance levels resort to periodic measurements of network paths. Typical metrics of interest are latency, bandwidth and loss rates. While the latency metric has been the focus of many research studies, the bandwidth metric has received comparatively little attention. In this paper, we report our bandwidth measurements between PlanetLab nodes and analyze various trends and insights from the data. For this work, we assessed the capabilities of several existing bandwidth measurement tools and describe the difficulties in choosing suitable tools as well as using them on PlanetLab.

1   Introduction

The lack of end-to-end QoS and multicast support in the underlying best effort networking infrastructure has spurred a trend towards providing application level intermediaries such as web-caches and service replicas to mitigate the performance issues. It is not only important to provide intermediary services but also to connect the end-clients to the intermediary that can meet the client QoS requirements and provide the best performance. For instance, web applications might want to select the nearest content cache while the online multiplayer game players might want to choose the least loaded game server. There also have been attempts to build overlays by connecting application level intermediaries for composable and personalized web and media services. Normally, such services have QoS requirements such as bandwidth and delay. Hence, building and maintaining such overlays requires periodic or on-demand measurement of end-to-end paths. Motivated by these trends, there have been significant research studies on active and passive network measurement techniques, and measurement studies from many large scale networks [1]. Clearly, periodic or on-demand measurement of all possible network paths will incur a high overhead and is inefficient. Thus a key concern is the development of scalable measurement and inference techniques which require minimum probing and yet provide the required measurement accuracy.

The primary network metrics of interest are end-to-end latency, bandwidth and loss rates while application level metrics are HTTP response times, media streaming rates, and so on. Many studies have focused on scalable network distance estimation, mainly using the triangular inequality heuristic [17]. However, similar triangular inequality does not apply to bandwidth, and hence it is much more difficult to identify the nodes that provide the maximal bandwidth without probing to each node.

C. Dovrolis (Ed.): PAM 2005, LNCS 3431, pp. 292–305, 2005.

Although there is a plethora of bandwidth measurement techniques, there have been only a few large-scale bandwidth or bottleneck capacity measurement studies. In fact, most of these studies are in conjunction with the validation of a new bandwidth measurement tool. New bandwidth measurement tools continue to be developed, aiming for better accuracy with faster measurements. In this paper, we present results from a large scale bandwidth measurement study on the PlanetLab infrastructure. Our goals are (i) to understand the bandwidth characteristics of network paths connecting PlanetLab nodes and (ii) to ultimately obtain insights into potential trends that will enable scalable bandwidth estimation. We primarily focus on the first goal in this paper. We do not develop a new bandwidth estimation tool nor evaluate and compare the accuracy of various tools. Rather, we assess the capabilities of a number of available tools from a PlanetLab deployment standpoint and report our findings in the hope that it will help other researchers to make an informed choice of a tool.

In the next section, we describe our methodology and the tools we assessed for this study, followed by an analysis of the data we collected. Section 3 concludes the paper.

2   Measurement Study

PlanetLab is an attractive platform for bandwidth measurement as it is an open, globally distributed network service platform with hundreds of nodes spanning over 25 countries. PlanetLab has gained the status of the de facto standard for conducting large scale Internet experiments. Although the interdomain connectivity of the PlanetLab hosts may not represent the global Internet [2], the characterization of PlanetLab topology is still of utmost importance for designing experiments on PlanetLab and drawing informed conclusions from the results. Several measurement studies have been conducted on PlanetLab topology, mostly focusing on the connectivity and the inter-node latency. In this paper, we study the bottleneck capacity between the PlanetLab nodes.

2.1   Methodology

Our methodology consisted of deploying the bandwidth[1] measurement tool on a selected set of responsive nodes on PlanetLab using standard PlanetLab tools and then executing a script to run the measurements. The collected data is then shipped back to a central node on our site for analysis.

We performed two sets of measurements at two different time periods. The first set (referred to as *Set 1* in rest of the paper) was measured and collected starting in August of 2004, and the second (referred to as *Set 2*) in January of 2005. Between the two measurements periods, PlanetLab went through a major version change. The second set of experiments were performed after the version 3 rollout on PlanetLab.

Although there are over 500 deployed nodes on PlanetLab, only a little over half the nodes consistently responded when we started the measurement process. A crucial first step was to select a tool; we describe the selection process below. Conducting pair-wise latency measurements for a few hundred nodes is a relatively quick process for which

---

[1] We use the terms bandwidth and capacity inter-changeably throughout the paper.

measurements can be run in parallel and finishes in the order of minutes. However, pair-wise capacity measurements for a few hundred nodes needs to be well coordinated because the capacity measurement tools often do not give accurate results when cross traffic is detected. Thus the measurement process for all pairs can take much longer and is of the order of days to even weeks.

There are a large number of bandwidth measurement/estimation tools available, with several new tools recently introduced. This in itself is an indication that accurate bandwidth measurement/estimation remains a hard problem even after many years of research and there is room for further improvements [9]. For the details of the various tools and their measurement accuracy, please use our bibliography or available survey articles [15,18]. For the purposes of our study, our goal was to find a reasonably accurate but low overhead tool that is easily deployable on the PlanetLab platform. Note that the purpose of this study is *not* to do an accuracy comparison of these tools. After some narrowing of the choices, we evaluated the following tools as described below. We merely present our experiences with different tools in the evaluation process.

We were hesitant to use per-hop capacity estimation tools as they generate excessive probing traffic overhead. Moreover, we could not build pathchar [10] or pchar [14] as they can not be built on newer Linux systems. Currently, PlanetLab runs kernel version 2.4.22, but pathchar supports up to 2.0.30 and pchar up to 2.3. When we tested Clink [5] on PlanetLab, the experiment were simply "hung" without making any progress. We suspect this is also because of a Linux version compatibility issue.

As for end-to-end capacity estimation tools, bprobe [3] works only on SGI Irix. SProbe [21, 22] is an attractive tool as it only requires to be run on the source machine, and hence can measure capacities to hosts where the user does not have account access. In addition, SProbe is included in the Scriptroute [23] tool that runs as a service on PlanetLab hosts. One key feature of SProbe is that when it detects cross traffic, instead of making a poor estimate of the capacity, it does not report any value. When we ran SProbe between PlanetLab hosts, less than 30% of the measurements returned the capacity estimate. The authors of the tool had a similar experience on their trials with Napster and Gnutella peers. As we have access to all the PlanetLab hosts, we can deploy and run pathrate [4]. Unless the network hosts are down or we could not login to the hosts for various reasons, we were able to measure capacity between PlanetLab nodes using pathrate. It was the only capacity estimation tool we could successfully run and obtain estimates on PlanetLab.

We also tested several available bandwidth estimation tools. Similar to bprobe, cprobe [3] does not run on Linux. One of the most popular tools is pathload [8]. When we tested pathload on PlanetLab nodes however, we ran into an invalid argument error on connect. This very issue was also recently brought up in PlanetLab user mailing list. We were able to run IGI (Initial Gap Increasing) [7] without any run-time errors. However, the tool showed poor accuracy with high variance in the estimation of the same pair on sequential attempts, and also reported unusually high estimates (ten times larger than the estimated capacity by pathrate). Spruce [24] has shown to be more accurate than pathload and IGI. However, Spruce requires the knowledge of the capacity of the path to estimate available bandwidth. We also tested pathChirp [19] and it ran successfully with reasonable accuracy in our first set of measurements performed in August 2004. However, after the version 3 rollout of PlanetLab, pathChirp, along with STAB [20],

Table 1. End-to-end capacity statistics

	Set 1	Set 2
Number of nodes	279	178
Measurement period	8/11/04~9/6/04	1/5/05~1/18/05
PlanetLab version	version 2	version 3
Number of pairs	12,006	21,861
Minimum capacity	0.1 Mbps	0.3 Mbps
Maximum capacity	1210.1 Mbps	682.9 Mbps
Average capacity	63.44 Mbps	64.03 Mbps
Median capacity	24.5 Mbps	91.4 Mbps
Standard deviation	119.22 Mbps	43.78 Mbps

developed by the same authors of pathChirp, failed to work on PlanetLab. After a few chirps, the tool stops running and hangs. We are communicating with the authors of pathChirp to resolve the issue.

In our future work, we are planning to test tools such as ABwE [16], CapProbe [11], pathneck [6], and MultiQ [12].

2.2   Measurement Analysis

For the first set of measurements, we show the capacity measurements from pathrate (version 2.4.0) as it was the only capacity estimation tool we were able to successfully run in a consistent manner. Each pathrate run on average took approximately 30 minutes. Pathrate returns two estimates, a high estimate and a low estimate, for bottleneck capacity between a pair of source and destination nodes. In the first experiment pathrate returned negative values for low capacity estimate in certain measurements. When we reported this to the authors of the pathrate tool, they kindly debugged the calculation error and the modified version (v2.4.1b) was used in the second set of measurements. To avoid this calculation error, we only report the high capacity estimate of the pathrate in this paper.

The first set of measurements was initiated on August 11th, 2004 and completed on September 6th, 2004. The second set was measured between January 5th, 2005 and January 18th, 2005. On our first attempt in August 2004, we tried measuring capacities between all PlanetLab nodes of the then nearly 400 nodes. However, many of the nodes did not respond consistently, and many of the pathrate capacity estimates were not returned. Ultimately, in the first set, we collected bottleneck capacity data on 12,006 network paths from 279 nodes. In the second set of experiments performed in January of 2005, we prefiltered 178 nodes (and no more than two nodes per site) that consistently responded. It could be one of the reasons why the experiments were finished in a shorter time compared with the first set of measurement experiments. In the second set of measurement we managed to collect data on 21,861 paths.

We first look at the statistics of the end-to-end capacity over all paths (source-destination node pairs) measured. It is important to note that given two nodes A and B, capacity measurements in both directions, i.e., source destination node pairs (A,B)

Table 2. End-to-end capacity distribution

Capacity (C)	Set 1		Set 2	
	Number of paths	Percentage (%)	Number of paths	Percentage (%)
$C < 20$ Mbps	4013	33.42	6733	30.8
20 Mbps $\leq C < 50$ Mbps	4246	35.37	1910	8.74
50 Mbps $\leq C < 80$ Mbps	674	5.61	1303	5.96
80 Mbps $\leq C < 120$ Mbps	2193	18.27	11744	53.72
120 Mbps $\leq C < 200$ Mbps	207	1.72	139	0.64
200 Mbps $\leq C < 500$ Mbps	392	3.27	21	0.096
500 Mbps $\leq C$	281	2.34	11	0.05

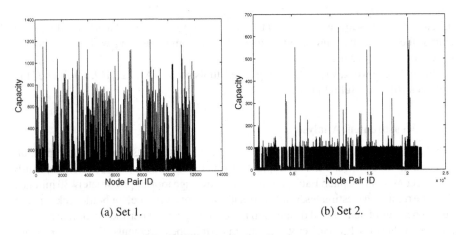

(a) Set 1.   (b) Set 2.

Fig. 1. Bandwidth capacity for all pairs measured

and (B,A) may not both be available. Table 1 shows that the average bandwidth between PlanetLab hosts is nearly 64 Mbps. Table 2 shows the distribution, Figure 1 visualizes this distribution (notice the different scaling of y-axis between two subfigures) and Figure 2 shows the cumulative distribution function (notice the different scaling of x-axis between two subfigures). On further analysis, we observed that when certain nodes were the source or the destination, the bandwidth measured was very low. In the first set of measurements for instance, for paths with freedom.ri.uni-tuebingen.de as the source, the average bandwidth was 4.61 Mbps. We noticed a similar behavior in the second set as when 200-102-209-152.paemt7001.t.brasiltelecom.net.br was the source, the average capacity was 0.42 Mbps and when it was the destination, the average capacity was 0.41 Mbps. On the other hand, when planetlab1.eurecom.fr was the destination, the average bandwidth was 3.85 Mbps, with the path from planetlab1.inria.fr having 199.2 Mbps of bandwidth. Without this measurement of 199.2 Mbps, the average bandwidth with planetlab1.eurecom.fr as the destination is 2.13 Mbps. We also noticed nodes with high average bandwidth. For instance, measurements from planet1.ottawa.canet4.nodes.planetlab.org showed an average of 508.46 Mbps.

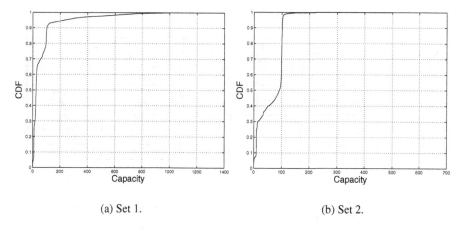

(a) Set 1.   (b) Set 2.

Fig. 2. Cumulative distribution function of bandwidth capacity

Table 3. Capacity by regions (Mbps)

Source	Destination									
	North America		South America		Asia		Europe		Oceania	
	Set 1	Set 2	Set 1	Set 2	Set 1	Set 2	Set 1	Set 2	Set 1	Set 2
North America	60.67	66.28	8.34	0.41	55.74	60.8	71.78	68.11	N/A	79.64
South America	7.94	0.42	106	N/A	N/A	0.43	6.1	0.41	N/A	0.4
Asia	69.13	55.15	N/A	0.41	73.5	62.36	72.28	42.94	N/A	99.52
Europe	69.94	66.43	30.67	0.4	61.38	47.54	74.82	66.69	N/A	13.15
Oceania	N/A	37.25	N/A	0.4	N/A	22.54	N/A	7.03	N/A	50.9

Some PlanetLab nodes have imposed outgoing bandwidth limit, ranging from 500 Kbps to 10 Mbps. We observed interesting interplay between the traffic shaper for bandwidth limiting and the pathrate probing scheme. In some cases we measured end-to-end capacity of 100 Mbps even though the source was bandwidth limited to 500 Kbps. We are further exploring this interaction.

The standard deviation for the second set is much smaller than the first set. We believe the prefiltering of the nodes for the second set is the main reason as the nodes that showed extremely low or high capacities in the first set were relatively unstable, and could have been removed from our second experiments. In the second set, we have limited the number of nodes per site to at most two nodes, and hence we have less number of high capacity local paths than the first set. We can also observe that in the second set, more than 99% of the paths show the capacity of less than 120 Mbps.

Table 3 shows the capacity measured region by region. We categorize each node into five regions: North America, South America, Asia, Europe, and Oceania. In our first measurement set we did not have any node from the Oceania region part of PlanetLab. On other entries of the table with N/A, no estimates were returned. There were only two nodes from Brasil in South American region in the second set, and as mentioned earlier,

Table 4. Asymmetry factor distribution

Asymmetry factor (α)	Set 1		Set 2	
	Number of pairs	Percentage (%)	Number of pairs	Percentage (%)
$\alpha < 0.01$	132	6.08	1843	21.49
$0.01 \leq \alpha < 0.05$	395	18.19	3237	37.74
$0.05 \leq \alpha < 0.1$	165	7.6	817	9.52
$0.1 \leq \alpha < 0.2$	243	11.19	880	10.26
$0.2 \leq \alpha < 0.5$	328	15.1	1111	12.95
$0.5 \leq \alpha$	909	41.85	870	10.14

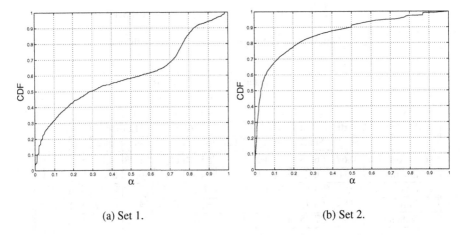

(a) Set 1.   (b) Set 2.

Fig. 3. Cumulative distribution function of asymmetry factor

capacities of the path to and from these nodes were very low. One might think that pairs between the same region will have higher capacity than those for inter-regions. The table however does not show any strong confirmation of that belief. It is also interesting to see that although the paths from Asia to Oceania show high capacity, the same cannot be said for the reverse direction.

We now investigate whether the paths between PlanetLab hosts are symmetric in terms of capacity. For the first set, among 12,006 measurements, only 2,172 pairs (4,344 measurements) reported capacity estimates in both directions, and for the second set, 8,758 pairs (17,516 measurements) out of 21,861 measurements returned estimates for both directions. To understand the path asymmetry, we define asymmetry factor between two nodes i and j, $\alpha_{i,j}$, as follows:

$$\alpha_{i,j} = \frac{|BW_{i,j} - BW_{j,i}|}{max(BW_{i,j}, BW_{j,i})}$$

where $BW_{i,j}$ is the bottleneck bandwidth from node i to node j.

Table 5. Comparison of End-to-end capacity statistics of common node pairs in Sets 1 and 2

	Set 1	Set 2
Number of common pairs	3,409	
Measurement period	8/11/04~9/6/04	1/5/05~1/18/05
PlanetLab version	version 2	version 3
Minimum capacity	0.1 Mbps	0.5 Mbps
Maximum capacity	1014.1 Mbps	152.6 Mbps
Average capacity	55.79 Mbps	71.09 Mbps
Median capacity	24.3 Mbps	97.3 Mbps
Standard deviation	109.94 Mbps	39.32 Mbps

When the capacity of the forward path equals the capacity of the reverse path (i.e., complete symmetry), the asymmetry factor is zero. A high asymmetry factor implies stronger asymmetry.

The distribution of the asymmetry factor for both sets are reported in Table 4 while the CDF is plotted in Figure 3. In the first set, only 132 pairs (6%) showed α of less than 0.01 and 692 pairs (32%) are less than 0.1. Although about 60% of the pairs have asymmetry factor below 0.5, there are a significant number of pairs with high asymmetry factor. We further investigated the reason for high asymmetry in 328 pairs that have α larger than 0.5. The PlanetLab imposed artificial bandwidth limit was reason for asymmetry in 189 of these pairs.

In the second set however, surprisingly large portion of the paths showed high symmetry. Nearly 60% has the asymmetry factor of less than 0.05. We believe that the main reason is, as shown in Table 2, more than half of the capacity estimates were between 80 and 120 Mbps.

Temporal Analysis. Since the measurements from Set 1 and Set 2 were done almost 5 months apart, the obvious question to ask is whether the data suggests significant changes in the PlanetLab infrastructure during this period. Note that we already know of two significant changes - the PlanetLab software version was upgraded to version 3 and the pathrate tool was upgraded to a new version. To answer the above question, we computed the common source-destination node pairs between the two sets and analyzed the bandwidth measurements. We found 128 common nodes in the two sets and 3,409 common source-destination node pair measurements.

The summary statistics of the measured capacity for these node pairs common to both measurement sets are given in Table 5. There are some interesting differences between the two sets, which could be caused by infrastructure changes, measurement errors or both. The average capacity between the measured node-pairs increased to 71 Mbps from 55 Mbps, as did the minimum measured bandwidth, implying an upgrade of the infrastructure on average. An interesting point to note is that the maximum capacity between any node pair decreased significantly from 1 Gbps to 152 Mbps. This could have been due to stricter bandwidth limits imposed on PlanetLab nodes. In the first set, the capacity between the nodes planetlab1.cse.nd.edu and planetlab2.cs.umd.edu were measured to be 1 Gbps, which in the second set is now close to 100 Mbps. We were unable to determine whether this is due to an infrastructure change, imposed bandwidth limit or measurement error. Diagnosing the causes for these measurement changes is future work.

While the stated goal of this work was not to verify accuracy of the pathrate tool (this has been done by other researchers in earlier work), we mentioned earlier that in some of the measurements in Set 1, the low estimate of bandwidth reported by pathrate were found to be negative and the authors of pathrate rectified this in the subsequent release. Although the negative values do not affect any of our presented results as we use the high estimate of the bandwidth, it is interesting to note that with the new version of pathrate, of the 3,409 measurements, no negative low estimates were observed in Set 2, while there were 93 negative measurements in Set 1.

The capacity distribution of the 3,409 common node pairs is given in Table 6. The biggest changes are in the paths with capacity between 20 Mbps and 50 Mbps and those with capacity between 80 Mbps and 120 Mbps. From the data presented it seems that significant number of paths were upgraded from the first band (20~50 Mbps) to the second band (80~120 Mbps) in the time between our measurements.

As mentioned earlier, given two nodes A and B in this common set, capacity measurements in both directions, i.e., source destination node pairs (A,B) and (B,A) may not both be available. Of the 3,409 source-destination node pairs common to sets 1 and 2, 661 node pairs (i.e., 1,322 measurements) had bandwidth measurements in both directions and hence the asymmetry metric could be computed for these. The asymmetry factor distribution is tabulated in Table 7. Again, the second set of experiments show a significantly reduced asymmetry than the first set.

Correlation Study. We now study the correlation between bandwidth and latency. Before we report the result of this study, we explain the motivation of attempting to relate the delay with bandwidth. As mentioned in Section 1, our ultimate goal is to gain insights into potential correlation that will enable scalable bandwidth estimation. For example, to find a node whose path from a given node has the largest capacity, instead of performing bandwidth estimates to all the nodes, can we do the probing to just a small number of nodes (five for instance)? Since measuring latency can be done with less probing overhead with quick turnaround time than measuring bandwidth, there already exist tools that perform scalable network distance estimation [26]. With these tools available and

Table 6. End-to-end capacity distribution of common node pairs in Sets 1 and 2

Capacity (C)	Set 1		Set 2	
	Number of paths	Percentage (%)	Number of paths	Percentage (%)
$C < 20$ Mbps	1041	30.54	909	26.66
20 Mbps $\leq C < 50$ Mbps	1491	43.74	103	3.02
50 Mbps $\leq C < 80$ Mbps	105	3.08	180	5.28
80 Mbps $\leq C < 120$ Mbps	587	17.22	2205	64.68
120 Mbps $\leq C < 200$ Mbps	37	1.09	12	0.35
200 Mbps $\leq C < 500$ Mbps	86	2.52	0	0.00
500 Mbps $\leq C$	62	1.82	0	0.00

Table 7. Asymmetry factor distribution of common node pairs in Sets 1 and 2

Asymmetry factor (α)	Set 1		Set 2	
	Number of pairs	Percentage (%)	Number of pairs	Percentage (%)
$\alpha < 0.01$	65	9.83	145	21.94
$0.01 \le \alpha < 0.05$	167	25.26	299	45.23
$0.05 \le \alpha < 0.1$	57	8.62	86	13.01
$0.1 \le \alpha < 0.2$	64	9.68	70	10.59
$0.2 \le \alpha < 0.5$	83	12.56	48	7.26
$0.5 \le \alpha$	225	34.04	13	1.97

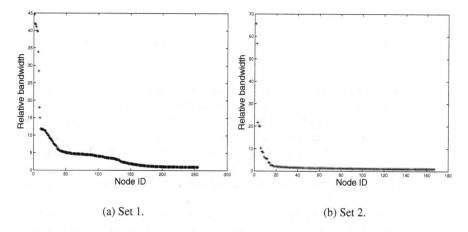

(a) Set 1.   (b) Set 2.

Fig. 4. Bandwidth/delay correlation

latency values easily in hand, if there is any relationship or trend between latency and bandwidth, we can scalably estimate network bandwidth without excessive bandwidth probing. That is the main motivation of this trend analysis. Note also that using capacity, instead of available bandwidth, is more appropriate as the values of available bandwidth vary with time, and unless the measurement of bandwidth and latency are done at the same period, the analysis could be meaningless.

For the latency measurement, we initially used the all pair ping data.[2] However, due to some missing ping data, there was little overlap between the available ping data and collected bandwidth measurement. Hence we also used the RTT measurements from pathrate. The resulting trends from ping and RTT measures from pathrate are quite similar, and hence we only present the results based upon the pathrate RTT latency.

We used two metrics for studying the bandwidth and latency correlation. The first metric, called relative bandwidth correlation metric, captures the ratio of maximum bandwidth path and bandwidth to the closest node. For a given host ($node_i$), using the

---

[2] We obtain this from http://www.pdos.lcs.mit.edu/~strib/pl_app/

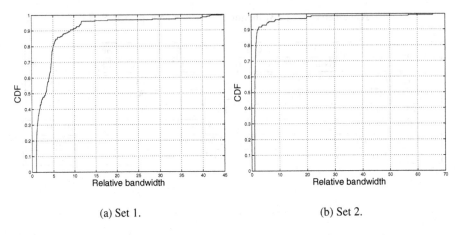

(a) Set 1.   (b) Set 2.

Fig. 5. Cumulative distribution function of bandwidth/delay correlation

latency data, we find the host that has the minimum latency ($node_{minLat}$). Similarly, using the capacity measurements, we find the host that provides the maximum bandwidth ($node_{maxBW}$). The relative bandwidth correlation penalty metric for $node_i$ is then defined as the ratio of the maximum bandwidth ($BW_{i,maxBW}$) and the bandwidth from $node_i$ to $node_{minLat}$ ($BW_{i,minLat}$). This metric takes values greater than or equal to 1.0. The closer this metric is to 1.0, we consider the correlation between latency and bandwidth to be stronger. We plot this metric and its cumulative distribution in Figures 4 and 5.

We see from the figure that latency and bandwidth are surprisingly quite correlated, especially in the second set. On further analysis with the CDF for the first set, we notice that for about 40% of the nodes, the bandwidth to the closest node is roughly 40% smaller than the actual maximum bandwidth. Our preliminary investigations reveal one of the primary reasons for this behavior is the imposed bandwidth limit. In some cases, the capacities to even the nearby nodes are quite less than the maximum bandwidth as they have bandwidth limits. For instance, in the Set 1, in lots of cases the node with highest capacity is planetlab1.ls.fi.upm.es which did not have any imposed bandwidth limit.

For the second set on the other hand, we see that about 90% of the closest nodes has the capacity that is nearly equal to the maximum bandwidth. We must remember however, that the roundtrip time measurement was performed by pathrate itself. Tools such as Netvigator [26] that perform network distance estimation typically uses traceroute or ping, and those tools may return different values. We can further test the correlation between bandwidth and latency using these tools.

We also used the Spearman rank correlation coefficient [13], which is commonly used instead of the Pearson's correlation coefficient, when the two sets of data come from different distributions. In Figure 6, we rank all the node pairs based on both bandwidth and latency. We plot the bandwidth rank versus the latency rank. If there is good correlation, we expect the points to be clustered along the $y = x$ diagonal line. However, using the Spearman coefficient we did not find any such correlation. The value of

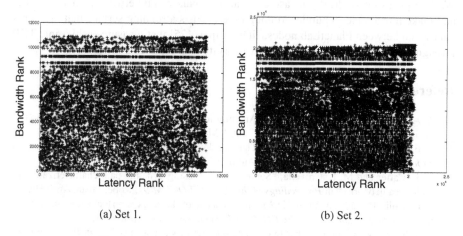

(a) Set 1.   (b) Set 2.

Fig. 6. Bandwidth/delay correlation using rank correlation

Spearman coefficient for sets 1 and 2 are 0.027 and 0.138 respectively based on the data presented in these figures. We believe that the degree of rank-order correlation is higher between closeby nodes and it decreases as distance between nodes increases. Hence, the low value of Spearman coefficient might be due to its computation over all the possible nodes. We plan to evaluate other rank-order correlation coefficients in this context.

3   Summary and Conclusions

In this paper, we presented a large scale measurement study of end-to-end capacity of paths between PlanetLab nodes. Our contributions are two-fold in that in addition to presenting the analysis of the data we collected from two sets of experiments, we also described the issues with the deployment of a number of bandwidth measurement tools on the PlanetLab platform. The measurement work is ongoing, but even with the data we have collected so far, there are a number of interesting conclusions one can draw. Foremost, we verified our intuition that network paths connecting PlanetLab nodes are highly heterogeneous in the capacity values and in planning PlanetLab experiments, one needs to take this into account. The capacity of paths may have an order of magnitude difference even when they are sourced from the same node and similarly for the same receiver. Paths between two nodes do not necessarily show capacity symmetry. According to PlanetLab policy, bandwidth limits on outgoing traffic have been implemented. But we observed violations of the policy, which could have been due to the inaccuracy of the tool itself and we are investigating this further. In attempting to draw insights for scalable bandwidth estimation, we studied the correlation between latency and bandwidth of a path. Our preliminary results are promising and we plan to investigate this further.

One of our future work includes modifying the SProbe tool to keep attempting measurements until a valid estimate is made. We can then measure the bandwidth to nodes outside PlanetLab as SProbe does not require user access to destination hosts.

We also plan to periodically measure the all pair bandwidth between planet-lab hosts and make it available to public. Although there exists a running service that measures bandwidth between PlanetLab nodes,[3] it uses Iperf [25] that measures achievable TCP throughput, which is not necessarily raw capacity or available bandwidth.

References

1. A. Akella, S. Seshan, and A. Shaikh, "An empirical evaluation of wide-area Internet bottlenecks," in *Proceedings of the ACM IMC 2003*, Miami, FL, October 2003.
2. S. Banerjee, M. Pias, and T. Griffin, "The interdomain connectivity of planetlab nodes," in *Proceedings of the PAM 2004*, Sophia-Antipolis, France, April 2004.
3. R. L. Carter and M. E. Crovella, "Server selection using dynamic path characterization in wide-area networks," in *Proceedings of the IEEE INFOCOM'97*, Kobe, Japan, April 1997.
4. C. Dovrolis, P. Ramanathan, and D. Moore, "What do packet dispersion techniques measure?" in *Proceedings of the IEEE INFOCOM 2001*, Anchorage, AK, April 2001.
5. A. B. Downey, "Using pathchar to estimate Internet link characteristics," in *Proceedings of the ACM SIGCOMM'99*, Cambridge, MA, September 1999.
6. N. Hu, L. Li, Z. M. Mao, P. Steenkiste, and J. Wang, "Locating internet bottlenecks: Algorithms, measurements, and implications," in *Proceedings of the ACM SIGCOMM 2004*, Portland, OR, August 2004.
7. N. Hu and P. Steenkiste, "Evaluation and characterization of available bandwidth probing techniques," *IEEE J. Select. Areas Commun.*, vol. 21, no. 6, August 2003.
8. M. Jain and C. Dovrolis, "Pathload: A measurement tool for end-to-end available bandwidth," in *Proceedings of the PAM 2002*, Fort Collins, CO, March 2002.
9. ———, "Ten fallacies and pitfalls on end-to-end available bandwidth estimation," in *Proceedings of the ACM IMC 2004*, Taormina, Italy, October 2004.
10. V. Jocobson. pathchar: A tool to infer characteristics of internet paths. [Online]. Available: ftp://ftp.ee.lbl.gov/pathchar
11. R. Kapoor, L.-J. Chen, L. Lao, M. Gerla, and M. Y. Sanadidi, "CapProbe: A simple and accurate capacity estimation technique," in *Proceedings of the ACM SIGCOMM 2004*, Portland, OR, August 2004.
12. S. Katti, D. Kitabi, C. Blake, E. Kohler, and J. Strauss, "MultiQ: Automated detection of multiple bottleneck capacities along a path," in *Proceedings of the ACM IMC 2004*, Taormina, Italy, October 2004.
13. E. L. Lehmann and H. J. M. D'Abrera, *Nonparametrics: Statistical Methods Based on Ranks, rev. ed.* Prentice-Hall, 1998.
14. B. A. Mah. pchar: A tool for measuring internet path characteristics. [Online]. Available: http://www.kitchenlab.org/www/bmah/Software/pchar
15. F. Montesino-Pouzols, "Comparative analysis of active bandwidth estimation tools," in *Proceedings of the PAM 2004*, Sophia-Antipolis, France, April 2004.
16. J. Navratil and R. L. Cottrell, "ABwE: A practical approach to available bandwidth estimation," in *Proceedings of the PAM 2003*, La Jolla, CA, April 2003.
17. T. S. E. Ng and H. Zhang, "Predicting Internet network distance with coordinates-based approaches," in *Proceedings of the IEEE INFOCOM 2002*, New York, NY, June 2002.
18. R. Prasad, C. Dovrolis, M. Murray, and kc claffy, "Bandwidth estimation: Metrics, measurement techniques, and tools," *IEEE Network*, vol. 17, no. 6, pp. 27–35, November/December 2003.

---

[3] http://www.planet-lab.org/logs/iperf/

19. V. Ribeiro, R. Riedi, R. Baraniuk, J. Navratil, and L. Cottrell, "pathChirp: Efficient available bandwidth estimation for network paths," in *Proceedings of the PAM 2003*, La Jolla, CA, April 2003.
20. V. J. Ribeiro, R. H. Riedi, and R. G. Baraniuk, "Locating available bandwidth bottlenecks," *IEEE Internet Comput.*, vol. 8, no. 6, pp. 34–41, September/October 2004.
21. S. Saroiu. SProbe: A fast tool for measuring bottleneck bandwidth in uncooperative environments. [Online]. Available: http://sprobe.cs.washington.edu/
22. S. Saroiu, P. K. Gummadi, and S. D. Gribble, "SProbe: Another tool for measuring bottleneck bandwidth," in *Work-in-Progress Report at the USITS 2001*, San Francisco, CA, March 2001.
23. N. Spring, D. Wetherall, and T. Anderson. Scriptroute: A facility for distributed internet debugging and measurement. [Online]. Available: http://www.scriptroute.org
24. J. Strauss, D. Katabi, and F. Kaashoek, "A measurement study of available bandwidth estimation tools," in *Proceedings of the ACM IMC 2003*, Miami, FL, October 2003.
25. A. Tirumala, F. Qin, J. Dugan, J. Ferguson, and K. Gibbs. Iperf: The TCP/UDP bandwidth measurement tool. [Online]. Available: http://dast.nlanr.net/Projects/Iperf/
26. Z. Xu, P. Sharma, S.-J. Lee, and S. Banerjee, "Netvigator: Scalable network proximity estimation," HP Laboratories, Tech. Rep., 2005.

Comparison of Public End-to-End Bandwidth Estimation Tools on High-Speed Links

Alok Shriram[1], Margaret Murray[2], Young Hyun[1], Nevil Brownlee[1], Andre Broido[1], Marina Fomenkov[1], and kc claffy[1]

[1] CAIDA, San Diego Supercomputer Center, University of California, San Diego
{alok, youngh, nevil, broido, marina, kc}@caida.org
[2] Texas Advanced Computing Center
marg@tacc.utexas.edu

Abstract. In this paper we present results of a series of bandwidth estimation experiments conducted on a high-speed testbed at the San Diego Supercomputer Center and on OC-48 and GigE paths in real world networks. We test and compare publicly available bandwidth estimation tools: *abing, pathchirp, pathload,* and *Spruce*. We also tested *Iperf* which measures achievable TCP throughput. In the lab we used two different sources of known and reproducible cross-traffic in a fully controlled environment. In real world networks we had a complete knowledge of link capacities and had access to SNMP counters for independent cross-traffic verification. We compare the accuracy and other operational characteristics of the tools and analyze factors impacting their performance.

1   Introduction

Application users on high-speed networks perceive the network as an end-to-end connection between resources of interest to them. Discovering the least congested end-to-end path to distributed resources is important for optimizing network utilization. Therefore, users and application designers need tools and methodologies to monitor network conditions and to rationalize their performance expectations.

Several network characteristics related to performance are measured in bits per second: capacity, available bandwidth, bulk transfer capacity, and achievable TCP throughput. Although these metrics appear similar, they are not, and knowing one of them does not generally imply that one can say anything about others. Prasad *et al.* [1] provide rigorous definitions of terms used in the field, survey underlying techniques and methodologies, and list open source measurement tools for each of the above metrics.

By definition [1], end-to-end capacity of a path is determined by the link with the minimum capacity (narrow link). End-to-end available bandwidth of a path is determined by the link with the minimum unused capacity (tight link). In this study our goal is to test and compare tools that claim to measure the available end-to-end bandwidth (Table 1). We did not test tools that measure end-to-end capacity.

Candidate bandwidth estimation tools face increasingly difficult measurement challenges as link speeds increase and network functionality grows more complex. Consider the issue of time precision: on faster links, intervals between packets decrease, rendering

C. Dovrolis (Ed.): PAM 2005, LNCS 3431, pp. 306–320, 2005.

Table 1. Available bandwidth estimation tools

Tool	Author	Methodology
abing	Navratil [2]	packet pair
cprobe	Carter [3]	packet trains
IGI	Hu [4]	SLoPS
netest	Jin [5]	unpublished
pathchirp	Ribeiro [6]	chirp train
pipechar	Jin [7]	unpublished
pathload	Jain [8]	SLoPS
Spruce	Strauss [9]	SLoPS

packet probe measurements more sensitive to timing errors. The nominal 1 μs resolution of UNIX timestamps is acceptable when measuring 120 μs gaps between 1500 B packets on 100 Mb/s links but insufficient to quantify packet interarrival time (IAT) variations of 12 μs gaps on GigE links. Available bandwidth measurements on high-speed links stress the limits of clock precision especially since additional timing errors may arise due to the NIC itself, the operating system, or the Network Time Protocol (designed to synchronize clocks of computers over a network) [10].

Several other problems may be introduced by network devices and configurations. Interrupt coalescence improves network packet processing efficiency, but breaks end-to-end tools that assume uniform per packet processing and timing [11]. Hidden Layer 2 store-and-forward devices distort an end-to-end tool's path hop count, resulting in calculation errors [12]. MTU mismatches impede measurements by artificially limiting path throughput. Modern routers that relegate probe traffic to a slower path or implement QoS mechanisms may also cause unanticipated complications for end-to-end probing tools. Concerted cooperative efforts of network operators, researchers and tool developers can resolve those (and many other) network issues and advance the field of bandwidth measurement.

While accurate end-to-end measurement is difficult, it is also important that bandwidth estimation tools be fast and relatively unintrusive. Otherwise, answers are wrong, arrive too late to be useful, or the end-to-end probe may itself interfere with the network resources that the user attempts to measure and exploit.

1.1 Related Work

Coccetti and Percacci [13] tested *Iperf*, *pathrate*, *pipechar* and *pathload* on a low speed ($\leq$ 4 Mb/s) 3 or 4 hop topology, with and without cross-traffic. They found that tool results depend strongly on the configuration of queues in the routers. They concluded that in a real network environment, interpreting tool results requires considerable care, especially if QoS features are present in the network.

Strauss *et al.* [9] introduced *Spruce*, a new tool to measure available bandwidth, and compared it to *IGI* and *pathload*. They used SNMP data to perform an absolute comparison on two end-to-end paths that both had their tight and narrow links at 100 Mb/s. They also compared the relative sensitivity of the tools on 400 paths in the RON and PlanetLab testbeds. The authors found that *IGI* performs poorly at higher loads, and that *Spruce* was more accurate than *pathload*.

Hu and Steenkiste [4] explored the packet pair mechanism for measuring available bandwidth and presented two measurement techniques, *IGI* and *PTR*. They tested their methods on 13 Internet paths with bottleneck link capacities of ≤ 100 Mb/s and compared the accuracies to *pathload* using the bulk data transmission rate (*Iperf*) as the benchmark. On some paths all three tools were in agreement with *Iperf*, while on others the results fluctuated. Since the authors did not have any information about true properties and state of the paths, the validity of their results is rather uncertain.

Finally, Ubik *et al.* [14] ran *ABwE* (a predecessor to *abing*) and *pathload* on two paths on the GEANT network, obtaining one month of hourly measurements. The measured paths consisted of at least 10 routers, all with GigE or OC-48 links. Measurements from both tools did not agree. These preliminary results await further analysis.

Our study takes one step further the testing and comparing publicly available tools for available bandwidth estimation. First, we considered and evaluated a larger number of tools than previous authors. Second, we conducted two series of reproducible laboratory tests in a fully controlled environment using two different sources of cross-traffic. Third, we experimented on high-speed paths in real networks where we had a complete knowledge of link capacities and had access to SNMP counters for independent cross-traffic verification. We compare the accuracy and other operational characteristics of the tools, and analyze factors impacting their performance.

2   Testing Methodology

From Table 1 we selected the following tools for this study: *abing, pathchirp, pathload,* and *Spruce*. For comparison we also included *Iperf* [15] which measures achievable TCP throughput. *Iperf* is widely used for end-to-end performance measurements and has become an unofficial standard [16] in the research networking community.

We were unable to test *cprobe* [3] because it only runs on an SGI Irix platform and we do not have one in our testbed. We did not include *netest* in this study since in our initial tests this tool inconsistently reported different metrics on different runs and different loads. We excluded *pipechar* [7] after tests on 100 Mb/s paths and *IGI* [4] after tests on 1 Gb/s paths since they were unresponsive to variations in cross-traffic.

2.1   Bandwidth Estimation Testbed

In collaboration with the CalNGI Network Performance Reference Lab [17], CAIDA researchers developed an isolated high-speed testbed that can be used as a reference center for testing bandwidth estimation tools. This resource allows us to test bandwidth estimation tools against known and reproducible cross-traffic scenarios and to look deeply into internal details of tools operation. We also attempt to offer remote access to the lab to tool developers wishing to further refine and enhance their tools.

In our current testbed configuration (Figure 1), all end hosts are connected to switches capable of handling jumbo MTUs (9000 B). Three routers in the testbed end-to-end path are each from a different manufacturer. Routers were configured with two separate network domains (both within private RFC1918 space) that route all packets across a single backbone. An OC48 link connects a Juniper M20 router with a Cisco GSR 12008

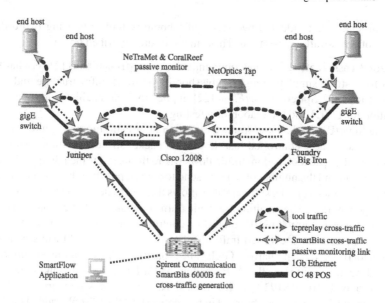

Fig. 1. Bandwidth Estimation Testbed. The end-to-end path being tested traverses three routers and includes OC48 and GigE links. Tool traffic occurs between designated end hosts in the upper part of this figure. Cross-traffic is injected either by additional end hosts behind the jumbo-MTU capable GigE switches or by the Spirent SmartBits 6000 box (lower part of figure). Passive monitors tap the path links as shown to provide independent measurement verification

router, and a GigE link connects the Cisco with a Foundry BigIron 10 router. We use jumbo MTUs (9000 B) throughout our OC48/GigE configuration in order to support traffic flow at full line speed [18].

Bandwidth estimation tools run on two designated end hosts each equipped with a 1.8 GHz Xeon processor, 512 MB memory, and an Intel PRO/1000 GigE NIC card installed on a 64b PCI-X 133 MHz bus. The operating system is the CAIDA reference FreeBSD version 4.8.

Our laboratory setup also includes dedicated hosts that run *CoralReef* [19] and *NeTraMet* [20] passive monitor software for independent verification of tool and cross-traffic levels and characteristics. Endace DAG 4.3 network monitoring interface cards on these hosts tap the OC-48 and GigE links under load. *CoralReef* can either analyze flow characteristics and packet IATs in real time or capture header data for subsequent analysis. The *NeTraMet* passive RTFM meter can collect packet size and IAT distributions in real time, separating tool traffic from cross-traffic.

2.2 Methods of Generating Cross-Traffic

The algorithms used by bandwidth estimating tools make assumptions about characteristics of the underlying cross-traffic. When these assumptions do not apply, tools cannot perform correctly. Therefore, test traffic must be as realistic as possible with respect to its packet IAT and size distributions.

In our study we conducted two series of laboratory tool tests using two different methods of cross-traffic generation. These methods are described below.

Synthetic Cross-Traffic. Spirent Communications SmartBits 6000B [21] is a hardware system for testing, simulating and troubleshooting network infrastructure and performance. It uses the Spirent *SmartFlow* [22] application that enables controlled traffic generation for L2/L3 and QoS laboratory testing.

Using SmartBits and *SmartFlow* we can generate pseudo-random, yet reproducible traffic with accurately controlled load levels and packet size distributions. This traffic generator models pseudo-random traffic flows where the user sets the number of flows in the overall load and the number of bytes to send to a given port/flow before moving on to the next one (burst size). The software also allows the user to define the L2 frame size for each component flow. The resulting synthetic traffic emulates realistic protocol headers. However, it does not imitate TCP congestion control and is not congestion-aware.

In our experiments we varied traffic load level from 100 to 900 Mb/s which corresponds to 10-90% of the narrow GigE link capacity. At each load level, *SmartFlow* generated nineteen different flows. Each flow had a burst size of 1 and consisted of either 64, 576, 1510 or 8192 byte L2 frames. The first three sizes correspond to the most common L2 frame sizes observed in real network traffic [23]. We added the jumbo packet component because high-speed links must employ jumbo MTUs in order to push traffic levels to line saturation. While [23] data suggest a tri-modal distribution of small/medium/large frames in approximately 60/20/20% proportions, we are not aware of equivalent published packet-size data for links where jumbo MTUs are enabled. We mixed the frames of four sizes in equal proportions.

Packet IATs (Figure 2(a)) ranged from 4 to more than 400 μs. We used passive monitors *CoralReef* and *NeTraMet* to verify the actual load level of generated traffic and found that it matched the requirements within 1-2%.

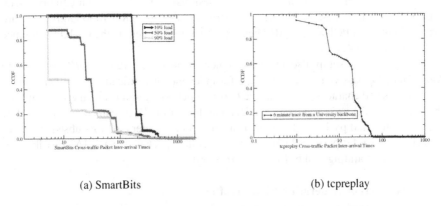

(a) SmartBits   (b) tcpreplay

Fig. 2. SmartBits and tcpreplay cross-traffic packet inter-arrival times

Playing Back Traces of the Real Traffic. We replayed previously captured and anonymized traffic traces on our laboratory end-to-end path using a tool *tcpreplay* [24]. This method of cross-traffic generation reproduces actual IAT and packet size distributions but is not congestion-aware. The playback tool operated on two additional end hosts (separate from the end hosts running bandwidth estimation tools) and injected the cross-traffic into the main end-to-end path via GigE switches.

We tested bandwidth estimation tools using two different traces as background cross-traffic:

- a 6-minute trace collected from a 1 Gb/s backbone link of a large university with approximately 300-345 Mb/s of cross-traffic load;
- a 6-minute trace collected from a 2.5 Gb/s backbone link of a major ISP showing approximately 100-200 Mb/s of cross-traffic load.

Neither trace contained any jumbo frames. Packet sizes exhibited a tri-modal distribution as in [23]. Packet IATs (Figure 2(b)) ranged from 1 to 60 μs.

We used CoralReef to continuously measure *tcpreplay* cross-traffic on the laboratory end-to-end path and recorded timestamps of packet arrivals and packet sizes. We converted this information into timestamped bandwidth readings and compared them to concurrent tool estimates. Both traces exhibited burstiness on microsecond time scales, but loads were fairly stable when aggregated over one-second time intervals.

3   Tool Evaluation Results

In this section we present tool measurements in laboratory tests using synthetic, non-congestion-aware cross-traffic with controlled traffic load (*SmartFlow*) and captured traffic traces with realistic workload characteristics (*tcpreplay*). In Section 4 we show results of experiments on real high-speed networks.

3.1   Comparison of Tool Accuracy

Experiments with Synthesized Cross-Traffic. We used the SmartBits 6000B device with the *SmartFlow* application to generate bi-directional traffic loads, varying from 10% to 90% of the 1 Gb/s end-to-end path capacity in 10% steps. We tested one tool at a time. In each experiment, the synthetic traffic load ran for six minutes. To avoid any edge effects, we delayed starting the tool for several seconds after initiating cross-traffic and ran the tool continuously for five minutes. Figure 3 shows the average and standard deviation of all available bandwidth values obtained during these 5 minute intervals for each tool at each given load.

Our end-to-end path includes three different routers with different settings. To check whether the sequence of routers in the path affects the tool measurements, we ran tests with synthesized cross-traffic in both directions. We observed only minor differences between directions. The variations are within the accuracy range of the tools and we suspect are due to different router buffer sizes.

We found that *abing* (Figure 3a) reports highly inaccurate results when available bandwidth drops below 600 Mb/s (60% on a GigE link). Note that this tool is currently deployed on the Internet End-to-End Performance Monitoring (IEPM) measurement

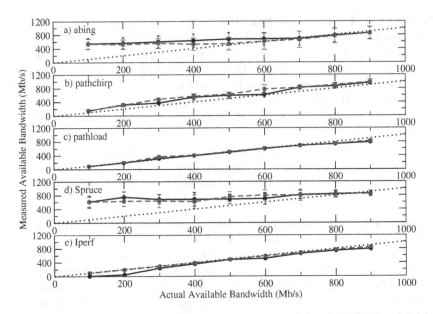

Fig. 3. Comparison of available bandwidth measurements on a 4-hop OC48/GigE path loaded with synthesized cross-traffic. For each experimental point, the x-coordinate is the actual available bandwidth of the path (equal to the GigE link capacity of 1000 Mb/s minus the generated load). The y-coordinate is the tool reading. Measurements of the end-to-end path in both directions are shown. The dash-dotted line shows expected value from SmartBits setting

infrastructure [25] where the MTU size is 1500 B, while our high-speed test lab uses a jumbo 9000 B MTU. We attempted to change *abing* settings to work with its maximum 8160 B probe packet size, but this change did not improve its accuracy.

We looked into further details of *abing* operating on an empty GigE path. The tool continuously sends back-to-back pairs of 1478 byte UDP packets with a 50 ms waiting interval between pairs. *abing* derives estimates of available bandwidth from the amount of delay introduced by the "network" between the paired packets. *abing* puts a timestamp into each packet, and the returned packet carries a receiver timestamp. Computing the packet IAT does not require clock synchronization since it is calculated as a difference between timestamps on the same host. Since these timestamps have a μs granularity, the IAT computed from them is also an integer number of μs. For back-to-back 1500 B packets on an empty 1 Gb/s link (12 Kbits transmitted at 1 ns per bit) the IAT is between 11 and 13 μs, depending on rounding error. However, we observed that every 20-30 packets the IAT becomes 244 μs. This jump may be a consequence of interrupt coalescence or a delay in some intermediate device such as a switch. The average IAT then changes to more than 20 μs yielding a bit rate of less than 600 Mb/s. This observation explains *abing* results: on an empty 1 Gb/s tight link it reports two discrete values of available bandwidth, the more frequent one of 890-960 Mb/s and occasional drops to 490-550 Mb/s. This oscillating behavior is clearly observed in time series of *abing* measurements (Figure 4) described below.

Another tool, *Spruce* (Figure 3d), uses a similar technique and, unsurprisingly, its results are impeded by the same phenomenon. *Spruce* sends 14 back-to-back 1500 B UDP packet pairs with a waiting interval of 160-1400 ms between pair probes (depending on some internal algorithm). In *Spruce* measurements, 244 μs gaps between packet pairs occur randomly between normal 12 μs gaps. Since the waiting time between pairs varies without pattern, the reported available bandwidth also varies without pattern in the 300-990 Mb/s range.

Results of our experiments with *abing* and *Spruce* on high-speed links caution that tools utilizing packet pair techniques must be aware of delay quantization possibly present in the studied network. Also, 1500-byte frames and microsecond timestamp resolution are simply not sensitive enough for probing high-speed paths.

In SmartBits tests, estimates of available bandwidth by *pathchirp* are 10-20% higher than the actual value determined from SmartBits settings (Figure 3b). This consistent overestimation persists even when there is no cross-traffic. On an empty 1 Gb/s path this tool yields values up to 1100 Mb/s. We have as yet no explanation for this behavior.

We found that results of *pathload* were the most accurate (Figure 3c). The discrepancy between its readings and actual available bandwidth was <10% in most cases.

The last tested tool, *Iperf*, estimates not the available bandwidth, but the achievable TCP throughput. We ran *Iperf* with the maximum buffer size of 227 KB and found it to be accurate within 15% or better (Figure 4e). Note that a smaller buffer size setting significantly reduces the *Iperf* throughput. This observation appears to contradict the usual rule of thumb that the optimal buffer size is the product of bandwidth and delay, which in our case would be $(10^9$ b/s) x $(10^{-4}$ s) $\sim$ 12.5 KB. Dovrolis *et al.* discuss this phenomenon in [26].

Experiments with Trace Playbacks. The second series of laboratory tests used previously recorded traces of real traffic. For these experiments we extracted six-minute samples from longer traces to use as a *tcpreplay* source. As in SmartBits experiments, in order to avoid edge effects we delayed the tool start for a few seconds after starting *tcpreplay* and ran each tool continuously for five minutes.

Figure 4 plots a time series of the actual available bandwidth, obtained by computing the throughput of the trace at a one-second aggregation interval and subtracting that from the link capacity of 1 Gb/s. Time is measured from the start of the trace. We then plot every value obtained by a given tool at the time it was returned.

As described in Section 2.2, we performed *tcpreplay* experiments with two different traces. We present tool measurements of the University backbone trace, which produced a load of about 300 Mb/s leaving about 700 Mb/s of available bandwidth. The tool behavior when using the ISP trace with a load of about 100 Mb/s was similar and is not shown here.

In tests with playback of real traces, *abing* and *Spruce* exhibit the same problems that plagued their performance in experiments with synthetic cross-traffic. Figure 4a shows that *abing* returned one of two values, neither of which was close to the expected available bandwidth. *Spruce* results (Figure 4d) continued to vary without pattern.

pathchirp measurements (Figure 4b) had a startup period of about 70 s when the tool returned only a constant value. The length of this period is related to the tool's measurement algorithm and depends on the number of chirps and chirp packet size

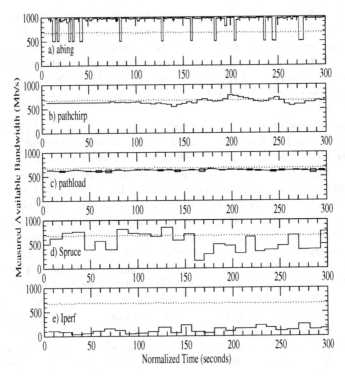

Fig. 4. Comparison of available bandwidth tool measurements on a 4-hop OC48/GigE path loaded with played back real traffic. The X-axis shows time from the beginning of trace playback. The Y-axis is the measured available bandwidth reported by each tool. The dotted line shows the actual available bandwidth that was very stable on a one-second aggregation scale

selected for the given tool run. After the startup phase, *pathchirp*'s values alternate within 15-20% of the actual available bandwidth.

The range reported by *pathload* (Figure 4c) slightly underestimates the available bandwidth by <16%.

Iperf reports surprisingly low results when run against *tcpreplay* traffic (Figure 4e). Two factors are causing this gross underestimation: packet drops requiring retransmission and a too long retransmission timeout of 1.2 s (default value). In the experiment shown, the host running *Iperf* and the host running *tcpreplay* were connected to the main end-to-end path via a switch. We checked the switch's MIB for discarded packets and discovered a packet loss of about 1% when the tool and cross-traffic streams merge. Although the loss appears small, it causes *Iperf* to halve its congestion window and triggers a significant number of retransmissions. The default retransmission timeout is so large it consumes up to 75% of the *Iperf* running time. Decreasing the retransmission timeout to 20 ms and/or connecting the *tcpreplay* host directly to the path bypassing the switch considerably improves *Iperf*'s performance. Note that we were able to reproduce the degraded *Iperf* performance in experiments with synthetic SmartBits traffic when we

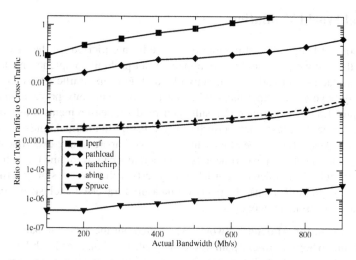

Fig. 5. Tool overhead vs. available bandwidth. Note that *pathchirp, abing,* and *Spruce* exhibit essentially zero overhead

flooded the path with a large number of small (64 B) packets. These experiments confirm that ultimately the TCP performance in the face of packet loss strongly depends on the OS retransmission timer.

3.2 Comparison of Tool Operational Characteristics

We considered several parameters that may potentially affect a user's decision regarding which tool to use: measurement time, intrusiveness, and overhead. We measured all these characteristics in experiments with SmartBits synthetic traffic where we can stabilize and control the load.

We define tool measurement time to be the average measurement time of all runs at a particular load level. On our 4-hop OC-48/GigE topology, the observed measurement durations were: 1.3 s for *abing*, 11 s for *Spruce*, 5.5 s for *pathchirp*, and 10 s for *Iperf* independent of load. The *pathload* measurement time generally increased when the available bandwidth decreased, and ranged between 7 and 22 s.

We define tool intrusiveness as the ratio of the average tool traffic rate to the available bandwidth, and tool overhead as the ratio of tool traffic rate to cross-traffic rate (Figure 5). *pathchirp, abing,* and *Spruce* have low overhead, each consuming less than 0.2% of the available bandwidth on the GigE link and introducing practically no additional traffic into the network as they measure. *pathload* intrusiveness is between 3 and 7%. Its overhead slightly increases with the available bandwidth (that is, when the cross-traffic actually decreases) and reaches 30% for the 10% load. As expected, *Iperf* is the most expensive tool both in terms of its intrusiveness (74-79%) and overhead costs. Since it attempts to occupy all available bandwidth, its traffic can easily exceed the existing cross-traffic.

4   Real World Validation

Comparisons of bandwidth estimation tools have been criticized for their lack of validation in the real world. Many factors impede if not prohibit comprehensive testing of tools on production networks. First, network conditions and traffic levels are variable and usually beyond the experimenters' control. This uncertainty prevents unambiguous interpretation of experimental results and renders measurements unreproducible. Second, a danger that tests may perturb or even disrupt the normal course of network operations makes network operators reluctant to participate in any experiments. Only close cooperation between experimenters and operators can overcome both obstacles.

We were able to complement our laboratory tests with two series of experiments in the real world. In both setups, the paths we measured traversed exclusively academic, research and government networks.

Experiments on the Abilene Network. We carried out the available bandwidth measurements on a 6 hop end-to-end path from Sunnyvale to Atlanta on the Abilene Network. Both end machines had a 1 Gb/s connection to the network and sourced no traffic except from running our tools. The rest of links in the path had either 2.5 or 10 Gb/s capacities.

We chose not to test *Spruce* on the Abilene Network since this tool performed poorly in our laboratory experiments[1]. We ran *pathload, pathchirp, abing*, and *Iperf* for 5 min each, in that order, back-to-back. We concurrently polled the SNMP 64-bit InOctect counters for all routers along the path every 10 s and hence knew the per-link utilization with 10 s resolution. We calculated the per-link available bandwidth as the difference between link capacity and utilization. The end-to-end available bandwidth is the minimum of per-link available bandwidths. During our experiments, the Abilene network did not have enough traffic on the backbone links to bring their available bandwidth below 1 Gb/s. Therefore, the end machines' 1Gb/s connections were both narrow and tight links in our topology.

Figure 6 shows our tool measurements and SNMP-derived available bandwidth. Measurements with *pathload, pathchirp*, and *Iperf* are reasonably accurate, while *abing* readings wildly fluctuate in the whole range between 0 and 1000 Mb/s.

The discrepancy between *Iperf* measurements and SNMP-derived values reflects tool design: *Iperf* generates large overhead (>70%) because it intentionally attempts to fill the tight link. Consequent readings of SNMP counters indicate how many bytes traversed an interface of a router during that time interval. They report total number of bytes without distinguishing tool traffic from cross-traffic. If a tool's overhead is high, then available bandwidth derived from SNMP data during this tool run is low. At the same time, since tools attempt to measure available bandwidth ignoring their own traffic, a high-overhead tool will report more available bandwidth than SNMP. Therefore, *Iperf* shows a correct value of achievable TCP throughput of ~950 Mb/s while concurrent SNMP counters account for *Iperf*'s own generated traffic, and thus yield less than 200 Mb/s of available bandwidth. A smaller discrepancy between *pathload* and SNMP results reflects *pathload*'s overhead (~10% per our lab tests).

---

[1] We tested *Spruce* in the other series of real network experiments, see subsection on SDSC-ORNL paths below

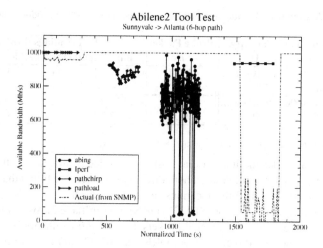

Fig. 6. Real world experiment conducted on the Abilene network. The dashed line shows the available bandwidth derived from SNMP measurements. See explanations in the text

Experiments on SDSC-ORNL Paths. In the second series of real-world experiments we tested *abing, pathchirp, pathload,* and *Spruce* between our host at SDSC (running CAIDA reference FreeBSD version 4.8) and a host at Oak Ridge National Lab (running Red Hat Linux release 9 with a 2.4.23 kernel and Web100 patch [27]). These experiments are of limited value since we did not have concurrent SNMP data for comparison with our results. However, we had complete information about link capacities along the paths which at least allows us to distinguish plausible results from impossible ones. We include these experiments since they present first insights into the interplay between the probing packet size and the path MTU.

The two paths we measured are highly asymmetric. The SDSC-ORNL path crosses CENIC and ESNet, has a narrow link capacity of 622 Mb/s (OC12) and MTU of 1500 bytes. The ORNL-SDSC path crosses Abilene and CENIC, has a narrow link capacity of 1 Gb/s and supports 9000-byte packets end-to-end. Both paths remained stable over the course of our experiments and included OC12, GigE, 10 GigE, and OC192 links. Under most traffic scenarios, it seems highly unlikely for the 10 Gb/s links to have less than 1 Gb/s of available bandwidth. Lacking true values of available bandwidth from SNMP counters for absolute calibration of tool results, we assume that the narrow link is also the tight link in both our paths.

We ran each tool using either 1500 or 9000 byte packets. *abing, pathchirp,* and *pathload* support large probe packet size as an option[2]. *Spruce* uses a hardcoded packet size of 1500 bytes; we had to trivially modify the code to increase the packet size to 9000 B. Table 2 summarizes our results while a detailed description is available in [28].

---

[2] The *abing* reflector has a hardcoded packet size of 1478 bytes.

Table 2. Summary of wide-area bandwidth measurements ("f"= produced no data)

Direction	Path Capacity, MTU	Probe Packet Size	Tool readings (Mb/s)			
			abing[a]	pathchirp	pathload	Spruce
SDSC to ORNL	622 Mb/s (OC12), 1500	1500 9000	178 / 241 f / 664	543 f	>324 409 – 424	296 0
ORNL to SDSC	1000 Mb/s (GigE), 9000	1500 9000	727 / 286 f / 778	807 816	>600 846	516 807

[a] Sender at SDSC for 1st value and at ORNL for 2nd value.

abing has a sender module on one host and a reflector module on the other host and measures available bandwidth in both directions at once. We found that its behavior changed when we switched the locations of sender and reflector. *abing* with 9000 B packets did not return results from SDSC to ORNL ("f" in Table 2). We could see that the ORNL host was receiving fragmented packets, but the *abing* reflector was not echoing packets. In the opposite direction, from ORNL to SDSC, *abing* with 9000 B packets overestimates the available bandwidth for the OC12 path (reports 664 Mb/s on 622 Mb/s capacity). Note that almost the factor of 3 difference in GigE path measurements with 1500 B packets (727 and 286 Mb/s) may be due to different network conditions since these tests occurred on different days.

pathchirp produced results on both paths when run with 1500 B packets and on the GigE path with 9000 B packets, but failed on the OC12 path with large packets. There does not appear to be any significant advantage to using large packets over small ones. Variations between consequent measurements with the same packet size are sometimes greater than the difference between using large and small packets.

In tests with 1500 B packets, on both paths *pathload* reports that results are limited by the maximum host sending rate. With 9000 B packets, this tool yielded available bandwidth estimates for both paths, but issued a warning "actual probing rate [does not equal] desired probing rate" for the OC12 path.

Spruce performed poorly in experiments with small packets from SDSC to ORNL, reporting wildly fluctuating values of available bandwidth. Tests with 9000 B packets in this direction always produced 0 Mb/s. However, in the ORNL to SDSC direction, its readings were more consistent and on par with other tools.

We suspect that fragmentation is responsible for most of the problems when probing packet size and path MTU mismatch. While using large packets to measure high-speed links is beneficial, more work is necessary to consistently support large packets and to reduce failures and inaccuracies stemming from fragmentation.

5   Conclusions and Future Work

Our study is the first comprehensive evaluation of publicly available tools for available bandwidth estimation on high-speed links. We conducted testing in the lab and over research networks. We found that *pathload* and *pathchirp* are the most accurate tools under conditions of our experiments.

Iperf performs well on high-speed links if run with its maximum buffer window size. Even small ($\sim$1%) but persistent amounts of packet loss seriously degrade its performance. Too conservative settings of the OS retransmission timer further exacerbate this problem.

Results of our experiments with *abing* and *Spruce* caution that tools utilizing packet pair techniques must be aware of delay quantization possibly present in the studied network. Also, 1500 byte frames and microsecond timestamp resolution are not sensitive enough for probing high-speed paths.

Despite the revealed problems, experimenting with available bandwidth estimating tools using large packets is worthwhile, considering the importance of using large packets on high-speed links.

We demonstrated how our testbed can be used to evaluate and compare end-to-end bandwidth estimation tools against reproducible cross-traffic in a fully controlled environment. Several bandwidth estimation tool developers have taken advantage of our offer of remote access to the testbed to conduct their own tests. We plan to use what we have learned from our testing methodology to conduct monitoring efforts on both research and commodity infrastructure.

Acknowledgments

We gratefully acknowledge access to the Spirent 6000 network performance tester and Foundry BigIron router in the CalNGI Network Performance Reference Lab created by Kevin Walsh. Many thanks to Cisco Systems for the GSR12008 router, Juniper Networks for the M20 router, and Endace, Ltd. for access to their DAG4.3GE network measurement card. Nathaniel Mendoza, Grant Duvall and Brendan White provided testbed configuration and troubleshooting assistance. Feedback from remote testbed users Jiri Navratil, Ravi Prasad, and Vinay Ribeiro was helpful in refining test procedures. Aaron Turner provided us with a lot of support on installing and running *tcpreplay*. We are grateful to Matthew J Zekauskas for invaluable assistance with running experiments on the Abilene network. This work was supported by DOE grant DE-FC02-01ER25466.

References

1. Prasad, R., Murray, M., claffy, k., Dovrolis, C.: Bandwidth Estimation: Metrics, Measurement Techniques, and Tools. In: IEEE Network. (2003)
2. Navratil, J.: ABwE: A Practical Approach to Available Bandwidth. In: PAM. (2003)
3. Carter, R., Crovella, M.: Measuring Bottleneck Link Speed in Packet-Switched Networks. Technical Report 96-006, Boston University (1996)
4. Hu, N., Steenkiste, P.: Evaluation and Characterization of Available Bandwidth Probing Techniques. IEEE JSAC Internet and WWW Measurement, Mapping, and Modeling (2003)
5. Jin, G.: netest-2 (2004) http://www-didc.lbl.gov/NCS/netest.html.
6. Ribeiro, V.: pathChirp: Efficient Available Bandwidth Estimation for Network Path. In: PAM. (2003)
7. Jin, G., Yang, G., Crowley, B.R., Agarwal, D.A.: Network Characterization Service (NCS). Technical report, LBNL (2001)
8. Jain, M., Dovrolis, C.: Pathload: an available bandwidth estimation tool. In: PAM. (2002)

9. Strauss, J., Katabi, D., Kaashoek, F.: A measurement study of available bandwidth estimation tools. In: IMW. (2003)

10. Pasztor, A., Veitch, D.: On the Scope of End-to-end Probing Methods. Communications Letters, IEEE **6(11)** (2002)

11. Prasad, R., Jain, M., Dovrolis, C.: Effects of Interrupt Coalescence on Network Measurements. In: PAM. (2004)

12. Prasad, R., Dovrolis, C., Mah, B.: The effect of layer-2 store-and-forward devices on per hop capacity estimation. In: IMW. (2002)

13. Coccetti, F., Percacci, R.: Bandwidth measurements and router queues. Technical Report INFN/Code-20 settembre 2002, Instituto Nazionale Di Fisica Nucleare, Trieste, Italy (2002) http://ipm.mib.infn.it/bandwidth-measurements-and-router-queues.pdf.

14. Ubik, S., Smotlacha, V., Simar, N.: Performance monitoring of high-speed networks from the NREN perspective. In: TERENA Networking Conference, Rhodes, Greece. (2004) http://staff.cesnet.cz/ ubik/publications/2004/terena2004.pdf.

15. NLANR: Iperf v1.7.0 (2004) http://dast.nlanr.net/Projects/Iperf.

16. Cottrell, L., Logg, C.: Overview of IEPM-BW Bandwidth Testing of Bulk Data Transfer. In: Sc2002: High Performance Networking and Computing. (2002)

17. San Diego Supercomputer Center : CalNGI Network Performance Reference Lab (NPRL) (2004) http://www.calngi.org/about/index.html.

18. Jorgenson, L.: Size Matters: Network Performance on Jumbo Packets. In: Joint Techs Workshop Columbus, OH. (2004)

19. Keys, K., Moore, D., Koga, R., Lagache, E., Tesch, M., Claffy, k.: The architecture of CoralReef: an Internet traffic monitoring software suite. In: Passive and Active Network Measurement (PAM) Workshop, Amsterdam, Netherlands. (2001)

20. Brownlee, N.: NeTraMet 5.0b3 (2004) http://www.caida.org/tools/measurement/netramet/.

21. Spirent Corp.: Smartbits 6000B (2004) http://spirentcom.com/analysis/view.cfm?P=141.

22. Spirent Corp.: Smartflow (2004) http://spirentcom.com/analysis/view.cfm?P=119.

23. NLANR: Passive Measurement Analysis Datacube (2004) http://pma.nlanr.net/Datacube/.

24. Turner, A.: tcpreplay 2.2.2 - a tool to replay saved tcpdump files at arbitrary speed (2004) http://tcpreplay.sourceforge.net/.

25. SLAC: Internet End-to-end Performance Monitoring - Bandwidth to the World (IEPM-BW) Project (2004) http://www-iepm.slac.stanford.edu/bw/.

26. Dovrolis, C., Prasad, R., Jain, M.: Socket Buffer Auto-Sizing for High-Performance Data Transfers. Journal of Grid Computing **1(4)** (2004)

27. Mathis, M.: The Web100 Project (2003) http://www.web100.org.

28. Hyun, Y.: Running Bandwidth Estimation Tools on Wide-Area Internet Paths (2004) http://www.caida.org/projects/bwest/reports/tool-comparison-supplement.xml.

Traffic Classification Using a Statistical Approach

Denis Zuev[1] and Andrew W. Moore[2]

[1] University of Oxford, Mathematical Institute
zuev@maths.ox.ac.uk*
[2] University of Cambridge, Computer Laboratory
andrew.moore@cl.cam.ac.uk**

Abstract. Accurate traffic classification is the keystone of numerous network activities. Our work capitalises on hand-classified network data, used as input to a supervised Bayes estimator. We illustrate the high level of accuracy achieved with a supervised NaïveBayes estimator; with the simplest estimator we are able to achieve better than 83% accuracy on both a per-byte and a per-packet basis.

1 Introduction

Traffic classification enables a variety of other applications and topics, including Quality of Service, security, monitoring, and intrusion-detection that are of use to researchers, accountants, network operators and end users. Capitalising on network traffic that had been previously hand-classified provides us with training and testing data-sets. We use a *supervised* Bayes algorithm to demonstrate an accuracy of better than 66% of flows and better than 83% for packets and bytes. Further, we require only the network protocol headers of unknown traffic for a successful classification stage.

While machine-learning has been used previously for network-traffic/flow classification e.g., [1], we consider our work to be the first that combines this technique with the use of accurate test and training data-sets.

2 Experiment

In order to perform analysis of data using the NaïveBayes technique, appropriate input data is needed. To do this, we capitalised on trace-data described and categorised in [2]. This classified data was further reduced, and split into 10 equal time intervals each containing around 25,000–65,000 objects (flows). To evaluate the performance of the NaïveBayes technique, each dataset was used as a training set in turn and evaluated against the remaining datasets, allowing computation of the average accuracy of classification.

Traffic Categories. Fundamental to classification work is the idea of classes of traffic. Throughout this work we use classes of traffic defined as a common group of user-centric applications. Other users of classification may have both simpler definitions,

---

* This work was completed when Denis Zuev was employed by Intel Research, Cambridge.
** Andrew Moore thanks the Intel Corporation for its generous support of his research fellowship.

C. Dovrolis (Ed.): PAM 2005, LNCS 3431, pp. 321–324, 2005.

e.g., Normal versus Malicious, or more complex definitions, e.g., the identification of specific applications or specific TCP implementations.

Described further in [2], we consider the following categories of traffic: BULK (e.g., ftp), DATABASE (i.e., postgres, etc.), INTERACTIVE (ssh, telnet), MAIL (smtp, etc.), SEVICES (X11, dns), WWW, P2P (e.g., KaZaA, . . .), ATTACK (virus and worm attacks), GAMES (Half-Life, . . .), MULTIMEDIA (Windows Media Player, . . .).

Importantly, the characteristics of the traffic within each category are not necessarily unique. For example, the BULK category which is made up of ftp traffic, consists of both the control channel, which transfers data in both directions, and the data channel consisting a simplex flow of data for each object transferred. The assignment of categories to applications is an artificial grouping that further illustrates that such arbitrary clustering of only-minimally-related traffic-types is possible with our approach.

Objects and Discriminators. Our central object for classification is the flow and for the work presented in this extended-abstract we have limited our definition of a flow to being a complete TCP flow — that is all the packets between two hosts for a specific tuple. We restrict ourselves to complete flows, those that start and end validly, e.g., with the first SYN, and the last FIN ACK.

As noted in Section 1, the application of a classification scheme requires the parameterisation of each object to be classified. Using these parameters, the classifier allocates an object to a class, due to their ability to allow discrimination between classes. We refer to these object-describing parameters as discriminators. In our research we have used 249 different discriminators to describe traffic flows, these include: flow duration statistics, TCP Port information, payload size statistics, fourier transform of the packet interarrival time discriminators — a complete list is given in [3].

3   Method

Machine Learned Classification. Here we briefly describe the machine learning (ML) approach we take, a trained NaïveBayes classifier, along with a number of the refinements we use. We would direct interested readers to [4] for one of many surveys of all ML techniques.

Several methods exist for classifying data and all of them fall into two broad classes: deterministic and probabilistic classification. As the name suggests, deterministic classification assigns data points to one of a number of mutually-exclusive classes. This is done by considering some metric that defines the distance between data points and by defining the class boundaries. On the other hand, the probabilistic method classifies data by assigning it with probabilities of belonging to each class of interest.

We believe that probabilistic classification of Internet traffic, and our approach in particular, is more suitable given the need to be robust to measurement error, to allow for supervised training with pre-classified traffic, to be able to identify similar characteristics of flows after their probabilistic class assignment. We believe that the method be tractable and understood, and be able to cope with the unstable-dynamic nature of Internet traffic and that the method allow identification of when the model requires retraining. Additionally, the method needs to be available in a number of implementations.

Naïve Bayesian Classifier. The main approach that is used in this work is the NaïveBayes technique described in [5]. Consider a collection of flows $\mathbf{x} = (x_1, \ldots, x_n)$, where each flow x_i is described by m discriminators $\{d_1^{(i)}, \ldots, d_m^{(i)}\}$ that can take either numeric or discrete values. In the context of the Internet traffic, $d_j^{(i)}$ is a discriminant of flow x_i, for example it may represent the mean interarrival time of packets in the flow x_i. In this paper, flows x_i belong to exactly one of the mutually-exclusive classes described in Section 2. The supervised Bayesian classification problem deals with building a statistical model that describes each class based on some training data, and where each new flow y receives a probability of getting classified into a particular class according to the Bayes rule below,

$$p(c_j \mid y) = \frac{p(c_j)f(y \mid c_j)}{\sum\limits_{c_j} p(c_j)f(y \mid c_j)} \tag{1}$$

where $p(c_j)$ denotes the probability of obtaining class c_j independently of the observed data, $f(y \mid c_j)$ is the distribution function (or the probability of y given c_j) and the denominator acts as a normalising constant.

The NaïveBayes technique that is considered in this paper assumes the independence of discriminators $d_1, \ldots, d_m$ as well as the simple Gaussian behaviour of them. The authors understand that these assumptions are not realistic in the context of the Internet traffic, but [5] suggest that this model sometimes outperforms certain more complex models.

4   Naïve Bayes Results

Our experiments have shown that the NaïveBayes technique classified on average 66.71% of the traffic correctly. Table 1 demonstrates the classification accuracy of this techinique for each class. It can be seen from this table, that SERVICES and BULK are very well classified, with around 90% of correctly-predicted flows. In comparision to other results, it could be concluded that most discriminator distributions are well separated in the Euclinean space.

Table 1. Average accuracy of classification of NaïveBayes technique by class and Probability that the predictive class is the real class

	WWW	MAIL	BULK	SERV	DB	INT	P2P	ATT	MMEDIA	GAMES
Accuracy (%)	65.97	56.85	89.26	91.19	20.20	22.83	45.59	58.08	59.45	1.39
Probability (%)	98.31	90.69	90.01	35.92	61.78	7.54	4.96	1.10	32.30	100.00

At this stage, it is important to note why certain classes performed very poorly. Classes such as GAMES and INTERACTIVE do not contain enough samples, therefore, NaïveBayes training on these classes is not accurate or realistic. ATTACK flows were often confused with the WWW flows, due to the similarity in discriminators.

Alongside accuracy we consider it important to define several other metrics describing the classification technique. Table 1 shows how traffic from different classes gets classified — clearly an important measure. However, if a network administrator were to use our tool they would be interested in finding out how much trust can be placed in the classification outcome. Table 1 also shows the average probability that the predicted flow class is in fact the real class, e.g., if flow x_i has been classified as WWW, a measure of trust gives us a probability that x_i is in reality WWW.

A further indication of how well the NaïveBayes technique performs is to analyse the volume of accurately-classified bytes and packets. The results obtained are: 83.98% and 83.93% of packets and bytes, respectively, were correctly classified by the NaïveBayes technique described above. In contrast port-based classification achieved an accuracy of 71.02% by packet and 69.07% by bytes (from [2]). Comparing results in this way highlights the significant improvement of our NaïveBayes technique over the port-based classification alone.

5   Conclusions and Further Work

We demonstrate that, in its simplest form, our probabilistic-classification is capable of 67% accuracy per-flow or better than 83% accuracy both per-byte and per-packet. We maintain that access to a full-payload trace, the only definitive way to characterise network applications, will be limited due to technical and legal restrictions. We illustrate how data gathered without those restrictions may be used as training input for a statistical classifier which in turn can provide accurate, albeit estimated, classification of header-only trace data.

References

1. Anthony McGregor et al.: Flow Clustering Using Machine Learning Techniques. In: Proceedings of the Fifth Passive and Active Measurement Workshop (PAM 2004). (2004)
2. Moore, A.W., Papagiannaki, K.: Toward the accurate identification of network applications. In: Passive & Active Measurement Workshop 2005 (PAM2005), Boston, MA (2005)
3. Moore, A., Zuev, D.: Discriminators for use in flow-based classification. Technical Report, Intel Research, Cambridge (2005)
4. Mitchell, T.: Machine Learning. McGraw Hill (1997)
5. Witten, I.H., Frank, E.: Data Mining. Morgan Kaufmann Publishers (2000)

Self-Learning IP Traffic Classification Based on Statistical Flow Characteristics

Sebastian Zander[1], Thuy Nguyen[1], and Grenville Armitage[1]

Centre for Advanced Internet Architectures (CAIA),
Swinburne University of Technology, Melbourne, Australia
{szander, tnguyen, garmitage}@swin.edu.au

Abstract. A number of key areas in IP network engineering, management and surveillance greatly benefit from the ability to dynamically identify traffic flows according to the applications responsible for their creation. Currently such classifications rely on selected packet header fields (e.g. destination port) or application layer protocol decoding. These methods have a number of shortfalls e.g. many applications can use unpredictable port numbers and protocol decoding requires high resource usage or is simply infeasible in case protocols are unknown or encrypted. We propose a framework for application classification using an unsupervised machine learning (ML) technique. Flows are automatically classified based on their statistical characteristics. We also propose a systematic approach to identify an optimal set of flow attributes to use and evaluate the effectiveness of our approach using captured traffic traces.

1 Introduction

Over recent years there has been a dramatic increase in the variety of applications used in the Internet. Besides the 'traditional' applications (e.g. email, web) new applications have gained strong momentum (e.g. gaming, P2P). The ability to dynamically classify flows according to their applications is highly beneficial in a number of areas such as trend analysis, network-based QoS mapping, application-based access control, lawful interception and intrusion detection.

The most common identification technique based on the inspection of 'known port numbers' suffers because many applications no longer use fixed, predictable port numbers. Some applications use ports registered with the Internet Assigned Numbers Authority (IANA) but many applications only utilise 'well known' default ports that do not guarantee an unambiguous identification. Applications can end up using non-standard ports because (i) non-privileged users often have to use ports above 1024, (ii) users may be deliberately trying to hide their existence or bypass port-based filters, or (iii) multiple servers are sharing a single IP address (host). Furthermore some applications (e.g. passive FTP) use dynamic ports unknowable in advance.

A more reliable technique involves stateful reconstruction of session and application information from packet contents. Although this avoids reliance on fixed port numbers, it imposes significant complexity and processing load on the

---

[1] Work supported by Cisco Systems, Inc under the University Research Program.

C. Dovrolis (Ed.): PAM 2005, LNCS 3431, pp. 325–328, 2005.

identification device, which must be kept up-to-date with extensive knowledge of application semantics, and must be powerful enough to perform concurrent analysis of a potentially large number of flows. This approach can be difficult or impossible when dealing with proprietary protocols or encrypted traffic. The authors of [1] propose signature-based methods to classify P2P traffic. Although these approaches are more efficient than stateful reconstruction and provide better classification than the port-based approach they are still protocol dependent.

Machine Learning (ML) automatically builds a classifier by learning the inherent structure of a dataset depending on the characteristics of the data. Classification in a high dimensional attributes space is a big challenge for humans and rule-based methods, but stochastic ML algorithms can easily perform this task. The use of stochastic ML for traffic classification was raised in [2], [3] and [4]. However, to the best of our knowledge no systematic approach for application classification and evaluation has been proposed and an understanding of possible achievements and limitations is still lacking. We propose a detailed framework for self-learning flow classification based on statistical flow properties that includes a systematic approach of identifying the optimal set of flow attributes that minimizes the processing cost, while maximizing the classification accuracy. We evaluate the effectiveness of our approach using traffic traces collected at different locations in the Internet.

2 Related Work

Previous work used a number of different parameters to describe network traffic (e.g. [1], [5], [6]). The idea of using stochastic ML techniques for flow classification was first introduced in the context of intrusion detection [2]. The authors of [7] use principal component analysis and density estimation to classify traffic into different applications. They use only two attributes and their evaluation is based on a fairly small dataset. In [3] the authors use nearest neighbour and linear discriminate analysis to separate different application types (QoS classes). This supervised learning approach requires an a-priori knowledge of the number of classes. Also, it is unclear how good the discrimination of flows is because in [3] the sets of attributes are averaged over all flows of certain applications in 24-hour periods. In [4] the authors use the Expectation Maximization (EM) algorithm to cluster flows into different application types using a fixed set of attributes. From their evaluation it is not clear what influence different attributes have and how good the clustering actually is.

3 ML-Based Flow Classification Approach and Evaluation

As initial input we use traffic traces or capture data from the network. First we classify packets into flows according to IP addresses, ports, and protocol and compute the flow characteristics. The flow characteristics and a model of the flow attributes are then used to learn the classes (1). Once the classes have been learned new flows can be classified (2). The results of the learning and classification can be exported for evaluation. The results of the classification would be used for e.g. QoS mapping, trend analysis etc. We define a flow as a bidirectional series of IP packets with the

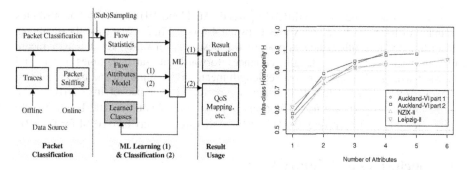

Fig. 1. ML-based flow classification   **Fig. 2.** Intra-class homogeneity

same source and destination address, port numbers and protocol (with a 60 second flow timeout). Our attribute set includes packet inter-arrival time and packet length mean and variance, flow size (bytes) and duration. Aside from duration all attributes are computed in both directions. We perform packet classification using NetMate [8], which supports flexible flow classification and can easily be extended with new flow characteristics. For the ML-based classification we use autoclass [9], an implementation of the Expectation Maximization (EM) algorithm [10]. EM is an unsupervised Bayesian classifier that automatically learns the 'natural' classes (also called clustering) inherent in a training dataset with unclassified cases. The resulting classifier can then be used to classify new cases (see [4], [9]).

For the evaluation we use the Auckland-VI, NZIX-II and Leipzig-II traces from NLANR [11] captured in different years at different locations. Because the learning process is slow we use 1,000 randomly sampled flows for eight destination ports (FTP data, Telnet, SMTP, DNS, HTTP, AOL Messenger, Napster, Half-Life), which results in a total of 8,000 flows. Finding the combination of attributes that provides the most contrasting application classes is a repeated process of (i) selecting a subset of attributes, (ii) learning the classes and (iii) evaluating the class structure.

We use sequential forward selection (SFS) to find the best attribute set because an exhaustive search is not feasible. The algorithm starts with every single attribute. The attribute that produces the best result is placed in a list of selected attributes SEL(1). Then all combinations of SEL(1) and a second attribute not in SEL(1) are tried. The combination that produces the best result becomes SEL(2). The process is repeated until no further improvement is achieved. To assess the quality of the resulting classes we compute the intra-class homogeneity H. We define C and A as the total numbers of classes and applications respectively. If N_{ac} is the number of flows of application a that fall into class c and N_c is the total number of flows in class c H_c is defined as:

$$H_c = \max(\frac{N_{ac}}{N_c} | 0 \le a \le A-1) \qquad (0<H\le1) \qquad (1)$$

For each trial H is the mean of H_c for $0 \le c \le C-1$ and the objective is to maximize H to achieve a good separation between different applications. For the evaluation we assume a flow's destination port defines the application. This may be incorrect (as stated initially) but we assume it is true for a majority of the flows. Unfortunately public available traces do not contain payload information usable for verification.

For each trace (and for two different parts of Auckland-VI) the best set of attributes found is different and the size varies between 4-6 (see Fig.2.). We rank the attributes according to how often they appear in the best set: forward packet length mean, forward/backward packet length variance, forward inter-arrival times mean and forward size (75%), backward packet length mean (50%), duration and backward size (25%). Clearly, packet length statistics are preferred over packet inter-arrival time statistics for the ports we use. The average maximum H is 0.87 ± 0.02 but H greatly differs for different ports (e.g. 0.98 ± 0.01 for Half-Life but only 0.74 ± 0.14 for HTTP).

4   Conclusions and Future Work

We have proposed a framework for ML-based flow classification based on statistical flow properties, identified a systematic approach of identifying an optimal set of flow attributes and evaluated the effectiveness of our approach. The results show that some separation of the applications can be achieved if the flow attributes are chosen properly. We plan to evaluate our approach with a larger number of flows and more applications (e.g. audio/video streaming). We hope to get traces that contain payload information usable for verifying the actual applications. We also plan to experiment with more attributes (e.g. idle time, burstiness) and possibly use payload information in a protocol independent way. Furthermore the precision of the resulting classifier and the classification performance has not yet been evaluated.

References

1. S. Sen, O. Spatscheck, D. Wang, "Accurate, Scalable In-Network Identification of P2P Traffic Using Application Signatures", WWW 2004, New York, USA, May 2004.
2. J. Frank, "Machine Learning and Intrusion Detection: Current and Future Directions", Proceedings of the National 17th Computer Security Conference, 1994.
3. M. Roughan, S. Sen, O. Spatscheck, N. Duffield, "Class-of-Service Mapping for QoS: A statistical signature-based approach to IP traffic classification", ACM SIGCOMM Internet Measurement Workshop 2004, Taormina, Sicily, Italy.
4. A. McGregor, M. Hall, P. Lorier, J. Brunskill, "Flow Clustering Using Machine Learning Techniques", Passive & Active Measurement Workshop 2004, France, April, 2004.
5. K. Lan, J. Heidemann, "On the correlation of Internet flow characteristics", Technical Report ISI-TR-574, USC/Information Sciences Institute, July, 2003.
6. K. Claffy, H.-W. Braun, G. Polyzos, "Internet Traffic Profiling", CAIDA, San Diego Supercomputer Center, http://www.caida.org/ outreach/papers/1994/itf/ , 1994.
7. T. Dunnigan, G. Ostrouchov, "Flow Characterization for Intrusion Detection", Oak Ridge National Laboratory, Tech Report, http://www.csm.ornl.gov/~ost/id/tm.ps, November 2000.
8. NetMate, http://sourceforge.net/projects/netmate-meter/ (as of January 2005).
9. P. Cheeseman, J. Stutz, "Bayesian Classification (Autoclass): Theory and Results", Advances in Knowledge Discovery and Data Mining, AAAI/MIT Press, USA, 1996.
10. A. Dempster, N. Laird, D. Rubin, "Maximum Likelihood from Incomplete Data via the EM Algorithm, Journal of Royal Statistical Society, Series B, Vol. 30, No. 1, 1977.
11. NLANR traces: http://pma.nlanr.net/Special/ (as of January 2005).

Measured Comparative Performance of TCP Stacks

Sam Jansen[1] and Anthony McGregor[1,2]

[1] Department of Computer Science, The University of Waikato
[2] National Laboratory for Applied Network Research (NLANR)*

Abstract. This extended abstract present findings on measured TCP performance of a range of network stacks. We have found that there are significant differences between the TCP implementations found in Linux, FreeBSD, OpenBSD and Windows XP.

1   Introduction

Implementations of a protocol vary in many respects, including how well they perform. There are several reasons for this variation. When implementing an Internet protocol, the programmer will refer to the protocol's specifications, for example its RFC. These specifications are normally written in English and may be ambiguous. Some aspects of the protocols behaviour may be left to the implementor. Even with a very tightly specified protocol, implementations do not always correctly meet the specification. This may be due to logic errors or misinterpretations of the specification. In some cases decisions are made that violate the specification to gain better performance.

In 1997 Paxson analysed TCP by writing a tool to automatically analyse a large amount of trace data he had available [1]. This was successful in finding implementation problems in a range of TCP variants of the time. However, the tool has a serious limitation: the code needs to be updated and hand crafted for every TCP implementation that is studied.

Previous work looking at TCP performance has looked at specific types of congestion control [2], sometimes under specific conditions such as a mobile ad hoc network [3], a lossy radio link [4] or ATM [5]. Studies comparing variants of TCP have also been performed, for example comparing New Reno and Vegas and Westwood+ [6]. Paxson's research involved TCP stacks from 1997: Solaris 2.4, NetBSD 1.0, Linux 1.0, Windows 95, Windows NT and others. TCP has evolved significantly in the past 7 years; this paper focuses on TCP implementations used in 2004 and 2005.

We hypothesise that todays TCP implementations will perform correctly under congestion regardless of their BSD lineage, in contrast to Paxson's findings.

* NLANR Measurement and Network Analysis Group (NLANR/MNA)is supported by the National Science Foundation (NSF) under cooperative agreement no. ANI-0129677.

C. Dovrolis (Ed.): PAM 2005, LNCS 3431, pp. 329–332, 2005.

Further, we believe that TCP implementations have diversified sufficiently that there are significant differences in measured performance between implementations, whether of BSD lineage or not. We use a test-bed network called the *WAND Emulation Network* which is described in the next section. Some measured TCP performance results are presented in Sect 3.

2   Emulation Network

The WAND Network Research Group has built a network of 24 machines dedicated to network testing. Machines are configured so there is a *control network* connecting the machines to the control machine and an *emulation network* which is configured by changing patch panels. Each machine has one Ethernet card connected to the control network, and one Ethernet card connected to the emulation network, which has four ports in the case of router machines. This allows arbitrary network topologies to be created between machines at a maximum speed of 100Mbit/s.

All machines are connected through one central switch to a control machine as well as having serial connections to the same machine. To simulate link delay and bandwidth limits, FreeBSD Dummynet [7] routers are used.

The control machine is able to install operating system images onto the machines on the emulation network in less than five minutes, making it possible to test a variety of different operating systems in a short time span. Images of Linux, FreeBSD, OpenBSD, Solaris and Windows XP are available.

It is possible to write scripts that run commands on the machines on the emulation network and send their output back to the control machine. This makes it possible to design, execute and record tests on the control machine.

3   TCP Performance

The following tests have been performed with the following operating systems: Linux, FreeBSD, OpenBSD and Windows XP with Service Pack 2.

3.1   Bidirectional Random Loss

This section presents a study of TCP performance under random loss in both directions; that is, both data and acknowledgement packets are dropped randomly using a uniform model. Random loss is interesting to study because Lakshman and Madhow [8] report that random loss is a simple model for transient congestion and of interest in the context of networks with multimedia traffic.

Figure 1 shows the topology used in this test. The bottleneck link is configured to have a propagation delay of 100ms and bandwidth limited at 2Mb/s. Router R1 drops packets randomly using Dummynet's *packet loss rate* option. The goodput over a single TCP stream from host H1 to H2 is measured. *Goodput* is the amount of data successfully read from the TCP socket by the application at the receiving end of a TCP connection. Hosts H3 and H4 are unused. Each test lasted 60 seconds.

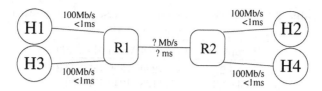

Fig. 1. Test network setup

Table 1. TCP performance during 5% bidirectional loss

TCP Implementation	Min	Mean	Max	SD
Linux 2.6.10	164.3	213.9	287.6	22.7
Linux 2.4.27	153.8	207.4	248.7	22.8
FreeBSD 5.3	136.7	176.2	225.0	17.1
FreeBSD 5.2.1	128.7	162.8	219.0	19.5
Windows XP SP2	89.9	137.3	191.0	21.6
OpenBSD 3.5	63.8	117.9	166.8	22.1

Table 1 shows recorded performance in kilobits per second under 5% random loss. The four numbers in the table are recorded goodput: minimum, mean, max and standard deviation. For each network stack, the test was run 100 times. All tests were run with kernel parameters with their defaults. Increasing the buffer sizes of any of the stacks studied made little difference, even though Windows XP defaults to only 8kB (compared to up to 64kB on other operating systems). While there is variation from run to run, it is small compared to the mean. Measurements with SACK turned on and off showed that SACK increases performance by just over 5% in this scenario.

3.2   Reverse Path Congestion

The test network topology is the same as presented in the previous section in Fig. 1. No artificial loss is added by routers R1 or R2 but host H4 sends data over a single TCP stream to host H3. The TCP stream from host H1 to host H2 is measured. The buffer sizes on routers R1 and R2 are set to 8 packets, the bottleneck link is set at 2Mb/s with 50ms delay. This allows R2 to be congested by the TCP stream from H4 to H3 which has the effect of congesting the acks of the measured TCP stream. H3 and H4 use Linux 2.4.27 while the operating system on H1 and H2 vary. The stacks are configured as in Sect. 3.1 apart from the size of the TCP socket buffers. The TCP socket buffer size for both receive and send buffers are set to 64kB for all network stacks in the test.

Table 2 shows the measured goodput at host H2 in kilobits per second. The variation between stacks is not as large as in the previous section but there is still a significant difference of 32% between the lowest and highest.

Table 2. TCP performance during reverse path congestion

TCP Implementation	Min	Mean	Max	SD
Linux 2.4.27	1220	1296	1375	33.3
FreeBSD 5.3	1128	1242	1366	52.8
FreeBSD 5.2.1	1099	1205	1289	48.6
Windows XP	906	1024	1152	58.5
OpenBSD 3.5	1273	1352	1438	40.4

4   Summary

This abstract shows that there is a large difference between the measured performance of the TCP stacks studied: Linux, FreeBSD, OpenBSD and Windows XP. During bidirectional random loss, the Linux TCP stack is able to obtain the most goodput by quite a long way. In this scenario OpenBSD is only able to achieve just over half the goodput that was measured with Linux 2.4 and 2.6 kernels while Windows XP achieves just 64% of the goodput measured with Linux. Windows XP is additionally limited by its default TCP window sizes, which are very small by today's standards.

Further analysis of these results is not presented because of the lack of space available.

References

1. Paxson, V.: Automated packet trace analysis of TCP implementations. In: SIG-COMM. (1997) 167–179
2. Fall, K., Floyd, S.: Comparison of Tahoe, Reno and SACK TCP (1995)
3. Holland, G., Vaidya, N.H.: Analysis of TCP performance over mobile ad hoc networks. In: Mobile Computing and Networking. (1999) 219–230
4. Kumar, A.: Comparative performance analysis of versions of TCP in a local network with a lossy link. IEEE/ACM Transactions on Networking **6** (1998) 485–498
5. Comer, D., Lin, J.: TCP buffering and performance over an ATM network (1995)
6. Grieco, L.A., Mascolo, S.: Performance evaluation and comparison of Westwood+, New Reno, and Vegas TCP congestion control. SIGCOMM Comput. Commun. Rev. **34** (2004) 25–38
7. Rizzo, L.: Dummynet: a simple approach to the evaluation of network protocols. ACM Computer Communication Review **27** (1997) 31–41
8. Lakshman, T.V., Madhow, U.: The performance of TCP/IP for networks with high bandwidth-delay products and random loss. IEEE/ACM Transactions on Networking **5** (1997)

Applying Principles of Active Available Bandwidth Algorithms to Passive TCP Traces

Marcia Zangrilli and Bruce B. Lowekamp

College of William and Mary, Williamsburg VA 23187, USA
{mazang, lowekamp}@cs.wm.edu

Abstract. While several algorithms have been created to actively measure the end-to-end available bandwidth of a network path, they require instrumentation at both ends of the path, and the traffic injected by these algorithms may affect the performance of other applications on the path. Our goal is to apply the self-induced congestion principle to passive traces of existing TCP traffic instead of actively probing the path. The primary challenge is that, unlike active algorithms, we have *no control over the traffic pattern* in the passive TCP traces. As part of the Wren bandwidth monitoring tool, we are developing techniques that use single-sided packet traces of existing application traffic to measure available bandwidth. In this paper, we describe our implementation of available bandwidth analysis using passive traces of TCP traffic and evaluate our approach using bursty traffic on a 100 Mb testbed.

1   Introduction

Available bandwidth is typically measured by actively injecting data probes into the network. The active approach often produces accurate measurements, but it may cause competition between application traffic and the measurement traffic, reducing the performance of useful applications. Most of these active algorithms rely on UDP traffic to probe the path for available bandwidth, however applications typically use TCP traffic. Because UDP and TCP traffic may be packet-shaped differently along the same path, measurements made with UDP traffic may not reflect the actual bandwidth available to TCP applications. Furthermore, these available bandwidth algorithms require instrumentation on both ends of the path, which may not always be possible.

Our goal is to use passive traces of existing TCP traffic instead of actively generating the traffic being used to measure available bandwidth. By monitoring the traffic that an application generates, we can calculate the available bandwidth even when the application has not generated sufficient traffic to saturate that path. Our available bandwidth measurements can be used by an application already generating traffic to determine if it can increase its sending rate, by network managers who are interested in observing traffic, capacity planning, SLA monitoring, etc., or by central monitoring systems[1, 2] that store measurements for future use or use by other applications.

C. Dovrolis (Ed.): PAM 2005, LNCS 3431, pp. 333–336, 2005.

Because our approach uses existing application traffic to measure available bandwidth, the monitored traffic is not an additional burden on the path and experiences the same packet shaping issues affecting applications. To achieve the necessary accuracy and avoid intrusiveness, our passive monitoring system uses the Wren packet trace facility [3] to collect kernel-level traces of application traffic and analyzes the traces in the user-level. Our trace facilities can be deployed on one or two end hosts or on a single packet capture box, which is an advantage over tools that must be deployed on both ends of the path.

This paper describes how to apply the self-induced congestion principle to passive traces of application traffic, a task complicated because we have no control over the application traffic pattern. We describe an algorithm for applying the self-induced congestion principle to passive, one-sided traces of TCP traffic and demonstrate that our algorithm produces measurements that are responsive to changes in available bandwidth.

2   Background

Available bandwidth describes what portion of the path is currently unused by traffic. More precisely, available bandwidth is determined by subtracting the utilization from the capacity of the network path [4, 5]. In practice, available bandwidth may also be affected by traffic shapers that allow some traffic to consume more or less bandwidth than other traffic can consume.

The basic principle of the self-induced congestion (SIC) technique is that if packets are sent at a rate greater than the available bandwidth, the queuing delays will have an increasing trend, and the rate the packets arrive at the receiver will be less than the sending rate. If the one-way delays are not increasing and the rate the packets arrive is the same as the sending rate of the packets, then the available bandwidth is greater than or equal to the sending rate. Tools that utilize this concept [6, 7, 8, 9] probe the network path for the largest sending rate that does not result in queuing delays with an increasing trend because this sending rate reflects the available bandwidth of the path.

Proposed improvements to the TCP protocol have set a precedent for measuring available bandwidth on a single end host. Paced Start (PaST)[10] incorporates the self-induced congestion principle into the TCP protocol to reduce the amount of time taken before transitioning into the congestion avoidance phase.

3   Passive One-Sided SIC Implementation

Our one-sided passive SIC implementation uses the timestamps of data and ACK packets on the sending host to calculate the round trip times (RTT) and the initial sending rates of the stream of packets. Our implementation is similar to the pathload [8], which uses trends in one-way delays to determine the available bandwidth.

We group packets together into streams and identify the trend in RTTs of each packet group. We impose the condition that grouped packets are the same size so that all packets we consider have experienced the same store-and-forward delays at the links along the path. Because congestion window size often determines the sending rate of the TCP application, we also ensure that all packets grouped together have the same congestion window size. The number of packets in each group is determined by the congestion window size. For each stream of packets, we calculate the RTTs of each packet, calculate the initial sending rate, and determine if there is an increasing trend in the RTTs. We group several streams together and try to identify the maximum value for the available bandwidth. For each group, the stream with the largest sending rate and no increasing trend determines the available bandwidth.

To emulate traffic generated by on-off applications, we created traffic generators that send 256K messages with a variable delay. The variable delay causes the throughput of the generators to oscillate.

Figure 1 presents the results of applying our passive SIC approach to one-sided traces of two traffic generators. In the left graph, the average throughput of the traffic generator is 65 Mbps on an uncongested LAN. The traffic generator was run on a 100 Mb testbed with varying amounts of cross traffic present. This graph shows distinct bands that demonstrate our algorithm can detect the changes in the amount of available bandwidth.

The second traffic generator was designed to send out bursts of messages with varying throughput. In this experiment, there is 20 Mbps of cross traffic present for the first 15 seconds and 40 Mbps of cross traffic present for the last 15 seconds. In the right graph in Fig. 1, the line represents the throughput of the traffic generator and the points are the measurements produced by our passive algorithm. Notice how the third peak does not reach the true available bandwidth, but our algorithm is still able to produce an accurate measurement. This graph shows that our SIC algorithm is able to measure the available bandwidth using

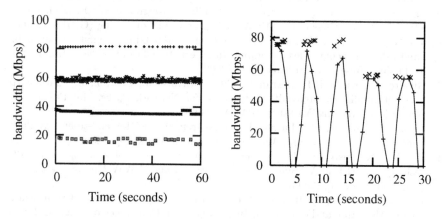

Fig. 1. The left graph shows how our SIC algorithm is responsive to changes in available bandwidth. The right graph demonstrates our SIC algorithm's ability to accurately measure the available bandwidth even when TCP throughput is ramping up

one-sided traces of application traffic with on-off communication patterns, even when the application traffic throughput is less than the available bandwidth.

4   Conclusion

We have described the implementation of a passive available bandwidth technique based on the self-induced congestion principle. Our preliminary evaluation of our one-sided passive SIC technique is quite promising and shows that we can obtain valid available bandwidth measurements in congested environments using bursty application traffic.

We are continuing to evaluate our one-sided passive SIC algorithms. We are interested in qualifying what types of traffic patterns are best suited for our algorithm to produce valid measurements, and performing more detailed analysis on the affects of traffic burstiness, bottlenecks, and delayed ACKs on the accuracy of our algorithm's measurements.

References

1. Adams, A., Mahdavi, J., Mathis, M., Paxson, V.: creating a scalable architecture for internet measurement. In: Proceedings of INET'98. (1998)
2. Wolski, R.: Forecasting network performance to support dynamic scheduling using the network weather service. In: Proceedings of the 6th High Performance Distributed Computing Conference (HPDC). (1997)
3. Zangrilli, M., Lowekamp, B.B.: Using passive traces of application traffic in a network monitoring system. In: High Performance Distributed Computing (HPDC 13). (2004)
4. Prasad, R., Murray, M., Dovrolis, C., Claffy, K.: Bandwidth estimation: Metrics, measurement techniques, and tools. In: IEEE Network. (2003)
5. Lowekamp, B.B., Tierney, B., Cottrell, L., Hughes-Jones, R., Kielmann, T., Swany, M.: Enabling network measurement portability through a hierarchy of characteristics. In: Proceedings of the 4th Workshop on Grid Computing (GRID). (2003)
6. Jin, G., Tierney, B.: Netest: A tool to measure the maximum burst size, available bandwidth and achievable throughput. In: International Conference on Information Technology Research and Education. (2003)
7. Ribeiro, V., Riedi, R.H., Baraniuk, R.G., Navratil, J., Cottrell, L.: pathChirp: Efficient Available Bandwidth Estimation for Network Paths. In: Passive and Active Measurement Workshop (PAM). (2003)
8. Jain, M., Dovrolis, C.: Pathload: a measurement tool for end-to-end available bandwidth. In: Passive and Active Measurements Workshop. (2002)
9. Hu, N., Steenkiste, P.: Evaluation and characterization of available bandwidth techniques. IEEE JSAC Special Issue in Internet and WWW Measurement, Mapping, and Modeling (2003)
10. Hu, N., Steenkiste, P.: Improving tcp startup performance using active measurements: Algorithm and evaluation. In: International Conference on Network Protocols (ICNP). (2003)

A Network Processor Based Passive Measurement Node

Ramaswamy Ramaswamy, Ning Weng, and Tilman Wolf

Department of Electrical and Computer Engineering,
University of Massachusetts Amherst, MA 01003
{rramaswa, nweng, wolf}@ecs.umass.edu

1   Introduction

The complexity of network systems and the heterogeneity of end systems will make networks increasingly difficult to manage. To understand the operational details of networks it is imperative that sufficient information on their behavior is available. This can be achieved through network measurement.

Passive network measurement systems typically collect packet traces that are then stored in trace databases. To extract information on the state of the network, the traces are searched and post-processed. In our work, we envision two extensions to this approach:

- **Distributed Measurement Nodes.** To provide a richer set of network management applications and traffic profiling capabilities, traffic is collected and correlated from multiple measurement nodes.
- **Preprocessing of Trace Data.** Scalability in distributed measurement is a key problem. The aggregate bandwidth of trace data from multiple measurement nodes can easily overwhelm a conventional database system. To alleviate this problem, we preprocess packet traces on the measurement node and perform simple statistics collection online.

The basic architecture of our measurement system is shown in Figure 1. In this paper, we discuss how to implement the packet capture and online preprocessing functions of this system on a network processor. Network processors are software programmable system-on-a-chip multiprocessors that are optimized for high bandwidth I/O and highly parallel processing of packets. We use the Intel IXP 2400 [1] network processor for our proposed measurement system. The IXP 2400 contains eight multi-threaded microengines for packet processing along with an XScale core processor to perform control plane related functions.

The measurement node performs three functions:

- **Packet Capture and Header Parsing:** Each packet is parsed to determine the sequence of headers that are present. This allows the consideration of nested protocol headers as well as different header sizes due to options.
- **Anonymization:** To ensure the privacy of network users, IP addresses are anonymized online on the network processor during trace collection.
- **Online Queries and Statistics Collection:** Packet traces can be preprocessed on the measurement node to reduce the load on the centralized

C. Dovrolis (Ed.): PAM 2005, LNCS 3431, pp. 337–340, 2005.

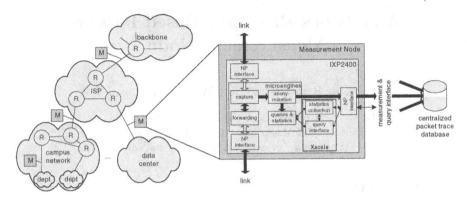

Fig. 1. Network Measurement Architecture

collection system. If traffic statistics match a particular query, a response is pushed from the measurement node.

2   Related Work

Traditionally, two approaches have been taken towards network measurement: active and passive.In the active approach (e.g., NLANR's AMP and Surveyor), a sender and/or receiver measure and record the traffic that they send/receive, obtaining end-to-end (e.g., path) characteristics.

In the passive approach, measurements are taken at a given point in a network and are typically used to characterize local properties of the network and its traffic. Traces of packets passing through a passive measurement point can be analyzed for traffic mix (e.g., protocol or application), or flow size and burstiness. The passive measurement projects that are most closely related to our proposed efforts here are Sprint's IPMON project [2], AT&T's Gigascope project [3] and NLANR's passive measurement efforts [4].

None of these efforts, however, leverage the use of network processors, which allow customized online queries. In this context the use of network processors is particularly crucial as complex centralized post-processing and storage of traffic traces can be off-loaded into the measurement node.

3   IP Address Anonymization

To ensure that no private information is revealed in a network trace, the IP source and destination addresses need to be anonymized. The main constraint on the anonymization algorithm is that it should be "prefix-preserving." Thus, some information on network-level characteristics of the measured traffic can be preserved across the anonymization step.

It is desirable to perform anonymization as early in the collection process as possible. By anonymizing header fields on the measurement node itself instead of external post-processing, it is less likely that unanonymized data is leaked. This requires the anonymization process to operate at a speed that can keep up with the link rates of the measurement node. This sort of *online* anonymization, however, cannot be achieved with current prefix-preserving anonymization algorithms ([5] and [6]), since they are computationally intensive.

We have developed a novel prefix-preserving anonymization algorithm, called TSA (top-hash subtree-replicated anonymization) [7], that addresses this problem by computing all necessary cryptographic functions offline. An IP address is anonymized by making a small number of accesses to a set of lookup tables in memory.

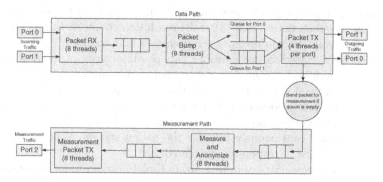

Fig. 2. Network Measurement Node on IXP2400

4   Measurement Node Prototype

The prototype implementation of the proposed measurement system is based on the IXP2400 network processor platform. The data flow and allocation of tasks to the underlying NP components is shown in Figure 2. The data path bumps incoming traffic from Port 0 to Port 1 and vice versa. Once a packet has successfully proceeded through the data path, it is enqued to the measurement part of the system. If this queue is full, the packet is dropped and no measurement tasks are performed on it. Thus, the measurement path is designed to have a minimal impact on the performance of the network processor in the data path. In the measurement path, packet headers are collected, IP addresses are anonymized, statistics are updated, and a "measurement packet" (which contains the packet headers and some meta data) is generated for each packet observed, and transmitted from Port 2.

The measurement system was simulated on the simulator for the IXP 2400 network processor. Simulation traffic consisted of unidirectional 60 byte TCP/IP packets over Ethernet. We were able to sustain a transmit rate of up to 1120Mbps (~900,000 packets per second) on the measurement port (Port 2). The measurement node was also tested on the Radisys ENP-2611 board [8], on a network access link of the University of Massachusetts. The node was observed to be functional at data rates of up to 140,000 packets per second.

5   Future Work

We are exploring an extension to the current measurement prototype that allows collection of online statistics and the implementation of queries to the measurement node. The key research question is what traffic statistics to collect and how this information can be accessed through the query interface. This issue is closely related to the capabilities of the underlying hardware.

We propose to implement simple counting functions on the microengines and leave more complex processing to the Xscale control processor. In particular, we consider collection of packet counts, layer 3 and 4 protocol distributions, counts of packets with special significance (e.g., TCP SYN packets). These statistics can be further extended to gather more information through (1) per-flow statistics, (2) window-based statistics, and (3) multi-resolution counters. In all three cases there is a tradeoff between memory requirements and accuracy.

For the query interface we consider two possible types of queries. Queries that "pull" information from the measurement node are comparable to those done on conventional packet trace collections. The query is sent to the system and the appropriate information is retrieved and sent in response. With the online operation of our system another type of query is possible. "Push" queries are such that they continuously monitor the packet stream. When a particular condition is matched, a response is triggered.

Finally, it is necessary to obtain accurate time information for timestamping. We are currently in the process of integrating a GPS receiver with the IXP2400 to operate an NTP-style clock synchronization mechanism on the Xscale control processor.

References

1. Intel Corp.: Intel Second Generation Network Processor. (2002) http://www.in-tel.com/design/network/products/npfamily/ixp2400.htm.
2. Fraleigh, C., Diot, C., Lyles, B., Moon, S.B., Owezarski, P., Papagiannaki, D., To-bagi, F.A.: Design and deployment of a passive monitoring infrastructure. In: Passive and Active Measurement Workshop, Amsterdam, Netherlands (2001)
3. Cranor, C., Gao, Y., Johnson, T., Shkapenyuk, V., Spatscheck, O.: Gigascope: High performance network monitoring with an SQL interface. In: Proc. of the 2002 ACM SIGMOD, Madison, WI (2002) 623
4. McGregor, T., Braun, H.W., Brown, J.: The NLANR network analysis infrastructure. IEEE Communications Magazine 38 (2000) 122–128
5. Minshall, G.: TCPDPRIV, (http://ita.ee.lbl.gov/html/contrib/tcpdpriv.html)
6. Xu, J., Fan, J., Ammar, M.H., Moon, S.B.: Prefix-preserving ip address anonymization: Measurement-based security evaluation and a new cryptography-based scheme. In: Proc. of 10th IEEE ICNP, Paris, France (2002) 280–289
7. Ramaswamy, R., Wolf, T.: High-speed prefix-preserving IP address anonymization for passive measurement systems. (under submission)
8. Radisys Corporation: ENP-2611 Product Data Sheet. (2004) http://www.radisys.com.

A Merged Inline Measurement Method for Capacity and Available Bandwidth

Cao Le Thanh Man, Go Hasegawa, and Masayuki Murata

Graduate School of Information Science and Technology, Osaka University
{mlt-cao, hasegawa, murata}@ist.osaka-u.ac.jp

Abstract. We have proposed a new TCP version, called ImTCP (Inline measurement TCP), in [1]. The ImTCP sender adjusts the transmission intervals of data packets, and then utilizes the arrival intervals of ACK packets for the *available bandwidth* estimation. This type of active measurement in a TCP connection (inline measurement) is preferred because it delivers measurement results that are as accurate as active measurement, even though no extra probe traffic is injected into the network. In the present research, we combine a new *capacity* measurement function with the currently used measurement method to enable simultaneous measurement of both capacity and available bandwidth in ImTCP. The capacity measurement algorithm is essentially based on the packet pair technique, but also consider the estimated available bandwidth values for data filtering or data calculation, so that this algorithm promises better measurement results than current packet-pair-based measurement algorithms.

Extended Abstract

The capacity of an end-to-end network path is the maximum possible throughput that the network path can provide. Traffic may reach this maximum throughput when there is no other traffic along the path. The available bandwidth indicates the unused bandwidth of a network path, which is the maximum throughput that newly injected traffic may reach without affecting the existing traffic. The two bandwidth-related values are obviously important with respect to adaptive control of the network. In addition, these two values are often both required at the same time. Although network transport protocols optimize link utilization according to capacity, congestion is also avoided by using the available bandwidth information. For routing or server selection in service overlay networks, information about both capacity and available bandwidth offers a better selection than either capacity or available bandwidth information alone. For example, when the available bandwidth fluctuates often and the transmission time is long, the capacity information may be a better criterion for the selection. However, when the available bandwidth appears steady during the transmission, then the available bandwidth should be used for the selection. Moreover, the billing policy of the Internet Service Provider should be based on both the capacity and the available bandwidth of the access link they are providing to the customer.

Several passive and active measurement approaches exist for capacity or available bandwidth. Active approaches are preferred because of their accuracy and speed. How-

C. Dovrolis (Ed.): PAM 2005, LNCS 3431, pp. 341–344, 2005.

ever, sending extra traffic onto the network is a common disadvantage that is shared by all active measurement tools.

We propose herein an active measurement method that does not add probe traffic to the network. The proposed method uses the concept of "plugging" the new measurement mechanism into an active TCP connection (*inline measurement*). Passive inline measurement appeared in TCP Westwood [2], in which the sender checks the ACK packet arrival intervals to infer the available bandwidth. We herein introduce ImTCP (Inline measurement TCP), a Reno-based TCP that deploys active inline measurement. The ImTCP sender not only observes the ACK packet arrival intervals, but also actively adjusts the transmission interval of data packets, just as active measurement tools use probe packets. When the corresponding ACK packets return, the sender utilizes the arrival intervals thereof to calculate the measurement values. The measurement algorithm in ImTCP combines the available bandwidth and capacity measurement algorithms. The available bandwidth measurement algorithm utilizes Self Loading Periodic streams (SLoPS) proposed in [3]. However, SLoPS is changed so that the algorithm can be applied to inline measurement. The available bandwidth algorithm is described in detail in [4]. The measured values of available bandwidth are then used to supplement the packet pair technique to deliver a better capacity estimation than traditional packet pair based techniques.

We insert a measurement program into the sender program of TCP Reno to create an ImTCP sender. The measurement program is located at the bottom of the TCP layer. When a new data packet is generated at the TCP layer and is ready to be transmitted, the packet is stored in an intermediate FIFO buffer. The measurement program decides the time at which to send the packets in the buffer. The program waits until the number of packets in the intermediate buffer is sufficient to form a packet stream for available bandwidth measurment and a packet pair for capacity measurement, in each RTT. After sending packets required for measurement, the program then passes all data packets immediately to the IP layer while waiting for the corresponding ACK packets. The measurement program does not require any special changes in the TCP receiver program, except that an ACK packet must be sent back for each received packet. Therefore, delayed ACKs must be disabled at the TCP receiver; otherwise ImTCP will not perform measurement properly.

The principle of the packet-pair-based measurement technique for capacity is that, if the packet pairs are transmitted in a back-to-back manner at the bottleneck link (the link of smallest capacity bandwidth in the network path) and the time interval until they reach the receiver remains unchanged, then the capacity of the bottleneck link C (which is also the capacity of the network path) is calculated as:

$$C = \frac{P}{Gap} \tag{1}$$

where P is the packet size and Gap is the arrival time dispersion of the two packets at the receiver. The packet pairs are referred to as the C-indicator. Their time dispersion indicates the exact capacity value. If the packet pair is cut by packets from other traffic, then its dispersion can not be used to calculate capacity via Equation (1).

Current packet-pair-based measurement techniques have various mechanisms for determining C-indicators from packet pair measurement results. Some tools assume a high frequency of appearance of the C-indicator, and so search for the C-indicator from a frequency histogram (Pathrate [5]) or a weighting function (Nettimer [6]). CapProbe [7] repeatedly sends packet pairs until it discovers a C-indicator, based on the transmission delay of the packets. However, as shown in the following equation, when the available bandwidth is small, the C-indicator does not appear frequently. Thus, current existing tools may not discover the correct capacity.

Let δ be the time space of the packet pair when it arrive at the bottleneck link. We then assume that the links before the bottleneck link do not have a noticeable effect on the time space, so that δ is the approximate time interval in which the sender sends the packets. During the time of δ, the average amount of cross traffic that arrives at the bottleneck link, which is denoted as L, is

$$L = \delta \cdot (C - A) \qquad (2)$$

where A is the available bandwidth at the time the packet pair is sent. We can see that when A is small, L is large, which means that the probability for a packet pair to pass the bottleneck link without being cut by the traffic of another packet is low. In other words, the available bandwidth of the path is an important factor in measuring the capacity. Based on the above observation, we develop a new capacity measurement algorithm, which exploits the advantage of awareness of the available bandwidth of ImTCP.

From Equation (2) we can estimate that the dispersion of the packet pair when leaving the bottleneck link is:

$$Gap = \frac{P + L}{C} = \frac{P + \delta \cdot (C - A)}{C}$$

Therefore, the capacity can be calculated as:

$$C = \frac{P - \delta \cdot A}{Gap - \delta} \qquad (3)$$

There is a problem with current capacity measurement tools when every packet pair that passes the bottleneck link is cut by other packets, due to either a heavy load or constant and aggressive cross traffic at the bottleneck link. In this case, CapProbe will spend an extremely long time searching for C-indicators, and Pathrate and Nettimer will deliver incorrect estimations. Equation (3) introduces some important prospects, including ways to overcome the above problem:

- We can calculate the capacity bandwidth without the existence of C-indicators, assuming that the available bandwidth value is known.
- The measurement does not require δ as the smallest value that the sender can create. Any two TCP data packets that are sent in an appropriately small interval can be exploited for the calculation. This is a very important advantage because more data can be collected for the capacity search.
- We can discuss the statistical confidence of the measurement results based on the value of the variance of the calculated data.

We present a simulation of packet pair measurement as an example explaining Equation (3). We perform a simulation of packet pair measurements over 50 seconds on a

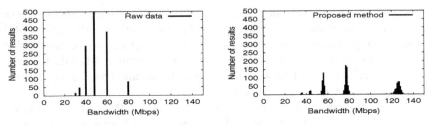

Fig. 1. Results calculated using Equation (1) and (3)

network path for which the available bandwidth is 15 Mbps during the time. The background traffic is made up of an UDP packet flow. The UDP packet size is 500 KB. The correct capacity of the path is 80 Mbps. In Figure 1, the "Raw data" graph shows the measurement results calculated using Equation (1), and the "Proposed method" graph shows the results obtained by using Equation (3). We can conclude that Equation (3) provides a better result for capacity because the calculated data concentrate at the correct value of capacity (80 Mbps).

References

1. Cao Man, Go Hasegawa and Masayuki Murata, "Available bandwidth measurement via TCP connection," in *Proceeding of the 2nd Workshop on End-to-End Monitoring Techniques and Services E2EMON*, Oct. 2004.
2. M.Gerla, B.Ng, M.Sanadidi, M.Valla, R.Wang, "TCP Westwood with adaptive bandwidth estimation to improve efficiency/friendliness tradeoffs," *To appear in Computer Communication Journal*.
3. M. Jain and C. Dovrolis, "End-to-end available bandwidth: Measurement methodology, dynamics, and relation with TCP throughput," in *Proceedings of ACM SIGCOMM 2002*, Aug. 2002.
4. Cao Man, Go Hasegawa and Masayuki Murata, "A new available bandwidth measurement technique for service overlay networks," in *Proceeding of 6th IFIP/IEEE International Conference on Management of Multimedia Networks and Services Conference, MMNS2003*, pp. 436–448, Sept. 2003.
5. C. Dovrolis and D. Moore, "What do packet dispersion techniques measure?," in *Proceedings of IEEE INFOCOM 2001*, pp. 22–26, Apr. 2001.
6. K. Lai and M. Baker, "Nettimer: A tool for measuring bottleneck link bandwidth," in *Proceedings of the USENIX Symposium on Internet Technologies and Systems*, Mar. 2001.
7. R. Kapoor, L. Chen, L. Lao, M. Gerla and M. Sanadidi, "Capprobe: a simple and accurate capacity estimation technique," in *Proceedings of the 2004 Conference on Applications, Technologies, Architectures, and Protocols for Computer Communications*, 2004.

Hopcount and E2E Delay: IPv6 Versus IPv4

Xiaoming Zhou and Piet Van Mieghem

Delft University of Technology,
Electrical Engineering, Mathematics and Computer Science,
P.O Box 5031, 2600 GA Delft, The Netherlands
{X.Zhou, P.VanMieghem}@ewi.tudelft.nl

Abstract. IPv6 provides an expanded address space to satisfy the future Internet requirements. In this paper we compare and analyze one-month measurements of the end-to-end IPv6 delay and hopcount between 26 testboxes of the RIPE TTM project with the corresponding parts in IPv4 network. By comparing IPv6 and IPv4 paths, we focus on problems that are only present in the IPv6 paths. In those poorly performing IPv6 paths, we run traceroute with the path maximum transmission unit (MTU) discovery to identify the problems and their causes.

1 Introduction

IPv6 is the next generation IP protocol to replace the current IPv4. IPv6 provides an expanded address space, and supports new Internet applications that require advanced features to provide services like real-time audio. However, IPv6 is still in its infancy and is rarely used. Because the network performance directly influences the user experience in many applications, such as online chatting and games, the poor IPv6 performance certainly limits its deployment. To qualify the IPv6 infrastructure, it is interesting to compare the IPv6 and IPv4 measurements under the current network situations. Specifically, for each source-destination pair, i.e. between 26 textboxes of RIPE NCC TTM project [1], we collect routing and one-way delay information using IPv4 and IPv6 versions of traceroute and delay measurements, and compare the routing and delay data on a path-by-path basis. By comparing IPv6 and IPv4 paths, we focus on problems only present in the IPv6 paths, and run traceroute with path MTU discovery to identify the causes.

2 Measurement Results

2.1 Statistical Results of Delays, IP Delay Variation and Hopcount

Statistical Results of Source-Destination Delays. Real-time applications will not perform well if the end-to-end delays between the communicating parties exceed a certain QoS delay threshold. For example, in case of VoIP, to maintain the high quality of voice, packets need to be received within about 150 millisecond

C. Dovrolis (Ed.): PAM 2005, LNCS 3431, pp. 345–348, 2005.

(ms). The importance of Internet delay for providing QoS triggered us to examine the congestion-free delay of each pair as a function of time. The congestion-free delay is computed as the minimum end-to-end IPv4 and IPv6 delay. We repeated the experiments to calculate the average delay of each pair. The delay can depend on the geographical distance. The results show that 37% of the IPv6 paths and 39% of the IPv4 paths have a minimal delay less than 10 ms, while 88% and 92% less than 50 ms, respectively. We also found that 25% of the IPv6 paths and 32% of the IPv4 paths have an average delay less than 10 ms, while 86% and 90% less than 50 ms, respectively.

IP Delay Variation. The one way IP delay variation (ΔD) is defined in RFC 3393. Low IP delay variation is important for applications requiring timely delivery of packets. For each source-destination pair, we compute ΔD for both IPv6 and IPv4 paths, from which we constructed the pdf (probability density function) of the IPDV. We categorize four main classes: Class 1 is a typical distribution. It is a symmetrical distribution with short tails. Class 1 has 97.5% of the delay variation smaller than +/- 20 ms. To isolate high quality connections, a subclass 1b is introduced, which contains plots with less than +/- 1 ms of delay variation. Class 1 is characteristic for a good quality in transmission. Class 2 is similar to Class 1 except that there are many variations exceeding 20 ms; Class 3 is a symmetric distribution with more than one peak, which is mainly caused by path switching. We observed that only about 18% of IPv6 traffic are of class 1b, while about 31% in IPv4; about 60.2% of IPv6 traffic are of class 1, while about 69% in IPv4; about 34.7% of IPv6 traffic are of class 2, while about 24.4% in IPv4; about 5.1% of IPv6 traffic are of class 3, while about 6.7% in IPv4. The experiments confirm that compared to IPv4, IPv6 paths suffer from a larger delay variation, which has a significant impact on the real-time application since more buffering in the end host is required.

Statistical Results of Hopcounts. The pdf of hopcounts (H) in Internet contributes to our understanding of the Internet's topological structure. All traceroute IP paths were converted to AS paths from the RIPE Whois database. In the traceroute data from the remaining boxes a total of 630 most dominant paths have been determined. From the pdf of the hopcount of those paths shown in Figure 1.a, we found that both IP hopcount and AS hopcount in IPv6 are alike their corresponding parts in IPv4. The interesting distinguishing factor between AS hopcounts and IP hopcounts lies in the ratio $\alpha = \frac{E[H]}{var[H]}$. For IPv6 and IPv4, we found approximately $\alpha \approx 1$ in the IP level, while $\alpha \approx 2$ in the AS level, respectively. These observations suggest that, to first order, the IP hopcount might be close to a Poisson random variable as explained in [2], while the AS hopcount behaves differently.

2.2   Comparison of IPv6 Delays and IPv4 Delays

For each source-destination pair, we compare the IPv6 and IPv4 delay data on a path-by-path basis. Figure 1.b shows the scatter plot of the IPv6 delays versus

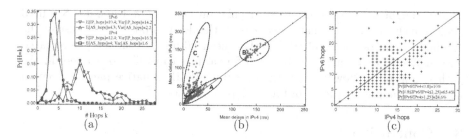

Fig. 1. (a) The hopcount distribution in the experiments; (b) Distribution of IPv6/IPv4 one-way delay; and (c) Distribution of IPv6/IPv4 hop

the IPv4 delays, where IPv6 delay is on the Y-axis while IPv4 delay on the X-axis. Each data point corresponds to a pair of peers. In Figure 1.b, following the idea from [3], the data points are approximately classified into three groups by R, the ratio of the IPv6 over the IPv4 one-way delay: group A for the European pairs with equal R ($0.8 \leq R \leq 1.25$) or small R ($R<0.8$); group B for the continent pairs (Europe-Japan, Europe-USA and USA-Japan) with equal or small R; and group C for the pairs with large R ($R>1.25$). The results indicate that compared with IPv4 paths about 54% of pairs are of group A, about 10% of group B, while about 36% of group C.

These poorly performing IPv6 paths (shown in the group C) consisted of several test-boxes located in different European counties (like UK, IT and NL). The large delay ratios might be a result of high level of IPv4 commitment and relatively low level of IPv6 responding in Europe. We repeat the experiments with the IP level hopcount. The results shown in Figure 1.c indicate that most IP level hopcounts are alike in IPv6 and IPv4.

2.3   Traceroute Results

For those 229 selected IPv6 paths whose IPv6:IPv4 delay ratios R are large, we run traceroute to identify specific problems and their causes. Many IPv6 networks use tunnels. Traceroute6 is one of the many tools used to obtain the quality of connectivity in a route. The experiments show that it is common for IPv6 paths to traverse different ASes than their IPv4 counterparts. The results also suggest that many problems lie in routing (e.g., 20 paths suffered routing loops, where 10 are native paths, while another 10 went through tunnels). The poor performance in IPv6 might be due to some poorly configured tunnels that disregard the underlying topologies. Tunnels are useful during the early stages of IPv6 deployment, but poorly configured tunnels lead to performance problems. In addition to the traceroute measurements, we use path MTU discovery to identify IPv6-in-IPv4 tunnels in those poorly performing IPv6 paths. The Tunnel discovery Tool allows us to detect an IPv6 tunnel by measuring the MTU over an entire path, since a drop in MTU at an intermediate router indicates a possible tunnel entry point. About 48.8% of those selected IPv6 paths went through native paths, while about 26.2% went through IPv6-in-IPv4 tunnels, about 21.3%

went throught Generic Routing Encapsulation tunnels; and about 3.8% used BSD tunnels. We expect that a decrease in the delay is possible because of the continuous improvements of IPv6 paths: the IPv6 in IPv4 tunnels are replaced with native IPv6 paths, and the IPv6 forwarding capability of routers in the path is improved. However, for those about 49% IPv6-native paths, we could not assert the precise causes of the poor performance.

3   Conclusion

Although IPv6 will replace IPv4 in the future, it is expected that IPv4 and IPv6 hosts will coexist for a substantial time during the steady migration from IPv6 to IPv4. To qualify the IPv6 infrastructure, it is interesting to compare the IPv6 and IPv4 measurements under the current network situations. Specifically, for each source-destination pair, we have collected the routing and delay information using both the IPv4 and the IPv6 versions of the traceroute and delay measurements, and have compared the delay data on a path-by-path basis. We have focused on problems that were only present in the IPv6 paths, and have run traceroute with path MTU discovery for identifying the causes. From our experiments, we can draw the following conclusions:

- Concerning the IP delay variation, our results suggest that compared to IPv4, IPv6 paths suffer from a larger delay variation, which has a significant impact on the real-time application since it might increase the cost of buffering in the end host;
- Compared with IPv4 paths, about 36% of the IPv6 paths are suffering from a significantly larger delay;
- The poorly performing IPv6 paths might be due to some badly configured tunnels that disregard the underlying topologies.

Acknowledgement. We are grateful to Henk Uijterwaal and Mark Santcroos for the use of the RIPE NCC TTM data.

References

1. RIPE Test Traffic Measurements, http://www.ripe.net/ttm
2. Van Mieghem, P., G. Hooghiemstra and R. W. van der Hofstad, "Scaling Law for the Hopcount", Delft University of Technology, report2000125, 2000
3. K.Cho, M.Luckie and B.Huffaker. Identifying IPv6 Network Problems in the Dual-Stack World. SIGCOMM'04 Network Troubleshooting Workshop, Portland, Oregon, September 2004

Scalable Coordination Techniques for Distributed Network Monitoring

Manish R. Sharma and John W. Byers

Dept. of Computer Science, Boston University,
Boston MA 02215, USA

Abstract. Emerging network monitoring infrastructures capture packet-level traces or keep per-flow statistics at a set of distributed vantage points. Today, distributed monitors in such an infrastructure do not co-ordinate monitoring effort, which both can lead to duplication of effort and can complicate subsequent data analysis. We argue that nodes in such a monitoring infrastructure, whether across the wide-area Internet, or across a sensor network, should coordinate effort to minimize resource consumption. We propose space-efficient data structures for use in gossip-based protocols to approximately summarize sets of monitored flows. With some fine-tuning of our methods, we can ensure that all flows observed by at least one monitor are monitored, and only a tiny fraction are monitored redundantly. Our preliminary results over a realistic ISP topology demonstrate the effectiveness of our techniques on monitoring tens of thousands of point-of-presence (PoP) level network flows. Our methods are competitive with optimal off-line coordination, but require significantly less space and network overhead than naive approaches.

1   Introduction

In monitoring applications ranging from wide-area traffic monitoring to event detection in sensor networks to surveillance by a set of pan/tilt/zoom cameras located at distributed vantage points, a growing challenge involves appropriate coordination of the activities of the individual monitors. In the applications above, monitors are *resource-constrained*, and thus it is essential to minimize the effort monitors expend on monitoring tasks. For example, when any of a set of monitors may perform a monitoring task equivalently well, a cost-saving strategy is to elect a single leader to perform the job. Of course, such a leader election process must be efficient, must be robust to losses and failures, and must err on the side of conservatism to ensure that all observable activities are monitored by at least one monitor. Avoiding duplication of effort has a secondary advantage for applications in which observed data is subsequently aggregated and processed, since complications associated with the presence of duplicate observations are avoided. In this work-in-progress paper, we consider the problem of minimizing duplication of effort in distributed event monitoring in which monitors are connected by a high-speed network. While we believe that both the statement of our problem and our methods are much more broadly applicable, we focus here exclusively on wide-area network traffic monitoring.

C. Dovrolis (Ed.): PAM 2005, LNCS 3431, pp. 349–352, 2005.

Wide-area Network Monitoring: Current technology to monitor network traffic by passively collecting flows or samples or logging packet headers either compromises router performance or incurs high costs due to costly measurement infrastructure. We argue that a brute force, non-adaptive approach to monitoring network traffic misses an opportunity to better manage resource consumption. Instead, we advocate a lower-cost alternative, i.e. developing scalable techniques to coordinate and distribute the monitoring effort. For example, if the monitoring effort can be distributed in such a way that each monitor monitors only a small subset of network flows, substantial savings can be achieved in terms of storage and processing overhead. Our work attempts to achieve the above goal without introducing too much control traffic overhead.

We assume a passive network monitoring infrastructure comprised of multiple monitoring systems, that coordinate to monitor network traffic traversing through them. Such systems are not expected to be ubiquitous or directly integrated into routers but are specially equipped with traffic capture and storage capabilities. We assume that: all monitors can communicate with all other monitors periodically, the monitors have sufficient memory to perform the monitoring functionality, and the monitors can compute the set of monitors on the route of a flow. We model the incoming traffic at monitors as a datastream consisting of items in the form of key-value pairs. Here, the key is taken to be a network flow at the Point of Presence (PoP) level, i.e. an ingress/egress pair, and the value is the size of each packet in bytes.

Problem Statement and Contribution: Let S denote a set of events, let $M = \{m_1, m_2, \ldots m_K\}$ be a set of monitors, and let $V_i \subseteq S$ be the set of events (flows) observable by m_i. Now let L_i denote the set of events monitored by m_i. Our objective is: Minimize $\sum_i |L_i|$ subject to $\bigcup_i L_i = S$ and $\forall i, L_i \subseteq V_i$.

In other words, monitors must monitor only flows they can observe, all observable flows must be monitored at least once, and the goal is to minimize duplication of effort. In the next sections, we describe our approach to this problem using Bloom filter-based summarization techniques to coordinate between a set of network monitors, quantify the cost, and present simulation results.

2   Coordination Algorithms

We now provide a brief overview of our data structures, algorithms and key results; full details and the analysis are in [5]. Each monitor locally maintains two data structures. The first is a lookup table of active flows marked either as actively monitored or monitored by someone else. The second is a counting Bloom filter, that approximately represents the set of active flows at a network monitor. Our approach starts by having each monitor represent the set of flows it is currently monitoring with a counting Bloom filter [1], a compact randomized data structure that supports lookup operations on keys. With a Bloom filter, lookups for inserted keys are always correct, but lookups for keys not present in the filter may yield a false positive, with a tunable false positive probability p. Full details are in [2]. Monitors use a simple gossiping protocol to periodically

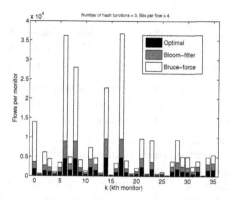

Fig. 1. Performance analysis of monitoring approaches

disseminate their Bloom filters to all other monitors in the system. Use of Bloom filters not only reduces the size of the summaries by orders of magnitude as compared to a full-fidelity representation but also keeps the network overhead of gossiping the summaries to other monitors manageable.

On the arrival of a new flow (whether new to the system or due to a route change), a monitor checks if it is the first monitor on the flow's route. If so, it inspects the Bloom filters of other monitors along the route. If the flow appears to be monitored elsewhere, it leaves the flow for the appropriate monitor. Otherwise, the monitor creates new state for the flow and maintains the state from that time onward. If the monitor is not first on the route, then it is not initially responsible for monitoring that flow. However, the approximate nature of the summaries makes them vulnerable to errors in the form of false positives, i.e., a flow is not actually monitored by a monitor but its summary reflects that it is.

The simple, but elegant solution to this potential problem is a central contribution of our work: we eliminate false positives by applying an idea of *self-inspection*: if a given monitor finds that an observable flow produces a false positive in its own Bloom filter, then it immediately starts to monitor that flow. The cost of this method is a small amount of redundant monitoring: in the event that two or more Bloom filters have a match on a given flow, they must all monitor that flow (redundantly). Our analysis in [5] shows that for a flow traversing j monitors using Bloom filters with false positive probability p, the expected number of monitors that will monitor the flow using our method is $(1-p)^j + pj$.

3   Experimental Results

We simulated deployment of our monitoring infrastructure over one of the PoP level topologies generated by Rocketfuel [4]. The topology consisted of 36 PoPs, producing 1296 origin-destination (OD) pairs. Next, we made use of inferred backbone link weights [3] to run Dijkstra's single-source shortest path algorithm at all PoPs to determine the route from one PoP to any other PoP. To create a plausible distribution of network flows between PoPs in our topology, for all PoP pairs, we compute a value l_{ij} that is the fraction of flows which originate at PoP i and terminate at PoP j. Using a gravity model, we take $l_{ij} \propto P_i \times P_j$, where P_i is the population of node i, and normalize $l_{ij} = \frac{P_i P_j}{\sum P_i P_j}$ to ensure $\sum_{i,j} l_{ij} = 1$. We simulated our proposed network monitoring technique with 50,000 network flows distributed amongst different PoP pairs using the gravity model, and compared the following three different monitoring approaches.

- **Brute force:** Each monitor monitors every flow that is visible to it.
- **Optimal:** Each monitor is given full information about the workload, and the flow is assigned to the least loaded monitor on its route.
- **Bloom filter:** Monitors have no prior information, flows arrive one by one, and our proposed methods do the online assignment of flows to monitors.

Figure 1 plots the number of flows monitored for each approach and at each of the 36 monitors. Our online Bloom filter approach is nearly as good as the offline optimal in terms of overall load reduction and load balance, and significantly improves worst-case load over the brute force approach. Using a simple back-of-the-envelope calculation (omitted for lack of space), we estimate that an unoptimized version of our approach affords more than a factor of two memory savings on average, and more than a factor of five at the worst case monitor in this scenario. Unlike brute force, our methods have an extra cost associated with data exchange to ensure continuous monitoring of all visible flows under route changes and to maintain load balance. In a naive all-pairs exchange of Bloom filters, the total aggregate traffic load is 3.24 MB (200 KB per pair) in our simulation setup.

Future work: Our ongoing work involves experimental evaluation, validation and refinement of our methods over large, realistic datasets. Along with additional evaluation, key considerations that we intend to further investigate in the full version of the paper are: further reducing network overhead when periodically exchanging summaries, refining load balancing mechanisms to improve their robustness, and specifying how data structure parameters can be set automatically.

References

1. Bloom, B. H.: Space/time trade-offs in hash coding with allowable errors. In *Comm. of the ACM*, volume 13(7), pages 422–426, 1970.
2. Broder, A. Z., and Mitzenmacher, M.: Network Applications of Bloom Filters: A Survey, In *Proc. of the 40th Annual Allerton Conference*, 2002.
3. Mahajan, R., Spring, N., Wetherall, D., and Anderson, T.: Inferring Link Weights using End-to-End Measurements. In *Proceedings of ACM SIGCOMM Internet Measurement Workshop*, November 2002.
4. Spring, N., Mahajan, R. and Wetherall, D.: Measuring ISP Topologies with Rocketfuel. In *Proceedings of ACM SIGCOMM*, August 2002.
5. Sharma M. R. and Byers J. W.: Scalable Coordination Techniques for Distributed Network Monitoring. *Tech. Report BUCS-TR-2005-001, Boston U.*, January 2004.

Evaluating the Accuracy of Captured Snapshots by Peer-to-Peer Crawlers

Daniel Stutzbach and Reza Rejaie

University of Oregon, Eugene OR 97403, USA
{agthorr, reza}@cs.uoregon.edu

Abstract. The increasing popularity of Peer-to-Peer (P2P) networks has led to growing interest in characterizing their topology and dynamics [1, 2, 3, 4], essential for proper design and effective evaluation. A common technique is to capture topology snapshots using a crawler. However, previous studies have not verified the accuracy of their captured snapshots. We present techniques to measure the inaccuracy of topology snapshots, quantify the effects of unreachable peers and crawling speed, and explore the impact of snapshot accuracy on derived characterizations.

1   Introduction

The accuracy of captured snapshots by P2P crawlers can be significantly affected by both the duration of a crawl and the ratio of unreachable peers. Determining the accuracy of captured snapshots of a P2P system is fundamentally difficult because a perfect reference snapshot for comparison is not available. The desired characterization of P2P systems determines the granularity and type of collected information in each snapshot, in the form of a tradeoff between the duration of a crawl and the completeness of the captured snapshot. For example, studying churn only requires a list of participating peers, and a crawler can gather this information from a subset of all peers with reasonable accuracy. In contrast, to study the overlay topology a captured snapshot should include all edges of the overlay; this requires the crawler to directly contact every peer, otherwise a connection between two unvisited peers would be missed.

To study snapshot accuracy, we developed a fast and efficient Gnutella crawler, called *Cruiser*, that is able to capture a complete snapshot of the Gnutella network in around 5 minutes with six off-the-shelf desktop PCs. Previous studies typically crawled their target P2P systems in 30 minutes to two hours (e.g., [5, 4]), despite crawling significantly smaller networks. Cruiser achieves this significant reduction in crawl time as follows: *(i)* it leverages several features of modern Gnutella, including its semi-structured topology and efficient new handshake mechanism; *(ii)* it substantially increases the degree of concurrency during the crawling process by deploying a master-slave architecture and allowing each slave crawler to contact hundreds of peers simultaneously. More details on the design and evaluation of Cruiser may be found in our tech report [6].

C. Dovrolis (Ed.): PAM 2005, LNCS 3431, pp. 353–357, 2005.

2   Modern Gnutella

We briefly describe the key features of modern Gnutella [7, 8] that are used by Cruiser.
The original Gnutella protocol had limited
scalability due to its flat overlay. To address
this limitation, most modern Gnutella clients
implement a two-tiered network structure by
dividing peers into two groups: *ultrapeers*
and *leaf* peers. As shown in Fig. 1, each ultrapeer neighbors with several other ultrapeers within a top-level overlay. The major-

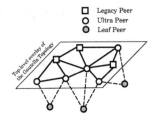

Fig. 1. Semi-Structured Topology
of Modern Gnutella

ity of the peers are leaves that are connected to the overlay through a few ultrapeers. Those peers that do not implement the ultrapeer feature can only reside
in the top-level overlay and do not accept any leaves. We refer to these peers as
legacy peers. We also refer to the legacy peers and ultrapeers collectively as the
top-level peers.

Also, modern Gnutella clients implement a special handshaking feature that
enables the crawler to quickly query a peer for a list of its current neighbors.
Previous crawlers relied on other features of the Gnutella protocol, namely Ping-
Pong messages, to retrieve this information, but these techniques were less efficient.

3   Accuracy of Captured Snapshots

We consider three effects that can impact the accuracy of topology snapshots.
First, we consider unreachable peers which, for one reason or another, cannot
be crawled. Second, we consider how much accuracy can be maintained while
cutting short the duration of crawls. Finally, we consider the impact of the
crawler's speed.

Unreachable Peers: A non-negligible subset of contacted peers in each crawl
time out (15–24%), prematurely drop (6–10%) or refuse TCP connections (5–
7%). Peers are unreachable when they have already left the system (i.e., departed), they are located behind a firewall (or NATed), or they receive SYN
packets at too high a rate (i.e., overloaded). Departed and firewalled peers
are noted in previous studies; however we find many unreachable peers are overloaded, refusing and accepting TCP connections sporadically over a short period
of time (i.e., within a single minute they alternate repeatedly). Unreachable ultrapeers can introduce the following errors in a captured snapshot: (i) including
departed peers, (ii) omitting branches between unreachable ultrapeers and their
leaves, and (iii) omitting branches between two unreachable top-level peers. To
minimize these errors, it is important to quantify what portion of unreachable
peers were departed versus firewalled or overloaded. Unfortunately, there is no
reliable test to firmly verify the status of unreachable peers among the three
possible scenarios, since overloaded, firewalled, and departed peers may or may

not reply to SYN packets. However, we found that repeatedly attempting to connect to peers which have timed out is unlikely to ever meet with success, even after attempting for several hours. This suggests that those peers, at least, are firewalled.

Impact of Crawling Duration: To examine the impact of crawl duration on the accuracy of captured snapshots, we modified Cruiser to stop the crawl after a specified period. Shorter crawls allow us to capture back-to-back snapshots more rapidly, which increases the granularity for studying churn. We performed two back-to-back crawls and repeated this process for different durations. We define δ_+ and δ_- as the number of new and missing peers in the second snapshot compared to the first one, respectively (normalized by the total number of peers in the first crawl). Figure 2(a) presents the sum $\delta = \delta_+ + \delta_-$ as well as the total number of discovered peers as a function of the crawl duration. During short crawls (the left side of the graph), δ is high because the captured snapshot is incomplete, and each crawl captures a different subset. As the duration of the crawl increases, δ decreases, indicating that the captured snapshot becomes more complete. Increasing the crawl length beyond four minutes does not decrease δ any further, and achieves a marginal increase in number of discovered peers. This figure reveals a few important points. First, there exists a "sweet spot" for the crawl duration beyond which crawling has diminishing returns if the goal is simply to capture the population. Second, the change of $\delta = 0.08$ is an upper-bound on the distortion due to the passage of time as Cruiser runs. Third, for sufficiently long crawls, Cruiser can capture a relatively accurate snapshot. The relatively flat values of delta for longer crawls suggest that a small but significant fraction of the network is unstable and turns over quickly. For shorter durations, the standard deviation of the peers discovered is small, since the size of the discovered topology is limited by the crawl's duration. For longer

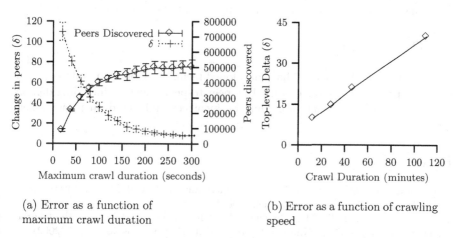

(a) Error as a function of
maximum crawl duration

(b) Error as a function of crawling
speed

Fig. 2. Effects of crawl speed and duration, generated by running two crawls back-to-back per x-value

durations, the standard deviation is larger and measures the actual variations in network size.

Impact of Crawling Speed: To examine the impact of crawling speed on the accuracy of captured snapshots, we decreased the speed of Cruiser by reducing the number of parallel connections that each slave process can open. Figure 2(b) depicts the error in between snapshots from back-to-back crawls as a function of crawl duration. The first snapshot was captured with the maximum speed and serves as a reference, whereas the speed (and thus duration) of the second snapshot has changed. The duration of the second snapshot is shown as the x value. This figure clearly demonstrates that the accuracy of snapshots decreases significantly for longer crawls.

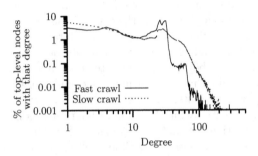

Fig. 3. Observed top-level degree distributions in a slow and a fast crawl

Impact of Snapshot Accuracy on Derived Characterization: To show the effect this error has on conclusions, in Fig. 3 we show the observed degree distribution of a fast crawl versus a crawl limited to 60 concurrent connections. The slow crawl distribution looks similar to that seen in [4][1], which lead to the conclusion that Gnutella has a two-piece power-law degree distribution. If we further limit the speed, the distribution begins to look like a single-piece power-law, the result reported by earlier studies [9, 5]. To a slow crawler, peers with long uptimes appear as high degree because many short-lived peers report them as neighbors. However, this is a misrepresentation since these short-lived peers are not all present at the same time.

4   Conclusion

In this extended abstract, we have developed techniques for examining the accuracy of topology snapshots captured by peer-to-peer crawlers, including demonstrating that earlier conclusions may be incorrect and based on measurement artifacts.

References

1. Bhagwan, R., Savage, S., Voelker, G.: Understanding Availability. In: International Workshop on Peer-to-Peer Systems. (2003)
2. Saroiu, S., Gummadi, P.K., Gribble, S.D.: Measuring and Analyzing the Characteristics of Napster and Gnutella Hosts. Multimedia Systems Journal **8** (2002)

---

[1] Their crawler was limited to 50 concurrent connections.

3. Liben-Nowell, D., Balakrishnan, H., Karger, D.: Analysis of the Evolution of Peer-to-Peer Systems. In: Principles of Distributed Computing, Monterey, CA (2002)
4. Ripeanu, M., Foster, I., Iamnitchi, A.: Mapping the Gnutella Network: Properties of Large-Scale Peer-to-Peer Systems and Implications for System Design. IEEE Internet Computing Journal **6** (2002)
5. clip2.com: Gnutella: To the Bandwidth Barrier and Beyond (2000)
6. Stutzbach, D., Rejaie, R.: Characterizing Today's Semi-Structured Gnutella Network. Technical Report CIS-TR-04-02, University of Oregon (2004)
7. Singla, A., Rohrs, C.: Ultrapeers: Another Step Towards Gnutella Scalability. Gnutella Developer's Forum (2002)
8. Lime Wire LLC: Crawler Compatability. Gnutella Developer's Forum (2003)
9. Adamic, L.A., Lukose, R.M., Huberman, B., Puniyani, A.R.: Search in Power-Law Networks. Physical Review E **64** (2001)

HOTS: An OWAMP-Compliant Hardware Packet Timestamper

Zhang Shu and Katsushi Kobayashi

National Institute of Information and Communications Technology, Japan
{zhang, ikob}@koganei.wide.ad.jp

1 Introduction

Accurate timestamps on both the sender and the receiver side are crucial for one-way delay (OWD) measurements. Traditionally, the methods of (i) peering with NTP servers, and (ii) connecting to a time source directly, have been used to maintain the accuracy of a measurement system clock. However, it has became clear that such methods suffer from errors to different extents.

In this paper, we introduce the hardware OWAMP [1] timestamper (HOTS), which generates extremely precise clock information for OWAMP test packets on both the sender and the receiver side. Compared with traditional methods, HOTS offers the following advantages: (i) the generated timestamp can be extremely precise because HOTS accepts an external 10-MHz signal as well as the 1PPS signal as input, and (ii) HOTS bypasses all of the software processing, thus minimizing possible errors. We also present the early results of OWD measurements that we made using this timestamper.

DAG [2] is a similar measurement instrument which also uses hardware to generate timestamps for a packet. However, this product is only designed to record the arrival timestamp of a packet and cannot be used to measure OWD.

2 Methodology

2.1 OWAMP Overview

The one-way active measurement protocol (OWAMP) is designed to measure one-way delay, jitter or packet loss. It consists of two inter-related protocols: OWAMP-Control and OWAMP-Test. OWAMP-Control is used to initiate, start and stop test sessions and to fetch their results, while OWAMP-Test is used to define the format of the test packets.

OWAMP test packets are transmitted in UDP datagrams. The header of the packets includes an 8-byte "Timestamp" field where the sender inserts the clock information when an OWAMP test packet is sent.

2.2 HOTS

Simply speaking, HOTS is a packet over SONET (POS) network interface card (NIC) which has a function to generate timestamps for outgoing or incoming

C. Dovrolis (Ed.): PAM 2005, LNCS 3431, pp. 358–361, 2005.

packets destined for a specific port. Because it uses a PCI bus, HOTS can be used on most of the PCs.

HOTS has three I/O ports: one bidirectional SC connector and two mini-BNC jacks. The SC connector is used to send and receive packets, just as a normal SC connector does. The two mini-BNC jacks are used to obtain clock information from an external source such as a GPS or CDMA receiver.

HOTS can accept two kinds of signal as input: a 1PPS signal and a 10-MHz signal. The precision of the generated timestamp depends on the accuracy of the provided signals. In our measurements, we used two kinds of GPS receiver to generate these signals: the HP 58503A and the TymServe 2100. Both of them can generate extremely precise clock information.

HOTS maintains two clock-related counters: C_1 and C_2, which respectively hold the seconds and the fractions of a second based on the two kinds of external signals. The counters are operated as follows.

1. When reset (or the interface becomes up), both counters are cleared to zero.
2. In operational mode, C_2 is incremented based on the external 10-MHz signal.
3. When the 1PPS signal is received, C_1 is incremented and C_2 is cleared to zero.

Sender Behavior. When HOTS is used on the sender side, it works as follows.

1. When receiving a packet from the upper layer (usually the driver program), HOTS checks whether this packet is a UDP datagram and is destined for a specific port.
2. If it is, HOTS generates a timestamp (t_s) based on the two clock counters, inserts the timestamp into the "Timestamp" field of the packet, recalculates the UDP checksum based on the original one and the new timestamp, and then sends the packet to the physical link.
3. If the packet is not a UDP packet or is not destined for a specific port, HOTS does nothing apart from sending the packet to the physical link.

Receiver Behavior. When used on the receiver side, HOTS behaves as follows.

1. When receiving a packet from the physical link, it checks whether this packet is a UDP datagram and is destined for a specific port.
2. If it is, HOTS generates a timestamp (t_r), and passes the timestamp to the upper layer as well as the received packet. How the timestamp is passed to the upper layer depends on the users. In our measurements, we directly recorded (in the driver program) the timestamp in the body of the OWAMP test packet for simplicity. We also cleared the UDP checksum so that the datagram would not be dropped in the UDP processing because of inconsistent UDP checksums.
3. If the packet is not a UDP datagram or the port number is not a specific one, HOTS simply passes the received packet to the upper layer as other NICs would.

The OWAMP program on the receiver side will receive the packet in the user-space by normal socket API and the OWD can be calculated by

$$D = t_r - t_s \qquad (1)$$

HOTS works with both IPv4 and IPv6 OWAMP test packets.

3   One-Way Delay Measurements

3.1   Measured Network

We made OWD measurements on the APAN-JP network, which is part of the Asia-Pacific Advanced Network [3]. The topology of the measured network is shown in Fig. 1. PC_1 and PC_2 were the two end hosts between which we sent and received OWAMP test packets. These hosts were located at our institute in Tokyo and a data center in Fukuoka, respectively. There were five routers between the two hosts. The major distances between the two hosts were the 30 km between our institute and downtown Tokyo, and the 900 km between Tokyo and Fukuoka.

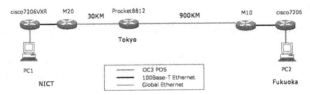

Fig. 1. Topology of the measured network

We periodically sent and received four kinds of test packet: packets of either 64 bytes or 1400 bytes in IPv4 or IPv6. All of these packets were sent once per second.

3.2   Measurement Results

Some early results for the IPv4 packets are shown in Fig. 2. From this graph, we can see that the OWD for the IPv4 64-byte packets was usually about 10.7ms, and the OWD for the IPv4 1400-byte packets was several milliseconds longer. For IPv6, the results were similar to those for IPv4 packets.

3.3   Adaptation for Other OWAMP Implementation

HOTS can be easily used in other OWAMP software, such as the implementation from Internet2 [4], to perform highly precise measurements with the following modifications.

- Specify the negotiated port numbers of the test packets on both the sender and the receiver side before transmitting a test packet.
- On the receiver side, use the hardware-generated timestamp when calculating the OWD.

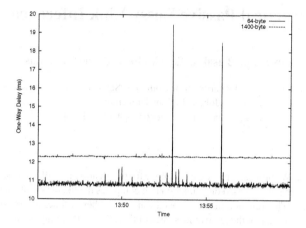

Fig. 2. Typical results for IPv4

4   Conclusion and Future Work

Highly precise OWD measurement is a challenge because of the difficulty of eliminating errors in process to obtain an accurate timestamp and other overheads. In this paper, we introduced HOTS - a hardware packet timestamper that we developed to measure OWD. HOTS can generate extremely precise clock information for OWAMP test packets provided an accurate time source such as a GPS or CDMA receiver. We presented the results of a preliminary OWD measurement that we did on the APAN-JP network to show its effectiveness.

References

1. S. Shalunov, B. Teitelbaum, A. Karp, J. W. Boote, and M. J. Zekauskas. (2004, Aug.) A one-way active measurement protocol (OWAMP). Internet draft. [Online]. Available: http://www.ietf.org/internet-drafts/draft-ietf-ippm-owdp-10.txt
2. DAG network monitoring interface cards. [Online]. Available: http://www.endace.com/networkMCards.htm
3. Asia-Pacific Advanced Network. [Online]. Available: http://www.apan.net
4. Internet2 OWAMP implementation. [Online]. Available: http://e2epi.internet2.edu/owamp

Practical Passive Lossy Link Inference

Alexandros Batsakis, Tanu Malik, and Andreas Terzis

Department of Computer Science,
Johns Hopkins University
{abat, tmalik, terzis}@cs.jhu.edu

Abstract. We propose a practical technique for the identification of lossy network links. Our scheme is based on a function that computes the likelihood of each link to be lossy. This function mainly depends on the number of times a link appears in lossy paths and on the relative loss rates of these paths. Preliminary simulation results show that our solution achieves accuracy comparable to statistical methods (e.g. Bayesian) at significantly lower running time.

1   Introduction and Related Work

Most loss inference techniques [1, 2, 3, 4, 5] attempt to deduce link loss rates from end-to-end measurements. *Active* techniques [1, 2] infer link loss by actively probing the network, while *passive* techniques [3, 4, 5] estimate packet loss by observing the evolution of application traffic. Passive measurements do not require coordination among end points, introduce no additional traffic, and hence, are easier to deploy and do not perturb the state of the network. Depending on the method used to infer packet loss, passive techniques can be further divided to *analytical* [3, 5] and *heuristic* [4]. Analytical techniques detect higher percentage of lossy links while techniques based on heuristics have an advantage on execution speed and resource consumption.

The insight behind this paper is that users care more about finding lossy links affecting the performance of their applications, than finding the exact loss rate of these links. Based on this, we present COBALT, a heuristics-based inference algorithm that detects with high probability the lossy links affecting applications' performance. COBALT assigns a *confidence level* to each link —the higher the confidence, the higher the probability that the link is lossy. The confidence level of a link depends on the lossy paths the link belongs to and how lossy are those paths compared to all the paths in the network.

Preliminary simulation results show COBALT's accuracy to be on par with the analytical methods while its running time is ten times faster. When compared to heuristic methods such as SCFS [4], COBALT infers 20% more truly lossy links, at the expense of higher percentage of false positives. However, the number of false positives decreases as the number of measurement points increases. The low running time of COBALT makes it possible to run the algorithm iteratively with smaller data sets. Each iteration of the algorithm provides new link confidence levels and previous results are incorporated using an exponential moving average. We sketch how this *real-time* variant of COBALT works and argue how it can track the variability in link loss characteristics.

C. Dovrolis (Ed.): PAM 2005, LNCS 3431, pp. 362–367, 2005.

2   Algorithm

Before presenting the details of our method, we briefly introduce the network model we used. In our model clients are connected to servers through a network whose topology is known a-priori. Clients connect and exchange data with the servers using any TCP-based protocol such as FTP or HTTP. We collect a trace of all the packets sent and received by each server. Using these traces, we calculate the loss rate of the path between the server and each client as the ratio of the retransmitted packets to the total number of packets sent by the server. This metric overestimates the actual loss rate, due to the retransmission strategy used by TCP, however our goal is not to estimate the exact link loss rates so this metric is used simply as an estimate of the path loss rate.

COBALT starts by separating lossy from non-lossy paths. Paths with loss rate higher than a user-configurable threshold T are labeled as *bad* while the remaining paths are labeled as *good*. The threshold T corresponds to a loss rate above which application performance is disrupted. The default value of T has been set to 1%.

Following path classification, links are categorized depending on the number of good paths they belong to. This approach is similar to the one followed in SCFS and is based on the intuition that lossy links dominate the end-to-end path loss rate. If a path contains a lossy link then the path's loss rate will be at least equal to the link's loss rate. Thus, a lossy link cannot be part of a good path. To make COBALT less susceptible to path loss rate estimation errors, we classify a link as good (non-lossy) only if it belongs to at least s good paths. To give a concrete example, if T is set to 2% and link l belongs to a path with estimated loss rate equal to 1.9%, it is difficult to draw conclusions about the probability of the link to be really lossy or not. If on the other hand l belongs to s paths with loss rate less than T we can assume with higher confidence that l is a good link. The parameter s, defined as the sensitivity of the algorithm, depends on the size of the network (i.e. number of hosts and paths) and is user configurable. Higher values of s give higher confidence that the identified links are truly lossy. At the same time the number of false positives increases because some good links might be classified as lossy if they don't participate in s paths.

After excluding the links found in good paths, COBALT computes the confidence levels for the remaining links. The confidence level $cfd(l)$ for a link l in a network N is computed as:

$$cfd(l) = K^{t(l)} \cdot \frac{avl(l)}{avp(N)} \tag{1}$$

In the formula above, $avl(l)$ is the average loss rate of all paths that l belongs to, while $avp(N)$ is the average loss rate among all bad paths in the network. $t(l)$ denotes the number of times l is found in lossy paths and finally K is a constant, with value greater or equal to one. Intuitively, a link is bad if it participates in paths whose loss rate is much higher than the average loss rate of all network paths. This effect is covered by the fraction in Equation 1. Second, we can have higher confidence that a link is bad if it belongs to many bad paths. This second effect is covered by the $K^{t(l)}$ term in Equation 1. If K is close to one, then $t(l)$ is of little significance in the computation of $cfd(l)$. The greater the value of K the higher the importance of $t(l)$. We use an exponential function of $t(l)$ so small differences in the number of lossy paths a link belongs to will create large difference in confidence level simplifying the final selection of the most problematic links. More details on how K should be selected will be discussed in future work.

As its last step, COBALT ranks the links by their confidence levels. The links with the highest confidence levels are the most likely to be problematic.

2.1 Real-Time Algorithm

While existing passive loss inference techniques can detect chronically lossy links, recent results [6] indicate that link characteristics tend to remain stable for only small period of times (approx. 20 minutes). Unfortunately, the running time of current inference methods makes them inappropriate for short timescale changes. Simply put, data analysis requires more time than the time in which link characteristics remain stable. For example, the running time of the Bayesian method in our experiments was 60 minutes for 30 seconds of simulated traffic.

In order to infer network characteristics in a timely manner, while minimizing the storage and computational needs, we propose a real-time variant of our original algorithm. Specifically, we compute the confidence level for each link in short time intervals where the amount of data is small. We then combine the new confidence with our previous knowledge of a link's state, the confidence level estimated at the previous interval, to infer the current conditions.

The online algorithm works similarly to its offline variant but uses an exponential moving average formula to compute the confidence level of a link l:

$$cfd_{t_{i+1}}(l) = (1 - w) \cdot cfd_{t_{i-1}}(l) + w \cdot cfd_{t_i}(l) \tag{2}$$

where $cfd_{t_{i-1}}(l)$ the confidence level computed by the previous measurements and $cfd_{t_i}(l)$ the confidence level computed by the most recent data. w is an aging variable ($0 \le w \le 1$) controlling the convergence time of the algorithm. If w is close to one, the algorithm will converge faster but it will be more susceptible to oscillations since new data will have more weight. An interesting point in our method is that the value of w might not be constant across all links or even for estimates made for the same link. Its exact value is a function of two parameters:

- The amount of time between two successive runs of the algorithm. If this interval is long, taking into account the temporal link characteristics, the significance of the previous confidence levels decreases, as they reflect an obsolete network image. Hence, in this case the value of w should approach its upper limit. In contrast, if the interval is small then the value of w should be close to zero.
- The number of packets received between t_{i-1} and t_i. Since our method is based on statistics, the larger the sample the more confident we are about the outcome of our analysis. Therefore, as the number of received packets increases, w should approach one.

3   Simulation Results

We used ns-2 to simulate our network of clients and servers. The simulated network topologies were created using BRITE's two-level hierarchical topologies [7]. The net-

Table 1. Comparison of four passive lossy link inference methods

Fraction of Lossy Link	5%				10%				20%			
Number of Bad Links	64				105				224			
	R	B	S	C	R	B	S	C	R	B	S	C
Correctly Identified	13	37	39	43	30	75	51	81	46	140	57	144
False Positives	20	12	4	14	62	23	9	24	100	45	15	54

work consists of 800 nodes and about 1400 links. We randomly chose 100 clients out of a pool of 250 to download a large file from the server using HTTP. We also picked a fraction f of the links to be lossy. We then randomly assigned a loss rate to each of the lossy links from a configurable range of loss rates. We used a bimodal loss model, where good links, have a loss rate between $0 - 0.5\%$, and bad links have loss rate between $1.0 - 3.0\%$. This model represents a challenging case for inference algorithms since the difference in loss rates between good and bad links is not significant. We further assume that packet losses are independent.

We compare COBALT to three other methods: Random Sampling [3], Bayesian with Gibbs Sampling [3] and the SCFS algorithm proposed by Duffield [4]. Given the difference of COBALT to previous techniques we need to redefine coverage and false positives in terms of the algorithm's parameters. In our evaluation, a correctly identified lossy link is one whose confidence level exceeds a threshold T_{lossy}. Thus, non-lossy links whose confidence level exceed this threshold count as false positives. The value of T_{lossy} used in the experimental evaluation is equal to K, the constant used in Equation 1.

For every experiment presented in this section, we ran the algorithm three times, each time with a different topology. The reported confidence level is the average of the confidence levels obtained over the three executions. We noticed little or no variation on the outcome over the three runs. We choose K to be $3/2$, the sensitivity s of the algorithm is 3, while T is equal to 0.01. For random sampling, the mean link loss rate is chosen over 500 iterations. If the mean exceeds the loss rate threshold of bad links, the link is said to be lossy. Similarly, for the Bayesian method, the "burn in" period in Gibbs sampling is chosen as 1000 iterations, and links are marked lossy if 99% of the samples found are above the loss rate threshold.

A comparison of COBALT with the three other methods is shown in Table 1. The results on this table are based on measurements from a single server. It is evident that COBALT provides the best coverage at the expense of a relatively high false positive rate compared to SCFS. SCFS has the lowest false positive rate but its coverage drops dramatically when lossy links are not rare. The Bayesian method finds about 70% of the truly lossy links with false positive rate close to 20%. Finally, random sampling fails to identify more than 30% of the lossy links, while at the same time the number of false positives is very high. Our findings about the Bayesian and random sampling methods are different from the results presented by Padmanabhan *et al* in [3]. This is because we used a different link loss model in which the loss margin between good and lossy links is narrow. Furthermore, Padmanabhan *et al* used random tree topologies while we use Internet-like topologies.

Figure 1(a) illustrates the performance of COBALT and Bayesian when multiple servers are used. In this scenario, traces from all the servers are combined and both algo-

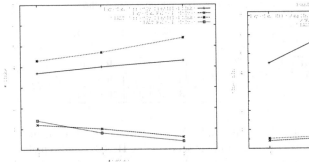 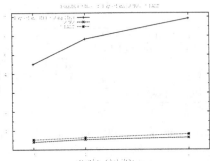

(a) Performance of COBALT and Bayesian as the number of measurement points increases

(b) Running time comparison of Bayesian with Gibbs Sampling, SCFS, COBALT in a network with 800 nodes and 1400 links

Fig. 1.

rithms run over the aggregate collected data. Notice that by using ten servers, randomly distributed across the network, COBALT improves its coverage and almost eliminates the false positives observed in the previous table.

Figure 1(b) shows the running time of Bayesian, SCFS and our approach as the fraction of lossy links increases. The execution time of the heuristics-based methods, SCFS and COBALT, is almost ten times faster than the Bayesian method with Gibbs sampling.

4   Future Work

We are currently evaluating COBALT across a wider range of simulated topologies in networks with thousands of nodes. We are also evaluating the real-time variant of the algorithm, in terms of its accuracy, execution time, and responsiveness to loss models that change over time.

The biggest challenge is to evaluate our method in the Internet. We are currently collecting and analyzing traces from two campus networks with thousands of users. This analysis will allow us to explore the loss patterns these users experience and provide insights about the actual lossy links in the Internet.

References

1. M. Rabbat, R. Nowak, and M. J. Coates, "Multiple Source, Multiple Destination Network Tomography," in *Proceedings of IEEE INFOCOM 2004*, Apr. 2004.
2. Ratul Mahajan, Neil Spring, David Wetherall, and Thomas Anderson, "User-level Internet Path Diagnosis," in *Proceedings of SOSP 2003*, Oct. 2003.
3. Venkat Padmanabhan, Lili Qiu, and Helen J. Wang, "Server-based Inference of Internet Link Lossines," in *Proceedings of IEEE INFOCOM 2003*, Apr. 2003.
4. Nick Duffield, "Simple Network Performance Tomography," in *Proceedings of Internet Measurement Conference (IMC) 2003*, Oct. 2003.

5. G. Liang R. Nowak R. Castro, M. J. Coates and B. Yu, "Internet Tomography: Recent Developments," *Statistical Science*, Mar. 2004.
6. Y. Zhang, N. Duffield, V. Paxson, and S. Shenker, "On the Constancy of Internet Path Properties," in *Proceedings ACM SIGCOMM Internet Measurement Workshop*, Nov. 2001.
7. Alberto Medina, Anukool Lakhina, Ibrahim Matta, and John Byers, "Brite: An approach to universal topology generation," in *Proceedings of MASCOTS 2001*, Aug. 2001.

Merging Network Measurement with Data Transport

Pavlos Papageorgiou and Michael Hicks

University of Maryland
pavlos@eng.umd.edu
mwh@cs.umd.edu

Abstract. The tasks of measurement and data transport are often treated independently, but we believe there are benefits to bringing them together. This paper proposes the simple idea of a transport agent to encapsulate useful data within probe packets in place of useless padding.

1   Introduction

Overlay networks have become a popular vehicle for introducing network services. Oftentimes, to drive its services, an overlay network infers characteristics of the network via application-layer probes. For example, nodes in RON [1], Detour [2], and Pastry [3] regularly ping their neighbors to check availability and/or measure latency. MediaNet [4] uses available bandwidth [5] estimations to determine along which paths to forward media streams.

For the most part, the measurement and transport aspects of overlay networks are treated independently. The measurement service is a *black box* used by the overlay to make decisions. But the fact that measurement traffic is *in addition* to transport traffic imposes an extra burden on the network. While not a problem for a single overlay under normal conditions, as congestion and/or the number of overlay networks in use increases, measurement traffic begins to influence the total traffic.

To reduce the overhead of measurement traffic, we propose the following simple idea: *merge the task of network measurement with the task of data transport.* In many cases, measurement traffic consists largely of *null padding* just meant to consume bandwidth for timing purposes. This is the case when measuring available bandwidth, for example [5]. To avoid this wasted bandwidth, the transport layer can replace null padding with user payloads available from other streams.

While others have proposed cooperative measurement services [6, 7, 8], or observed that network characteristics can be inferred passively [9], no one has proposed merging the tasks of measurement and transport. In this extended abstract, we outline the design and preliminary implementation of a *transport agent* that provides TCP and UDP-like transport along with an enhanced API for sending measurement probes.

2   Probe-Aware Transport Agent

The two goals of our probe-aware transport agent are: (i) to minimize the bandwidth that measurement tools consume and (ii) to allow probe traffic to be

C. Dovrolis (Ed.): PAM 2005, LNCS 3431, pp. 368–371, 2005.

responsive to congestion conditions. It is desirable that the API and the end-to-end semantics of user traffic (TCP and UDP) remain intact so that no changes are required to existing applications. Only measurement tools should be required to use the socket API extensions to set encapsulation and dispatch policy for their probes.

The challenge is how to maintain the same measurement accuracy while decreasing the probe bandwidth. The critical observation is that the *null padding*, which dominates probe packets, can be reused without sacrificing the tool's accuracy. The actual pad bytes are irrelevant to the measurement algorithm which means that probe packets can encapsulate user traffic if it is available. This approach satisfies the first goal. To address the second goal, we observe that probing schemes usually do not care about the absolute timing of probe packets but only about the relative timing between packets. Therefore, it should be possible to briefly delay certain probe packets without degrading the tool's accuracy. For example, the transport agent can delay the first packet of a packet train as long as it preserves the inter-packet timing and records the actual departure times.

There is an important trade-off between bandwidth efficiency and the timeliness of probe/user packet transmission. The bandwidth optimization achieved depends directly on how often probes encapsulate user traffic. Congestion conditions will increase this frequency, since it is more likely that user traffic will be buffered and available when probe packets are sent. Thus, probes consume less bandwidth as congestion increases. To improve the optimization under non-congested conditions, we can briefly delay certain probe packets, as directed by the measurement tool, when no user data is available. Conversely, we can delay a user packet when the application would allow it. For example, TCP already delays data to send it in larger chunks.

Socket API Extensions: Our probe-aware transport API extends the BSD-style socket API to define additional flags that affect the way packets are sent, either per-packet or per-session; these flags are presented in Table 1.

Probe packets are sent with the PAD_PKT [x] flag enabled. This states that the provided data should be sent with an additional x bytes of padding. Thus, if a probe tool wants to send d bytes of data (i.e., the byte content it wants delivered to the peer tool, such as control information) with x bytes of padding, it would pass only the d bytes to send, along with the flag PAD_PKT [x]. The transport agent will transport a packet of size $d + x$ bytes, and will attempt to use the x bytes portion of the packet to encapsulate user traffic. When the packet is actually sent, it is timestamped by the transport agent, and timestamped again when it is received. A separate function is used by tools to acquire the times.

The DELAY [t] flag can be used to delay the packet up to t ms, to increase the chances of encapsulating user data; otherwise the packet is queued for immediate departure. Note that this flag can be applied to either user or probe traffic. For example, to send a probe that waits up to 100 ms would require flags PAD_PKT[x] | DELAY [100].

Table 1. Probe-Aware Transport API flags

<table>
<tr><td colspan="2" align="center">send flags</td></tr>
<tr><td>PAD_PKT [s]</td><td>Probe packet that requires s bytes padding.</td></tr>
<tr><td>DELAY [t]</td><td>Packet can be delayed up to t ms.</td></tr>
<tr><td>PKT_FOLLOWS</td><td>This packet is not the last packet of a train (others follow).</td></tr>
<tr><td colspan="2" align="center">per-session flags</td></tr>
<tr><td>SINGLE_PKT</td><td>Packets should not be encapsulated.</td></tr>
<tr><td>WAIT_CONGESTION</td><td>Packets under congestion control.</td></tr>
</table>

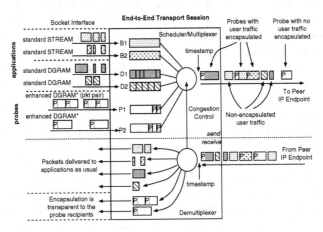

Fig. 1. Overview of a probe-aware *transport agent*

Finally, the PKT_FOLLOWS flag is used to indicate that the packet is part of a train, and should not be sent until all packets are available (i.e., a subsequent packet is submitted without this flag). Thus, the entire train may be delayed (by the first packet), but all packets in the train are sent back-to-back.

We also provide two session-level flags. If for some reason encapsulation should be avoided, users can establish sessions using the flag SINGLE_PACKET. Probe packets sent in such a session will not encapsulate other packets, and user packets will not be encapsulated by probes. Thus, our transport API semantics reverts to the standard semantics when this flag is set.

Additionally, we provide the flag WAIT_CONGESTION to subject sessions to congestion control. By default, STREAM sessions have this flag enabled (to conform to TCP semantics), but DGRAM sessions can specify it as well. This allows probe packets, which are often sent as datagrams, to be accounted for in the congestion window. However, the implementation is non-standard in that we must consider the DELAY and PKT_FOLLOWS flags when doing congestion accounting.

Figure 1 depicts our preliminary *probe-aware transport agent* which exchanges traffic between two IP endpoints. The transport agent would sit normally on top of IP; our current implementation tunnels over UDP. Internally, the transport agent multiplexes user and probe streams into one packet stream with uniform congestion control, as in the Congestion Manager [10], and encapsulates user traffic in probe packets, unless explicitly disallowed by SINGLE_PACKET flags.

For example, consider a measurement tool (session P2 in Figure 1) that periodically sends a probe packet consisting of 10 bytes of control data and 990 bytes of padding. Before the transport agent sends the packet out, it attempts to find user data to fill the 990 available bytes from candidate TCP sessions $B1$, $B2$ and UDP packet streams $D1$, $D2$. If any of them have bytes waiting in their buffers, then up to 990 bytes of user traffic will be encapsulated in the probe.

We have run preliminary experiments with real traffic on Emulab[1] that demonstrated bandwidth savings up to 95% during congestion conditions, i.e., most of the probe traffic piggy-backed on top of user traffic during that period. We are in the process of completing a fully functional implementation and modifying a number of measurement tools to run on top of our transport agent. We intend to run wide-area experiments with real traffic on PlanetLab[2] and continue our performance measurements in the controlled setting of Emulab.

References

1. Andersen, D.G., Balakrishnan, H., Kaashoek, K., Morris, R.: Resilient overlay networks. In: SOSP. (2001)
2. Savage, S., Collins, A., Hoffman, E., Snell, J., Anderson, T.: The end-to-end effects of Internet path selection. In: SIGCOMM. (1999)
3. Rowstron, A., Druschel, P.: Pastry: Scalable, distributed object location and routing for large-scale peer-to-peer systems. In: IFIP/ACM International Conference on Distributed Systems Platforms (Middleware). (2001) 329–350
4. Hicks, M., Nagarajan, A., van Renesse, R.: User-specified adaptive scheduling in a streaming media network. In: IEEE Conference on Open Architectures and Network Programming. (2003)
5. Prasad, R., Murray, M., Dovrolis, C., Claffy, K.: Bandwidth estimation: metrics, measurement, techniques, and tools. IEEE Network **17** (2003)
6. Nakao, A., Peterson, L., Bavier, A.: A Routing Underlay for Overlay Networks. In: SIGCOMM. (2003)
7. Srinivasan, S., Zegura, E.: Network Measurement as a Cooperative Enterprise. In: International Workshop on Peer-to-Peer Systems. (2002)
8. Danalis, A., Dovrolis, C.: ANEMOS: An Automomous NEtwork MOnitoring System. In: Passive and Active Measurement Workshop. (2003)
9. Paxson, V.: Automated Packet Trace Analysis of TCP Implementations. In: ACM SIGCOMM. (1997)
10. Balakrishnan, H., Rahul, H.S., Seshan, S.: An Integrated Congestion Management Architecture for Internet Hosts. In: ACM SIGCOMM. (1999)

---

[1] http://www.emulab.net

[2] http://www.planet-lab.org

Author Index

Lecture Notes in Computer Science

For information about Vols. 1–3345

please contact your bookseller or Springer

Vol. 3395: J. Grabowski, B. Nielsen (Eds.), Formal Approaches to Software Testing. X, 225 pages. 2005.

Vol. 3394: D. Kudenko, D. Kazakov, E. Alonso (Eds.), Adaptive Agents and Multi-Agent Systems III. VIII, 313 pages. 2005. (Subseries LNAI).

Vol. 3393: H.-J. Kreowski, U. Montanari, F. Orejas, G. Rozenberg, G. Taentzer (Eds.), Formal Methods in Software and Systems Modeling. XXVII, 413 pages. 2005.

Vol. 3391: C. Kim (Ed.), Information Networking. XVII, 936 pages. 2005.

Vol. 3390: R. Choren, A. Garcia, C. Lucena, A. Romanovsky (Eds.), Software Engineering for Multi-Agent Systems III. XII, 291 pages. 2005.

Vol. 3389: P. Van Roy (Ed.), Multiparadigm Programming in Mozart/OZ. XV, 329 pages. 2005.

Vol. 3388: J. Lagergren (Ed.), Comparative Genomics. VII, 133 pages. 2005. (Subseries LNBI).

Vol. 3387: J. Cardoso, A. Sheth (Eds.), Semantic Web Services and Web Process Composition. VIII, 147 pages. 2005.

Vol. 3386: S. Vaudenay (Ed.), Public Key Cryptography - PKC 2005. IX, 436 pages. 2005.

Vol. 3385: R. Cousot (Ed.), Verification, Model Checking, and Abstract Interpretation. XII, 483 pages. 2005.

Vol. 3383: J. Pach (Ed.), Graph Drawing. XII, 536 pages. 2005.

Vol. 3382: J. Odell, P. Giorgini, J.P. Müller (Eds.), Agent-Oriented Software Engineering V. X, 239 pages. 2005.

Vol. 3381: P. Vojtáš, M. Bieliková, B. Charron-Bost, O. Sýkora (Eds.), SOFSEM 2005: Theory and Practice of Computer Science. XV, 448 pages. 2005.

Vol. 3380: C. Priami, Transactions on Computational Systems Biology I. IX, 111 pages. 2005. (Subseries LNBI).

Vol. 3379: M. Hemmje, C. Niederee, T. Risse (Eds.), From Integrated Publication and Information Systems to Information and Knowledge Environments. XXIV, 321 pages. 2005.

Vol. 3378: J. Kilian (Ed.), Theory of Cryptography. XII, 621 pages. 2005.

Vol. 3377: B. Goethals, A. Siebes (Eds.), Knowledge Discovery in Inductive Databases. VII, 190 pages. 2005.

Vol. 3376: A. Menezes (Ed.), Topics in Cryptology – CT-RSA 2005. X, 385 pages. 2005.

Vol. 3375: M.A. Marsan, G. Bianchi, M. Listanti, M. Meo (Eds.), Quality of Service in Multiservice IP Networks. XIII, 656 pages. 2005.

Vol. 3374: D. Weyns, H.V.D. Parunak, F. Michel (Eds.), Environments for Multi-Agent Systems. X, 279 pages. 2005. (Subseries LNAI).

Vol. 3372: C. Bussler, V. Tannen, I. Fundulaki (Eds.), Semantic Web and Databases. X, 227 pages. 2005.

Vol. 3371: M.W. Barley, N. Kasabov (Eds.), Intelligent Agents and Multi-Agent Systems. X, 329 pages. 2005. (Subseries LNAI).

Vol. 3370: A. Konagaya, K. Satou (Eds.), Grid Computing in Life Science. X, 188 pages. 2005. (Subseries LNBI).

Vol. 3369: V.R. Benjamins, P. Casanovas, J. Breuker, A. Gangemi (Eds.), Law and the Semantic Web. XII, 249 pages. 2005. (Subseries LNAI).

Vol. 3368: L. Paletta, J.K. Tsotsos, E. Rome, G.W. Humphreys (Eds.), Attention and Performance in Computational Vision. VIII, 231 pages. 2005.

Vol. 3367: W.S. Ng, B.C. Ooi, A. Ouksel, C. Sartori (Eds.), Databases, Information Systems, and Peer-to-Peer Computing. X, 231 pages. 2005.

Vol. 3366: I. Rahwan, P. Moraitis, C. Reed (Eds.), Argumentation in Multi-Agent Systems. XII, 263 pages. 2005. (Subseries LNAI).

Vol. 3365: G. Mauri, G. Păun, M.J. Pérez-Jiménez, G. Rozenberg, A. Salomaa (Eds.), Membrane Computing. IX, 415 pages. 2005.

Vol. 3363: T. Eiter, L. Libkin (Eds.), Database Theory - ICDT 2005. XI, 413 pages. 2004.

Vol. 3362: G. Barthe, L. Burdy, M. Huisman, J.-L. Lanet, T. Muntean (Eds.), Construction and Analysis of Safe, Secure, and Interoperable Smart Devices. IX, 257 pages. 2005.

Vol. 3361: S. Bengio, H. Bourlard (Eds.), Machine Learning for Multimodal Interaction. XII, 362 pages. 2005.

Vol. 3360: S. Spaccapietra, E. Bertino, S. Jajodia, R. King, D. McLeod, M.E. Orlowska, L. Strous (Eds.), Journal on Data Semantics II. XI, 223 pages. 2005.

Vol. 3359: G. Grieser, Y. Tanaka (Eds.), Intuitive Human Interfaces for Organizing and Accessing Intellectual Assets. XIV, 257 pages. 2005. (Subseries LNAI).

Vol. 3358: J. Cao, L.T. Yang, M. Guo, F. Lau (Eds.), Parallel and Distributed Processing and Applications. XXIV, 1058 pages. 2004.

Vol. 3357: H. Handschuh, M.A. Hasan (Eds.), Selected Areas in Cryptography. XI, 354 pages. 2004.

Vol. 3356: G. Das, V.P. Gulati (Eds.), Intelligent Information Technology. XII, 428 pages. 2004.

Vol. 3355: R. Murray-Smith, R. Shorten (Eds.), Switching and Learning in Feedback Systems. X, 343 pages. 2005.

Vol. 3354: M. Margenstern (Ed.), Machines, Computations, and Universality. VIII, 329 pages. 2005.

Vol. 3353: J. Hromkovič, M. Nagl, B. Westfechtel (Eds.), Graph-Theoretic Concepts in Computer Science. XI, 404 pages. 2004.

Vol. 3352: C. Blundo, S. Cimato (Eds.), Security in Communication Networks. XI, 381 pages. 2005.

Vol. 3351: G. Persiano, R. Solis-Oba (Eds.), Approximation and Online Algorithms. VIII, 295 pages. 2005.

Vol. 3350: M. Hermenegildo, D. Cabeza (Eds.), Practical Aspects of Declarative Languages. VIII, 269 pages. 2005.

Vol. 3349: B.M. Chapman (Ed.), Shared Memory Parallel Programming with Open MP. X, 149 pages. 2005.

Vol. 3348: A. Canteaut, K. Viswanathan (Eds.), Progress in Cryptology - INDOCRYPT 2004. XIV, 431 pages. 2004.

Vol. 3347: R.K. Ghosh, H. Mohanty (Eds.), Distributed Computing and Internet Technology. XX, 472 pages. 2004.

Vol. 3346: R.H. Bordini, M. Dastani, J. Dix, A.E.F. Seghrouchni (Eds.), Programming Multi-Agent Systems. XIV, 249 pages. 2005. (Subseries LNAI).